ELEMENTS OF DIFFERENTIAL GEOMETRY

ELEMENTS OF DIFFERENTIAL GEOMETRY

RICHARD S. MILLMAN

Michigan Technological University
Houghton, Michigan

GEORGE D. PARKER

Southern Illinois University
Carbondale, Illinois

Prentice-Hall Inc., Upper Saddle River, NJ 07458

Library of Congress Cataloging in Publication Data

MILLMAN, RICHARD S (date)
Elements of differential geometry.

Bibliography: p. 248
Includes index.
1. Geometry, Differential. I. Parker, George D., (date) joint author. II. Title.
QA641.M52 516'.3604 76-28497
ISBN 0-13-264143-7

Printed in the United States of America

10 9 8

Prentice-Hall International (UK) Limited, London
Prentice-Hall of Australia Pty. Limited, Sydney
Prentice-Hall Canada Inc., Toronto
Prentice-Hall Hispanoamericana, S.A., Mexico
Prentice-Hall of India Private Limited, New Delhi
Prentice-Hall of Japan, Inc., Tokyo
Pearson Education Asia Pte. Ltd., Singapore
Editoria Prentice-Hall do Brasil, Ltda., Rio De Janeiro

To

Phillip, Steven, Robert, and Cheryl,
who do not understand geometry yet,

and

Margie,
who did once.

Contents

Preface

The intent of this book is to provide an elementary and geometric introduction to differential geometry. We adopt an approach that is elementary enough for undergraduates but still both conveys the spirit of modern differential geometry and prepares the student for a more advanced course on the subject. In particular, we have given significant emphasis to global considerations while providing the basic classical results on curves and surfaces. In these global results we have been guided in part by the beautiful article of Chern [1967]. This book is an outgrowth of courses taught by the first author at Ithaca College and by both authors at Southern Illinois University over a period of six years.

In differential geometry there is calculus and there is geometry, neither of which should be slighted. All too often the geometry is hidden in either machinery, abstraction, or symbolism. Furthermore, an unfortunate thing has happened to the subject in the last ten years—it has been relegated to the graduate curriculum. There is no question that differential geometry (or any subject for that matter) can be done more efficiently if the student has the background that every second year graduate student has, but to wait until graduate school to teach the subject is doing an injustice to both the student and the subject. We hope that this book will aid in the return of differential geometry to the undergraduate curriculum and that it can be taught by non-specialists, as well as by specialists.

A traditional undergraduate course in classical differential geometry usually hides the geometry in a myriad of symbols such as $\Gamma_{ij}{}^{k}$ or $R_i{}^j{}_{kl}$. We have minimized this difficulty by making use of linear algebra throughout. Modern vector space terminology can and should be used effectively to make the material more comprehensible to the students. Furthermore, we find that the students are very excited about using the linear algebra they have just learned in another course. They view differential geometry as an application of linear algebra, which it is. After the students understand the geometric content of a result (by use of linear algebra) the result is then expressed in classical notation. This enables the reader to consult texts in classical differential geometry without having to develop a whole new vocabulary.

In the advanced approach, which is usually restricted to the graduate level, the geometry is hidden in the machinery and abstraction. We have avoided topics which require building complicated machinery, such as differential forms or cohomology. We have resisted the temptation to prove the most general results possible if these would require a lot of machinery to state or to prove. In particular, for the first six chapters we stay in $\mathbf{R}^3$. In order to avoid the trap of abstraction we include many examples—not only of the definitions but also of the theorems because it is usually much easier to follow a proof with a firm example in mind.

When first introduced to differential geometry, individuals often have trouble reconciling the two views of the subject—classical and modern (manifolds). It is sometimes hard to tell what the connection (no pun intended) is between the two approaches. We attempt to remedy this situation in the last chapter. There we explain how the earlier material motivates the definitions of manifold, tangent space, Riemannian metric, etc. and give a brief introduction to the modern terminology. We hope that having finished this chapter the reader will have an easier transition to books such as Hicks [1965], Boothby [1975], Warner [1971], Matsushima [1972], or Kobayashi and Nomizu [1963].

There is only one way to learn mathematics and that is to get your hands dirty working problems. Differential geometry is no exception. We have included more than 350 problems. The easier problems are usually at the beginning of the problem sets. Those whose results are used in the text are marked with an asterisk (*). There is one exception to this rule—in Chapters 1–6 we have not starred any problems which would be needed in Chapter 7. It should be noted that an asterisk does not denote a hard problem: we have reserved the dagger (†) for this.

In Chapter 1 we present the material which is the necessary background for the book. Most of it is review and may be emphasized as needed. We have found that our students are generally weak in describing lines and planes in vector notation. The material on eigenvalues and linear transformations is not used until Chapter 4 and may be delayed until then.

Chapter 2 develops the basic local theory of space curves. The Frenet-Serret Theorem which expresses the derivative of a geometrically chosen basis of $\mathbf{R}^3$ in terms of itself is proved, and many important corollaries are derived. We include the Fundamental Theorem of Curves, showing the dependence of differential geometry on the theory of ordinary differential equations. Several classical topics are covered in the exercises, including sphere curves, contact, Bertrand curves, evolutes, and involutes.

In Chapter 3 we give our first sampling of global theorems. Here we stay in the plane and cover the Rotation Index Theorem of Hopf, the Isoperimetric Inequality, the Four-Vertex Theorem, and curves of constant width. We feel that this is a basic chapter because global differential geometry is a very important and popular subject which is all too often slighted in a first course. It is certainly easier to prove only local results. However, this really cheats the student who thereby misses one of the most fascinating parts of the subject and is robbed of the insight gained by examining results in the plane.

Chapter 4 presents the basics of local surface theory and serves as the motivation for the ideas of Chapter 7. We study surfaces in $\mathbf{R}^3$ and their first and second fundamental forms. Geodesics and their length-minimizing properties are discussed. Next comes an investigation of curvature (both Gaussian and mean), its relationship to the curvature of curves on the surface, and Gauss's *Theorema Egregium*. The chapter ends with optional material on isometries, the Fundamental Theorem of Surfaces, and surfaces of constant curvature. Again, several classical topics are covered in the exercises, including ruled surfaces, developable surfaces, and asymptotic curves.

In Chapter 5 we return to global notions, this time for space curves. Using the concepts of geodesics on the sphere and integration on a surface, we prove Fenchel's Theorem about the total curvature of closed space curves and the Fary-Milnor Theorem about the total curvature of a knot. We also include the nonstandard topic of total torsion of a closed space curve.

Chapter 6 gives various global results for surfaces. We start off with Meusnier's Theorem which states that a compact surface, all of whose points are umbilics, is a sphere. We then go on to the Gauss-Bonnet Formula, which relates the curvature of a region with the curvature of its boundary, and the Gauss-Bonnet Theorem, which gives the total curvature of a compact surface. As applications of the Gauss-Bonnet Theorem we prove theorems due to Jacobi, Hadamard, and Poincaré.

Motivated by the first six chapters, we introduce in Chapter 7 the basic definitions of manifold theory and Riemannian geometry. This chapter is written in an open-ended fashion, referring the reader to other books for details.

A brief historical summary and bibliography are given in the appendix. References to the bibliography are in the form of a name followed by a date in brackets, such as Chern [1967].

The first four chapters require only a knowledge of calculus and finite-dimensional vector spaces (mainly bases and linear independence). Chapter 5 has no additional prerequisites, but the material is more difficult. For Chapter 6 it would be helpful if the reader is familiar with the topology of $\mathbf{R}^3$. In particular, the reader should be familiar with the notion of the limit of a sequence of points in $\mathbf{R}^3$. We define compact as closed and bounded. In Chapter 7 it would be useful if the reader knew some metric space terminology, such as open sets and continuity. A knowledge of the inverse and implicit function theorems would also make the going a bit easier.

Several different courses can be made from this book. A one quarter undergraduate course could be based on Chapters 1, 2, and 3 or 4; a one semester course on 1, 2, 3, 4 (or more, depending on the pace of the course); a two quarter course could cover the first six chapters; the entire book could be used for a year course if the last chapter is supplemented. A graduate course could start with Chapters 2, 4, and 7 and then go on to a book like Hicks [1965], Boothby [1975], Matsushima [1972], or Warner [1971].

The following table shows the dependence of the various chapters.

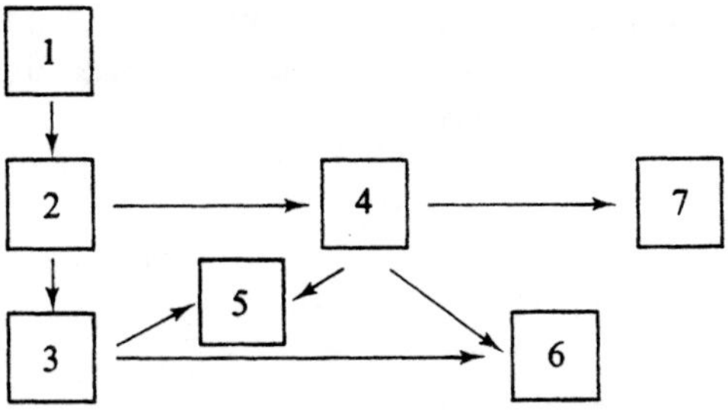

We would like to thank Sharon Champion for her magnificent job typing the manuscript, from the first set of class notes to the final draft. Her patience and attention to detail through the jth rewrite of the ith author ($i = 1$ or 2, $1 \leq j \leq \infty$) were truly amazing. We thank Marjorie Parker for her help with the proofreading. Finally, we thank Barbara Blum of the Prentice-Hall editorial staff for her help in making this book a reality.

Carbondale, Illinois

RICHARD S. MILLMAN
GEORGE D. PARKER

ELEMENTS OF DIFFERENTIAL GEOMETRY

1

Preliminaries

Differential geometry has two primary tools—linear algebra and calculus. In this text we assume that the reader is familiar with both. We shall recall the basic facts of linear algebra and vector calculus and urge the reader to review with care anything which is unfamiliar.

1–1. VECTOR SPACES

DEFINITION. A *real vector space* is a set V, whose elements are called *vectors*, together with two binary operations $+: V \times V \longrightarrow V$ and $\cdot: \mathbf{R} \times V \longrightarrow V$, called *addition* and *scalar multiplication*, which satisfy the following eight axioms for all $\mathbf{u}, \mathbf{v}, \mathbf{w} \in V$ and $r, s \in \mathbf{R}$:

(a) $\mathbf{u} + \mathbf{v} = \mathbf{v} + \mathbf{u}$;
(b) $\mathbf{u} + (\mathbf{v} + \mathbf{w}) = (\mathbf{u} + \mathbf{v}) + \mathbf{w}$;
(c) there is an element $\mathbf{0}$ in V such that $\mathbf{0} + \mathbf{u} = \mathbf{u}$;
(d) $(rs)\cdot\mathbf{u} = r\cdot(s\cdot\mathbf{u})$;
(e) $(r + s)\cdot\mathbf{u} = r\cdot\mathbf{u} + s\cdot\mathbf{u}$;
(f) $r\cdot(\mathbf{u} + \mathbf{v}) = r\cdot\mathbf{u} + r\cdot\mathbf{v}$;
(g) $0\cdot\mathbf{u} = \mathbf{0}$; and
(h) $1\cdot\mathbf{u} = \mathbf{u}$.

We shall normally omit the multiplication symbol $\cdot$

EXAMPLE 1.1. Ordinary three-dimensional space, $\mathbf{R}^3$, is a real vector space.

EXAMPLE 1.2. The set of all polynomials with real coefficients, $\mathbf{R}[x]$, is a real vector space.

DEFINITION. A set $\{\mathbf{v}_i \mid i \in I\} \subset V$ is *linearly independent* if whenever a finite linear combination $\sum a^i \mathbf{v}_i$ is zero, then each a^i must also be zero. If it is possible to find a finite linear combination $\sum a^i \mathbf{v}_i = \mathbf{0}$ with some $a^k \neq 0$, then the set is *linearly dependent*.

DEFINITION. A subset $S \subset V$ *spans* V if for each vector $\mathbf{v} \in V$ there are vectors $\mathbf{v}_1, \mathbf{v}_2, \ldots, \mathbf{v}_r$ in S and real numbers $a^1, a^2, \ldots, a^r$ such that $\mathbf{v} = \sum a^i \mathbf{v}_i$. (The number of elements used ("r") may depend on $\mathbf{v}$.)

DEFINITION. A *basis* of a vector space V is a linearly independent spanning set.

THEOREM 1.3. If V is a vector space, then V has a basis. Any two bases have the same number of elements, or all have infinitely many elements. This number is called the *dimension* of V.

EXAMPLE 1.4. $\mathbf{R}^3$ has dimension 3.

EXAMPLE 1.5. $\mathbf{R}[x]$ has infinite dimension. $\{1, x, x^2, \ldots, x^n, \ldots\}$ is a basis.

If $\{\mathbf{v}_i \mid i \in I\}$ is a basis of V, then every vector $\mathbf{v} \in V$ can be uniquely written as a finite sum $\mathbf{v} = \sum a^i \mathbf{v}_i$. The numbers a^i are called the *components* of $\mathbf{v}$ with respect to the given basis.

DEFINITION. An *inner product* on a vector space V is a function $\langle\ ,\ \rangle$: $V \times V \longrightarrow \mathbf{R}$ such that for all $\mathbf{u}, \mathbf{v}, \mathbf{w} \in V$ and $r, s \in \mathbf{R}$:
(a) $\langle \mathbf{u}, \mathbf{v} \rangle = \langle \mathbf{v}, \mathbf{u} \rangle$;
(b) $\langle \mathbf{u}, r\mathbf{v} + s\mathbf{w} \rangle = r\langle \mathbf{u}, \mathbf{v} \rangle + s\langle \mathbf{u}, \mathbf{w} \rangle$; and
(c) $\langle \mathbf{u}, \mathbf{u} \rangle \geq 0$ with equality if and only if $\mathbf{u} = \mathbf{0}$.

EXAMPLE 1.6. In $\mathbf{R}^3$ we may use the ordinary dot product:

$$\langle (a^1, a^2, a^3), (b^1, b^2, b^3) \rangle = a^1b^1 + a^2b^2 + a^3b^3.$$

EXAMPLE 1.7. In $\mathbf{R}[x]$ we may set $\langle p(x), q(x) \rangle = \int_{-1}^{1} p(x)q(x)\,dx$. See Problem 1.4.

DEFINITION. If V has an inner product and $\mathbf{v} \in V$, the *length* of $\mathbf{v}$ is $|\mathbf{v}| = \sqrt{\langle \mathbf{v}, \mathbf{v} \rangle}$.

DEFINITION. Let $\mathsf{T}: V \longrightarrow V$ be a linear transformation. A real number λ is an *eigenvalue* of T if there is a nonzero vector $\mathbf{v}$ such that $\mathsf{T}(\mathbf{v}) = \lambda\mathbf{v}$. $\mathbf{v}$ is called an *eigenvector* of T corresponding to λ.

If $(T^i{}_j)$ represents T, the eigenvalues of T are the real solutions of the polynomial equation $\det(T^i{}_j - x\delta^i{}_j) = 0$. Then if the dimension of V is n, there are at most n eigenvalues (counting multiplicity). There may be fewer, since some of the solutions might be complex but not real. Once the eigenvalues are known, the eigenvectors are found by solving appropriate linear equations.

EXAMPLE 2.1. Let $\mathsf{T}: \mathbf{R}^2 \longrightarrow \mathbf{R}^2$ be represented with respect to the standard basis $\{(1, 0)^t, (0, 1)^t\}$ by the matrix $\begin{pmatrix} 3 & 4 \\ 4 & -3 \end{pmatrix}$.

$$\begin{aligned}\begin{vmatrix} 3-\lambda & 4 \\ 4 & -3-\lambda \end{vmatrix} &= (3-\lambda)(-3-\lambda) - 16 \\ &= -9 + \lambda^2 - 16 \\ &= \lambda^2 - 25 \\ &= (\lambda - 5)(\lambda + 5).\end{aligned}$$

The eigenvalues are therefore 5 and -5. $\begin{pmatrix} 3 & 4 \\ 4 & -3 \end{pmatrix}\begin{pmatrix} x \\ y \end{pmatrix} = 5\begin{pmatrix} x \\ y \end{pmatrix}$ has $\begin{pmatrix} 2 \\ 1 \end{pmatrix}$ as one solution. $\begin{pmatrix} 3 & 4 \\ 4 & -3 \end{pmatrix}\begin{pmatrix} x \\ y \end{pmatrix} = -5\begin{pmatrix} x \\ y \end{pmatrix}$ has $\begin{pmatrix} 1 \\ -2 \end{pmatrix}$ as one solution. Hence $\begin{pmatrix} 2 \\ 1 \end{pmatrix}$ is an eigenvector corresponding to 5 and $\begin{pmatrix} 1 \\ -2 \end{pmatrix}$ is one corresponding to -5.

PROBLEMS

2.1. In Example 2.1 we have T represented by $\begin{pmatrix} 3 & 4 \\ 4 & -3 \end{pmatrix}$ with respect to the basis $\{(1, 0)^t, (0, 1)^t\}$ of $\mathbf{R}^2$.
(a) Represent T with respect to the basis $\{(1, 1)^t, (1, -1)^t\}$.
(b) Represent T with respect to the basis $\{(2, 1)^t, (1, -2)^t\}$ of eigenvectors.

2.2. Find the eigenvalues and eigenvectors of $\mathsf{T}: \mathbf{R}^2 \longrightarrow \mathbf{R}^2$ given by the matrix $\begin{pmatrix} -5 & 3 \\ -6 & 4 \end{pmatrix}$.

1–3. ORIENTATION AND CROSS PRODUCTS

Let $\{\mathbf{u}_1, \mathbf{u}_2, \ldots, \mathbf{u}_n\}$ and $\{\mathbf{v}_1, \mathbf{v}_2, \ldots, \mathbf{v}_n\}$ be two *ordered* bases of V and define a matrix $(a^i{}_j)$ by $\mathbf{v}_j = \sum a^i{}_j \mathbf{u}_i$.

Definition. The ordered bases $\{\mathbf{u}_1, \mathbf{u}_2, \ldots, \mathbf{u}_n\}$ and $\{\mathbf{v}_1, \mathbf{v}_2, \ldots, \mathbf{v}_n\}$ *give the same orientation* to V if $\det(a^i{}_j) > 0$. They give *opposite orientations* if $\det(a^i{}_j) < 0$.

Example 3.1. Let $\{(1, 0, 0), (0, 1, 0), (0, 0, 1)\}$ and $\{(1, 1, 0), (1, 0, -1), (2, 1, 3)\}$ be two ordered bases of $\mathbf{R}^3$. Since $(a^i{}_j) = \begin{pmatrix} 1 & 1 & 2 \\ 1 & 0 & 1 \\ 0 & -1 & 3 \end{pmatrix}$ has determinant -4, these ordered bases give opposite orientations.

Example 3.2. $\{(1, 0, 0), (0, 1, 0), (0, 0, 1)\}$ and $\{(1, 1, 0), (2, 1, 3), (1, 0, -1)\}$ give the same orientation.

The basis $\{(1, 0, 0), (0, 1, 0), (0, 0, 1)\}$ of $\mathbf{R}^3$ will be denoted $\{\mathbf{e}_1, \mathbf{e}_2, \mathbf{e}_3\}$. Its orientation will be called *right handed.*

Definition. If $\mathbf{u} = \sum a^i \mathbf{e}_i$ and $\mathbf{v} = \sum b^i \mathbf{e}_i$ are vectors in $\mathbf{R}^3$, the *cross* (or *vector*) *product* of $\mathbf{u}$ and $\mathbf{v}$ is

$$\mathbf{u} \times \mathbf{v} = (a^2b^3 - a^3b^2)\mathbf{e}_1 + (a^3b^1 - a^1b^3)\mathbf{e}_2 + (a^1b^2 - a^2b^1)\mathbf{e}_3.$$

By abuse of notation this may be written as

$$\mathbf{u} \times \mathbf{v} = \det \begin{pmatrix} \mathbf{e}_1 & \mathbf{e}_2 & \mathbf{e}_3 \\ a^1 & a^2 & a^3 \\ b^1 & b^2 & b^3 \end{pmatrix}.$$

Lemma 3.3. Let $\mathbf{u}, \mathbf{v}, \mathbf{w} \in \mathbf{R}^3$ and $r \in \mathbf{R}$. Then

(a) $\mathbf{u} \times \mathbf{v} = -\mathbf{v} \times \mathbf{u}$;
(b) $(r\mathbf{u}) \times \mathbf{v} = r(\mathbf{u} \times \mathbf{v})$;
(c) $\mathbf{u} \times \mathbf{v} = \mathbf{0}$ if and only if $\mathbf{u}$ and $\mathbf{v}$ are dependent;
(d) $(\mathbf{u} + \mathbf{v}) \times \mathbf{w} = (\mathbf{u} \times \mathbf{w}) + (\mathbf{v} \times \mathbf{w})$;
(e) $\mathbf{u} \times \mathbf{v}$ is perpendicular to both $\mathbf{u}$ and $\mathbf{v}$ under the usual dot product of $\mathbf{R}^3$ (Example 1.6);
(f) $|\mathbf{u} \times \mathbf{v}| = |\mathbf{u}||\mathbf{v}| \sin\theta$, where θ is the angle between $\mathbf{u}$ and $\mathbf{v}$;
(g) $\{\mathbf{u}, \mathbf{v}, \mathbf{u} \times \mathbf{v}\}$ gives a right handed orientation to $\mathbf{R}^3$ if $\{\mathbf{u}, \mathbf{v}\}$ is linearly independent.

Proof: Problem 3.1. ∎

LEMMA 3.4. Let $\mathbf{u}, \mathbf{v}, \mathbf{w} \in \mathbf{R}^3$. Then $\langle(\mathbf{u} \times \mathbf{v}), \mathbf{w}\rangle = \langle\mathbf{u}, (\mathbf{v} \times \mathbf{w})\rangle$.

Proof: Problem 3.2. ▌

DEFINITION. The *mixed* (or *triple*) *scalar product* of $\mathbf{u}, \mathbf{v}, \mathbf{w}$ is

$$[\mathbf{u}, \mathbf{v}, \mathbf{w}] \equiv \langle(\mathbf{u} \times \mathbf{v}), \mathbf{w}\rangle.$$

It is important to remember that $[\mathbf{u}, \mathbf{v}, \mathbf{w}]$ is always a number and *not* a vector.

LEMMA 3.5. $|[\mathbf{u}, \mathbf{v}, \mathbf{w}]|$ is the volume of the parallelopiped spanned by $\mathbf{u}, \mathbf{v}, \mathbf{w}$. (See Figure 1.1.)

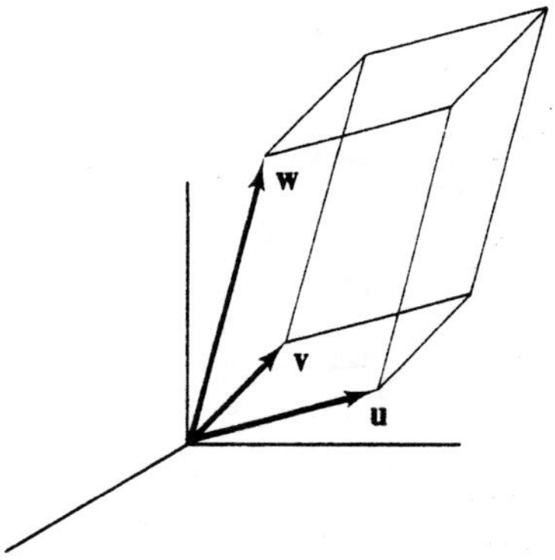

FIGURE 1.1

Proof: Problem 3.3. ▌

PROBLEMS

***3.1.** Prove Lemma 3.3.

***3.2.** Prove Lemma 3.4.

3.3. Prove Lemma 3.5.

3.4. Prove $\mathbf{a} \times (\mathbf{b} \times \mathbf{c}) = \langle\mathbf{a}, \mathbf{c}\rangle\mathbf{b} - \langle\mathbf{a}, \mathbf{b}\rangle\mathbf{c}$.

***3.5.** Let $\{\mathbf{a}, \mathbf{b}, \mathbf{c}\}$ be an orthonormal basis of $\mathbf{R}^3$ with $\mathbf{a} \times \mathbf{b} = \mathbf{c}$. Prove $\mathbf{b} \times \mathbf{c} = \mathbf{a}$ and $\mathbf{c} \times \mathbf{a} = \mathbf{b}$.

3.6. Prove $(\mathbf{a} \times \mathbf{b}) \times (\mathbf{c} \times \mathbf{d}) = [\mathbf{a}, \mathbf{b}, \mathbf{d}]\mathbf{c} - [\mathbf{a}, \mathbf{b}, \mathbf{c}]\mathbf{d}$.

3.7. Let $\mathbf{u}_1, \mathbf{u}_2, \mathbf{u}_3, \mathbf{v}_1, \mathbf{v}_2, \mathbf{v}_3$ be six vectors in $\mathbf{R}^3$. Let $a_{ij} = \langle\mathbf{u}_i, \mathbf{v}_j\rangle$. Prove $\det(a_{ij}) = [\mathbf{u}_1, \mathbf{u}_2, \mathbf{u}_3][\mathbf{v}_1, \mathbf{v}_2, \mathbf{v}_3]$.

1–4. LINES, PLANES, AND SPHERES

We shall recall the vector equations of lines, planes, and spheres in $\mathbf{R}^3$. See a standard calculus text such as Thomas [1968] for more details. Geometrically, a straight line is determined by a point on the line and a vector parallel to it. A plane is determined by a point on the plane and a vector perpendicular to the plane. A sphere is determined by its center and its radius.

Definition. The *line* through $\mathbf{x}_0 \in \mathbf{R}^3$ and parallel to $\mathbf{v} \neq \mathbf{0}$ has equation $\boldsymbol{\alpha}(t) = \mathbf{x}_0 + t\mathbf{v}$. (See Figure 1.2.)

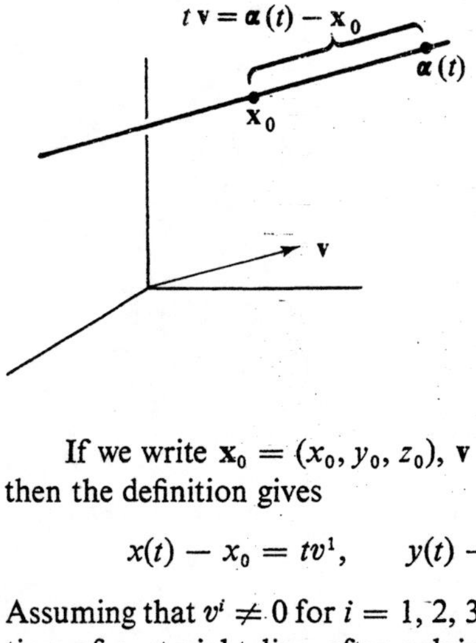

FIGURE 1.2

If we write $\mathbf{x}_0 = (x_0, y_0, z_0)$, $\mathbf{v} = (v^1, v^2, v^3)$ and $\boldsymbol{\alpha}(t) = (x(t), y(t), z(t))$, then the definition gives

$$x(t) - x_0 = tv^1, \qquad y(t) - y_0 = tv^2, \qquad z(t) - z_0 = tv^3.$$

Assuming that $v^i \neq 0$ for $i = 1, 2, 3$, these equations yield the classical definition of a straight line after solving each equality for t and setting these quantities equal:

$$\frac{x - x_0}{v^1} = \frac{y - y_0}{v^2} = \frac{z - z_0}{v^3},$$

where we have suppressed the t from the notation as is common classically.

If $\mathbf{x}_1$ and $\mathbf{x}_2$ are distinct points in $\mathbf{R}^3$ both of which lie on a line l, then the vector $\mathbf{x}_2 - \mathbf{x}_1$ is parallel to l. This observation proves:

Lemma 4.1. The line through $\mathbf{x}_1$ and $\mathbf{x}_2$ in $\mathbf{R}^3$ has equation

$$\boldsymbol{\alpha}(t) = \mathbf{x}_1 + t(\mathbf{x}_2 - \mathbf{x}_1).$$

(See Figure 1.3.)

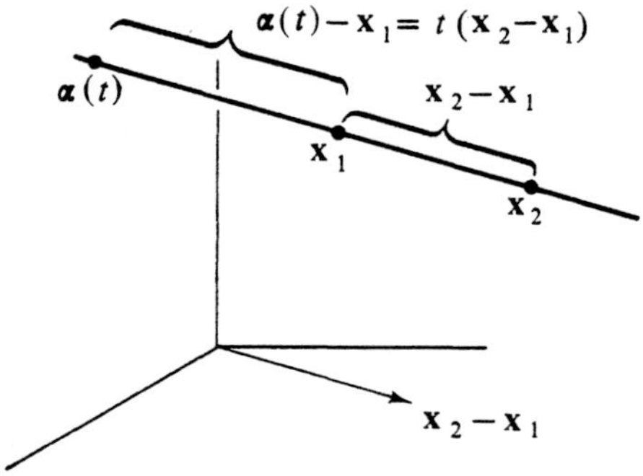

FIGURE 1.3

DEFINITION. The *plane* through $\mathbf{x}_0$ perpendicular to $\mathbf{n} \neq \mathbf{0}$ has equation

$$\langle \mathbf{x} - \mathbf{x}_0, \mathbf{n} \rangle = 0.$$

(See Figure 1.4.)

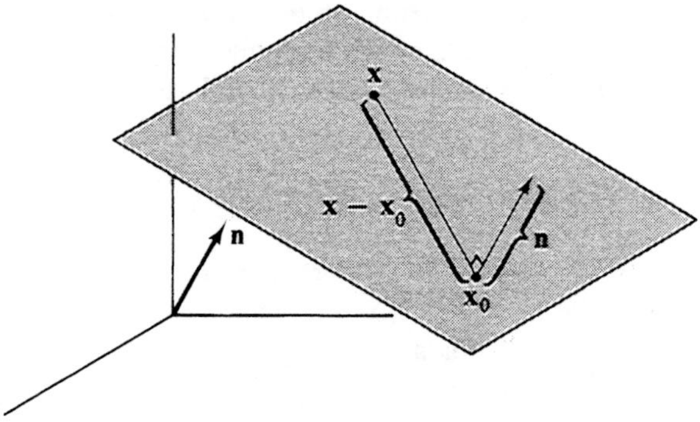

FIGURE 1.4

The following lemma is clear from Figure 1.5 since $\mathbf{u} \times \mathbf{v}$ is perpendicular to the desired plane.

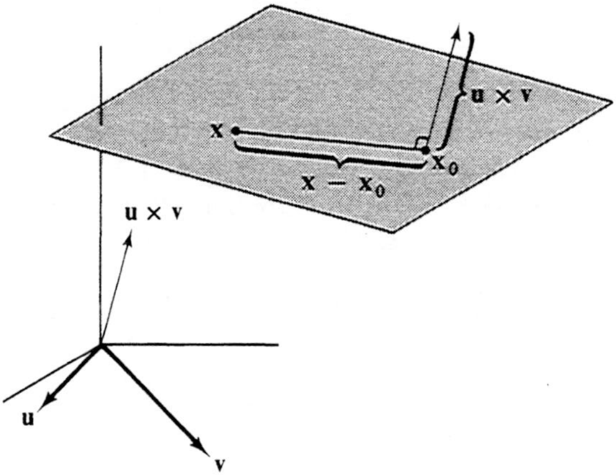

FIGURE 1.5

LEMMA 4.2. If $\{\mathbf{u}, \mathbf{v}\}$ is linearly independent, the plane through $\mathbf{x}_0$ parallel to both $\mathbf{u}$ and $\mathbf{v}$ has equation $\langle \mathbf{x} - \mathbf{x}_0, \mathbf{u} \times \mathbf{v} \rangle = 0$.

DEFINITION. The *sphere* in $\mathbf{R}^3$ with center $\mathbf{m}$ and radius $r > 0$ has equation $\langle \mathbf{x} - \mathbf{m}, \mathbf{x} - \mathbf{m} \rangle = r^2$.

If $\mathbf{m} = (a, b, c)$ and $\mathbf{x} = (x, y, z)$, then this definition is the familiar formula

$$(x - a)^2 + (y - b)^2 + (z - c)^2 = r^2.$$

PROBLEMS

4.1. Let $\mathbf{a} = (2, 1, -3)$, $\mathbf{b} = (1, 0, 1)$, $\mathbf{c} = (0, -1, 3)$. What is the equation of
(a) the line through $\mathbf{a}$ parallel to $\mathbf{b}$;
(b) the line through $\mathbf{b}$ and $\mathbf{c}$;
(c) the plane through $\mathbf{b}$ perpendicular to $\mathbf{a}$;
(d) the plane through $\mathbf{c}$ parallel to $\mathbf{a}$ and $\mathbf{b}$;
(e) the sphere with center $\mathbf{a}$ and radius 2?

1–5. VECTOR CALCULUS

If abstract vector spaces are unfamiliar, the reader may assume that $V = \mathbf{R}^3$ below. Let $\mathbf{f} : \mathbf{R} \longrightarrow V$, where V is a real vector space. If $\{\mathbf{v}_1, \mathbf{v}_2, \ldots, \mathbf{v}_n\}$ is a basis of V, then $\mathbf{f}(t) = \sum f^i(t)\mathbf{v}_i$. If the component functions $f^i(t)$ are differentiable or integrable, we may differentiate or integrate $\mathbf{f}$ componentwise:

$$\frac{d\mathbf{f}}{dt} = \sum \frac{df^i}{dt}\mathbf{v}_i$$

and

$$\int_a^b \mathbf{f}(t)\, dt = \sum \left(\int_a^b f^i(t)\, dt\right)\mathbf{v}_i.$$

We should check that these definitions do not depend on the choice of basis for V. However, this is a simple consequence of the linear properties of $d(\ \ )/dt$ and $\int_a^b (\ \ )\, dt$ and is left to the reader.

Similarly if $\mathbf{f}$ is a vector-valued function of several variables, we may take partial derivatives or multiple integrals.

LEMMA 5.1. Let $\mathbf{f}, \mathbf{g} : \mathbf{R} \longrightarrow V$ and suppose that V has an inner product $\langle\ ,\ \rangle$. Then

$$\frac{d}{dt}\langle \mathbf{f}, \mathbf{g} \rangle = \left\langle \frac{d\mathbf{f}}{dt}, \mathbf{g} \right\rangle + \left\langle \mathbf{f}, \frac{d\mathbf{g}}{dt} \right\rangle.$$

In particular, if $|\mathbf{f}|$ is constant then $d\mathbf{f}/dt$ is perpendicular to $\mathbf{f}$.

Proof: Problem 5.1. ▮

LEMMA 5.2. Let $\mathbf{f}, \mathbf{g} : \mathbf{R} \longrightarrow \mathbf{R}^3$. Then

$$\frac{d}{dt}(\mathbf{f} \times \mathbf{g}) = \frac{d\mathbf{f}}{dt} \times \mathbf{g} + \mathbf{f} \times \frac{d\mathbf{g}}{dt}.$$

Proof: Problem 5.2. ▮

DEFINITION. $f: \mathbf{R} \longrightarrow \mathbf{R}$ is of *class* C^k if all derivatives up through order k exist and are continuous. $f: \mathbf{R}^n \longrightarrow \mathbf{R}$ is of *class* C^k if all its (mixed) partial derivatives of order k or less exist and are continuous. A vector-valued function is of *class* C^k if all its components with respect to a given basis are of class C^k.

Note that if f is of class C^k it is also of class C^{k-1}.

Rather than continually worrying about what class a differentiable function belongs to, we shall usually assume it is of class C^3. We shall point out those cases where higher class is needed or lower class is sufficient.

Finally, before we start our study of curves, we want to remind you what form the chain rule for differentiation takes. Suppose that $\mathbf{x}$ is a function of several variables $u^1, u^2, \ldots, u^n$ and that the u^i are functions of variables $v^1, v^2, \ldots, v^m$. Then

$$\frac{\partial \mathbf{x}}{\partial v^\alpha} = \sum_{i=1}^{n} \frac{\partial \mathbf{x}}{\partial u^i} \frac{\partial u^i}{\partial v^\alpha}, \qquad \alpha = 1, 2, \ldots, m. \tag{5-1}$$

Note that we are writing the coefficients on the right of the vectors instead of the left as would be usual in linear algebra. This is done so that Equations (5-1) look more like the chain rule from calculus.

Special cases that we shall often use arise when $n = m = 2$ as in Equation (5-2), or $n = 2$ and $m = 1$, as in Equation (5-3).

$$\frac{\partial \mathbf{x}}{\partial v^\alpha} = \frac{\partial \mathbf{x}}{\partial u^1} \frac{\partial u^1}{\partial v^\alpha} + \frac{\partial \mathbf{x}}{\partial u^2} \frac{\partial u^2}{\partial v^\alpha}, \qquad \alpha = 1, 2 \tag{5-2}$$

$$\frac{d\mathbf{x}}{dt} = \frac{\partial \mathbf{x}}{\partial u^1} \frac{du^1}{dt} + \frac{\partial \mathbf{x}}{\partial u^2} \frac{du^2}{dt}. \tag{5-3}$$

PROBLEMS

***5.1.** Prove Lemma 5.1.

***5.2.** Prove Lemma 5.2.

Set A—Hyperbolic Functions

For some of the examples and problems in this book it will be necessary to use certain transcendental functions which behave in many ways like the trigonometric functions. Since they are not always covered in a calculus course we briefly cover them in this problem set. The *hyperbolic sine function* is the function $\sinh : \mathbf{R} \longrightarrow \mathbf{R}$ given by $\sinh(t) = (e^t - e^{-t})/2$. The *hyperbolic cosine function* is the function $\cosh : \mathbf{R} \longrightarrow \mathbf{R}$ given by $\cosh(t) = (e^t + e^{-t})/2$. For aid in reading, sinh is pronounced "cinch" while "cosh" rhymes with "gosh."

5.3. Prove that for any $t \in \mathbf{R}$, $\cosh^2(t) - \sinh^2(t) = 1$. (Since $x^2 - y^2 = 1$ is a hyperbola, this gives the origin of the term "hyperbolic function." Also, you should think about the analogy between this equation and $\cos^2 \theta + \sin^2 \theta = 1$.)

5.4. Prove that $\cosh'(t) = \sinh(t)$, and $\sinh'(t) = \cosh(t)$ for all $t \in \mathbf{R}$.

5.5. Using Problem 5.4 show that sinh is one-to-one. Show that cosh is one-to-one when restricted to $[0, \infty)$. Show also that sinh is onto the reals whereas $\cosh(\mathbf{R}) = [1, \infty)$. (This means we may define the appropriate inverse function.)

5.6. Show that the general solution to the differential equation $f'' = a^2 f$ is given by $f(t) = A \cosh(at) + B \sinh(at)$ where A and B are real constants. (If you cannot show this is the general solution, at least show that it satisfies the differential equation.)

5.7. Compute $\int dx/\sqrt{1 + x^2}$ by means of the substitution $x = \sinh(t)$.

2

Local Curve Theory

We shall begin our study of differential geometry with an investigation of curves in three-dimensional Euclidean space, $\mathbf{R}^3$. This is a good place to start for four reasons: (1) since curves in $\mathbf{R}^3$ are easy to draw and visualize, we can develop some geometric insight into the subject by looking at examples; (2) it is a very complete subject—the Frenet-Serret apparatus of a curve completely determines the local geometry of curves and gives a complete set of invariants for the problem of determining whether two curves are the same; (3) the theory of curves will introduce us to some techniques that are the mainstay of modern differential geometry (e.g., linear algebra); (4) we shall base our study of surfaces in Chapter 4 on the behavior of curves on the surfaces.

The history of the theory of curves (and of all differential geometry) is a fascinating one. Suffice it to say at this point that the many results in the theory of curves in $\mathbf{R}^3$, which we discuss in this chapter, were initiated by G. Monge (1746–1818) and his school (Meusnier, Lancret, and Dupin). Our approach is due to G. Darboux (1842–1917) whose idea of moving frames unified a great deal of the classical theory of curves. He accomplished this in 1887–1896. It is interesting to note that the approach to the theory of surfaces that we will take in Chapter 4 is that of K. Gauss in 1827. This anomaly of dates is not due to the fact that the curve theory is more difficult than the theory of surfaces (just the opposite is true) but rather to the pervasive genius of Gauss. The history of the theory of curves is discussed more completely in the historical notes at the end of this book.

There are two ways to think of curves. The first is as a geometric set of points, or locus. When this is the case, we refer to a *geometric curve*, or the *geometric shape* of the curve. Intuitively we are thinking of a curve as the path traced out by a particle moving in $\mathbf{R}^3$. The second way of thinking of a curve is as a function of some parameter, say t. Intuitively it is not always enough to know where a particle went—we also want to know when it got there. (The parameter t is often thought of as time.) It is necessary to view curves the second way if we are to apply the techniques of calculus to describe the geometric behavior of a curve. This means that we must pay careful attention to how the curve is parametrized (e.g., if you change the parameter you also change the velocity vector field to the curve). However, we are also interested in geometric properties of curves (e.g., arc length, tangent vector field). These should not depend on the way a curve is parametrized as a function but only on the image set of the function, that is, only on the geometric shape. Thus we shall ask whether our constructions and descriptions depend upon the parametrizations.

In Section 2-1 we define and give examples of parametrized curves. In Section 2-2 we introduce a particularly useful parametrization—that by arc length. Section 2-3 develops the Frenet-Serret apparatus which is the basic tool in the study of curves. It consists of three vector fields along the given curve (the tangent $\mathbf{T}$, the normal $\mathbf{N}$, and the binormal $\mathbf{B}$) and two scalar-valued functions (the curvature κ and the torsion τ). The Frenet-Serret Theorem is proved in the fourth section. This theorem expresses the derivatives of $\mathbf{T}$, $\mathbf{N}$, and $\mathbf{B}$ in terms of $\mathbf{T}$, $\mathbf{N}$, and $\mathbf{B}$. We then make several applications. There is also a long collection of problems at the end of this section, many of which deal with topics from the classical differential geometry of curves. Section 2-5 gives the Fundamental Theorem of Curves and shows that the Frenet-Serret apparatus does completely determine the geometry of the curve. Finally, in Section 2-6 we develop the necessary techniques for computing the Frenet-Serret apparatus for curves which are not parametrized by arc length.

2–1. BASIC DEFINITIONS AND EXAMPLES

Our study of curves will be restricted to a certain class of curves in $\mathbf{R}^3$. Not only do we want a curve to be described by a differentiable function so that we may use calculus to describe the geometry, we also want to avoid certain pathologies and technicalities. If $d\boldsymbol{\alpha}/dt = \mathbf{0}$ on an interval, then $\boldsymbol{\alpha}(t)$ is constant in that interval, which is geometrically very uninteresting. If

$d\boldsymbol{\alpha}/dt$ is $\mathbf{0}$ at some point, then the graph of $\boldsymbol{\alpha}$ can have a sharp corner, which is geometrically unappealing. (Consider the graph of $\boldsymbol{\alpha}(t) = (t^2, t^3, 0)$.) Because of these considerations we will only work with regular curves.

DEFINITION. A *regular curve* in $\mathbf{R}^3$ is a function $\boldsymbol{\alpha} : (a, b) \longrightarrow \mathbf{R}^3$ which is of class C^k for some $k \geq 1$ and for which $d\boldsymbol{\alpha}/dt \neq \mathbf{0}$ for all $t \in (a, b)$.

In this text, a regular curve will be assumed to be of class C^3 unless stated otherwise.

Note that from this point of view the curve is the function and *not* the image set (geometric curve). Two different curves may have the same image set (see Examples 1.2 and 1.4 below). A regular curve need not be one-to-one, but as Problem 1.9 shows, it cannot intersect itself too often.

Given a regular curve $\boldsymbol{\alpha}(t)$, we can define some vector fields along $\boldsymbol{\alpha}$. This means that for each t we will have a 3-vector $\mathbf{v}(t)$. The reader should think of the tail of $\mathbf{v}(t)$ to be at the point $\boldsymbol{\alpha}(t)$. The mapping $t \longrightarrow \mathbf{v}(t)$ is a vector-valued function and so we may use the material of Section 1-5.

DEFINITION. The *velocity vector* of a regular curve $\boldsymbol{\alpha}(t)$ at $t = t_0$ is the derivative $d\boldsymbol{\alpha}/dt$ evaluated at $t = t_0$. The *velocity vector field* is the vector-valued function $d\boldsymbol{\alpha}/dt$. The *speed* of $\boldsymbol{\alpha}(t)$ at $t = t_0$ is the length of the velocity vector at $t = t_0$, $|(d\boldsymbol{\alpha}/dt)(t_0)|$.

If we view the curve as the path of a moving particle, the velocity vector at $t = t_0$ points in the direction that the particle is moving at time t_0. (See Figure 2.1.) The regularity condition says that the speed is always nonzero—the particle never stops moving, even instantaneously.

DEFINITION. The *tangent vector field* to a regular curve $\boldsymbol{\alpha}(t)$ is the vector-valued function $\mathbf{T}(t) = (d\boldsymbol{\alpha}/dt)/|d\boldsymbol{\alpha}/dt|$.

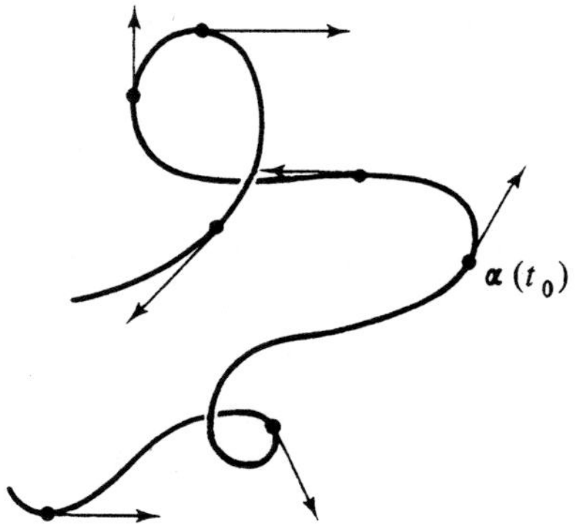

FIGURE 2.1

Note that we are able to define $\mathbf{T}$ (i.e., divide by $|d\boldsymbol{\alpha}/dt|$) precisely because of the regularity condition. $\mathbf{T}$ is the unit vector in the direction of the velocity vector.

We shall see later in this section that $\mathbf{T}$ is a geometric quantity: it depends only on the image set of $\boldsymbol{\alpha}$ and not the particular way this set is parametrized.

For each value of t, say t_0, there is a unique straight line through $\boldsymbol{\alpha}(t_0)$ parallel to $\mathbf{T}(t_0)$. This line is a linear approximation of the curve near $\boldsymbol{\alpha}(t_0)$. (This is an example of one of the basic techniques in differential geometry: an object of study (a curve) is replaced by a linear approximation (a tangent line). This is done because linear mathematics is so much better understood than nonlinear mathematics.) More formally:

Definition. The *tangent line* to a regular curve $\boldsymbol{\alpha}$ at the point $t = t_0$ is the straight line

$$l = \{\mathbf{w} \in \mathbf{R}^3 \,|\, \mathbf{w} = \boldsymbol{\alpha}(t_0) + \lambda \mathbf{T}(t_0), \lambda \in \mathbf{R}\}.$$

Note that the tangent line is a subset of $\mathbf{R}^3$ which contains the point $\boldsymbol{\alpha}(t_0)$ and actually is a straight line. Intuitively it is the line that most nearly approximates the curve near $\boldsymbol{\alpha}(t_0)$. (See Figure 2.2.)

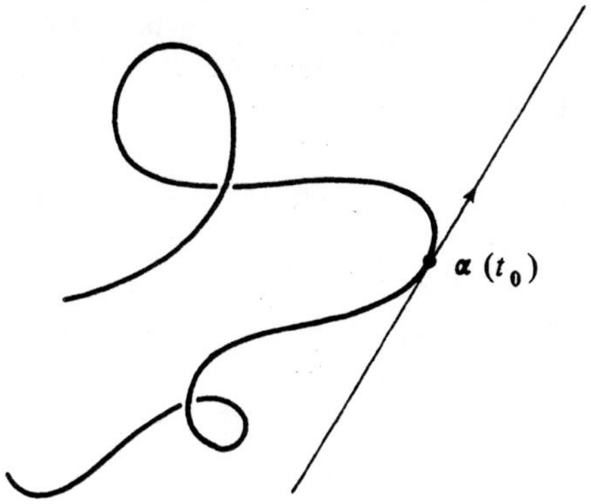

FIGURE 2.2

Since $d\boldsymbol{\alpha}/dt \neq \mathbf{0}$ and $d\boldsymbol{\alpha}/dt = |d\boldsymbol{\alpha}/dt|\mathbf{T}$, the tangent line at $t = t_0$ is also given by

$$\left\{\mathbf{w} \in \mathbf{R}^3 \,|\, \mathbf{w} = \boldsymbol{\alpha}(t_0) + \mu \frac{d\boldsymbol{\alpha}}{dt}(t_0), \mu \in \mathbf{R}\right\}.$$

Example 1.1. Let $\mathbf{u}$ and $\mathbf{v}$ be fixed vectors in $\mathbf{R}^3$. Then the curve $\boldsymbol{\alpha}: \mathbf{R} \longrightarrow \mathbf{R}^3$ given by $\boldsymbol{\alpha}(t) = \mathbf{u} + t\mathbf{v}$ is a regular curve if and only if $\mathbf{v} \neq \mathbf{0}$. In this case it is a straight line and $d\boldsymbol{\alpha}/dt = \mathbf{v}$. The tangent line at each point is the given straight line and $\mathbf{T} = \mathbf{v}/|\mathbf{v}|$.

Example 1.2. Let $\boldsymbol{\alpha}: \mathbf{R} \longrightarrow \mathbf{R}^3$ be given by $\boldsymbol{\alpha}(t) = (t, 0, 0)$. This is a special case of Example 1.1.

EXAMPLE 1.3. Let $\boldsymbol{\alpha} : \mathbf{R} \longrightarrow \mathbf{R}^3$ be given by $\boldsymbol{\alpha}(t) = (t^3, 0, 0)$. $d\boldsymbol{\alpha}/dt = (3t^2, 0, 0)$, which is zero at $t = 0$. $\boldsymbol{\alpha}$ is not a regular curve, even though its image is the same as the curve in Example 1.2!

EXAMPLE 1.4. Let $\boldsymbol{\alpha} : \mathbf{R} \longrightarrow \mathbf{R}^3$ be given by $\boldsymbol{\alpha}(t) = (t^3 + t, 0, 0)$.

$$\frac{d\boldsymbol{\alpha}}{dt} = (3t^2 + 1, 0, 0) \neq \mathbf{0}.$$

This is a regular curve whose image set is the same as the curve in Example 1.2.

EXAMPLE 1.5. Let $g: \mathbf{R} \longrightarrow \mathbf{R}$ be a differentiable function. Let $\boldsymbol{\alpha}(t) = (t, g(t), 0)$. Then $d\boldsymbol{\alpha}/dt = (1, g'(t), 0) \neq \mathbf{0}$ and $\boldsymbol{\alpha}$ is a regular curve. $\boldsymbol{\alpha}(t)$ is the graph of g except for the extra (third) coordinate. The tangent line at $t = t_0$ is $\{(t_0 + \lambda, g(t_0) + \lambda g'(t_0), 0) \mid \lambda \in \mathbf{R}\}$. In terms of t, y, z coordinates this line is $z = 0$, $y - g(t_0) = (t - t_0)g'(t_0)$, which should be familiar from Calculus I as the equation of the tangent line to the graph of $g(t)$ at $t = t_0$. (Remember $t = x$ in this example.)

EXAMPLE 1.6. Let $\boldsymbol{\alpha} : \mathbf{R} \longrightarrow \mathbf{R}^3$ be given by $\boldsymbol{\alpha}(t) = (r \cos t, r \sin t, ht)$, where $h > 0$ and $r > 0$. This is called a *right circular helix*. (If $h < 0$ it would be a *left circular helix*.) Circular refers to the fact that the projection in the (x, y) plane is a circle. Since $d\boldsymbol{\alpha}/dt = (-r \sin t, r \cos t, h) \neq \mathbf{0}$, $\boldsymbol{\alpha}$ is a regular curve. At $t = t_0$, $\mathbf{T} = (-r \sin t_0, r \cos t_0, h)/\sqrt{r^2 + h^2}$. (See Figure 2.3.)

FIGURE 2.3

In Examples 1.2 and 1.3 we saw a situation where the same image set (the x-axis) was given two different parametrizations, one of which was not regular. We wish to know what parametrizations can be used to describe a given image curve.

DEFINITION. A *reparametrization* of a curve $\boldsymbol{\alpha} : (a, b) \longrightarrow \mathbf{R}^3$ is a one-to-one onto function $g : (c, d) \longrightarrow (a, b)$ such that both g and its inverse $h : (a, b) \longrightarrow (c, d)$ are of class C^k for some $k \geq 1$.

What we have in mind is the new curve $\boldsymbol{\beta} = \boldsymbol{\alpha} \circ g$. If r denotes the variable in the interval (c, d), then $d\boldsymbol{\beta}/dr = (d\boldsymbol{\alpha}/dt)(dg/dr)$ by the chain rule. $\boldsymbol{\beta}$ is thus regular if $\boldsymbol{\alpha}$ is regular *and* $dg/dr \neq 0$. But $g(h(t)) = t$ so that by the chain rule $(dg/dr)(dh/dt) = 1$ and $dg/dr \neq 0$. Thus the composition of a regular curve with a reparametrization yields a regular curve. Note that if $\boldsymbol{\alpha}$ is of class C^m and g is of class C^k, then $\boldsymbol{\beta}$ is of class C^n with $n = \min(k, m)$. (See Figure 2.4.)

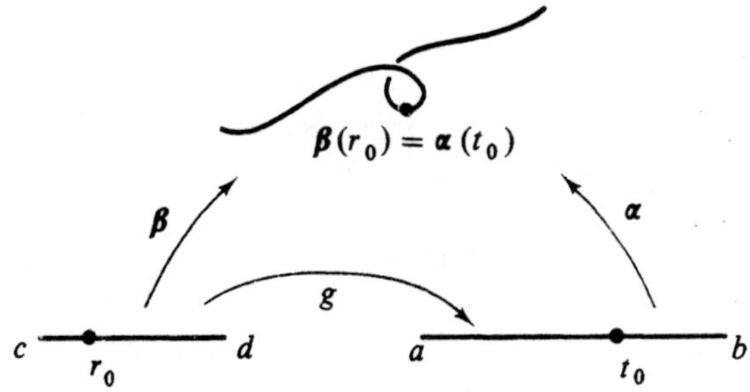

FIGURE 2.4

The image of a curve and any of its reparametrizations are the same. This means that any quantity which stays the same when we change parameters (i.e., make a reparametrization) is a quantity which depends only on the geometric shape of the curve. Briefly, the quantity is a geometric invariant. We will show below that the tangent line to a regular curve is such a geometric invariant after we give some examples.

EXAMPLE 1.7. Let $g: (0, 1) \longrightarrow (1, 2)$ be given by $g(r) = 1 + r^2$. g is one-to-one and onto with inverse $h(t) = \sqrt{t - 1}$. g is infinitely differentiable on $(0, 1)$ and so is h on $(1, 2)$. Thus g is a reparametrization of any regular curve on $(1, 2)$.

EXAMPLE 1.8. Let $g: \mathbf{R} \longrightarrow \mathbf{R}$ be given by $g(r) = r^3$. g is one-to-one, onto, and infinitely differentiable. However, $h(t) = t^{1/3}$ is the inverse and $h'(0)$ does not exist, so that h is not C^1. This is one reason why Example 1.3 was not a regular curve.

Now we shall show that the tangent vector field is a geometric property of the image set of a regular curve and does not depend on the parametrization. This means that the tangent line to a curve is a geometric property also.

PROPOSITION 1.9. Let $\boldsymbol{\alpha}: (a, b) \longrightarrow \mathbf{R}^3$ be a regular curve and let

$$g: (c, d) \longrightarrow (a, b)$$

be a reparametrization. Set $\boldsymbol{\beta} = \boldsymbol{\alpha} \circ g$. If $t_0 = g(r_0)$, the tangent vector field $\mathbf{T}$ of $\boldsymbol{\alpha}$ at t_0 and the tangent vector field $\mathbf{S}$ of $\boldsymbol{\beta}$ at r_0 satisfy $\mathbf{S} = \pm\mathbf{T}$.

Proof:

$$\mathbf{S} = \frac{\dfrac{d\boldsymbol{\beta}}{dr}}{\left|\dfrac{d\boldsymbol{\beta}}{dr}\right|} = \frac{\dfrac{d\boldsymbol{\alpha}}{dt}\dfrac{dg}{dr}}{\left|\dfrac{d\boldsymbol{\alpha}}{dt}\right|\left|\dfrac{dg}{dr}\right|}$$

$$= \frac{\dfrac{d\boldsymbol{\alpha}}{dt}}{\left|\dfrac{d\boldsymbol{\alpha}}{dt}\right|}\frac{\dfrac{dg}{dr}}{\left|\dfrac{dg}{dr}\right|} = (\mathbf{T})(\pm 1) = \pm\mathbf{T}. \quad \blacksquare$$

Note that $\mathbf{S} = \mathbf{T}$ if $dg/dr > 0$ (g is increasing) and $\mathbf{S} = -\mathbf{T}$ if $dg/dr < 0$ (g is decreasing). Geometrically the difference is whether $\boldsymbol{\alpha}$ and $\boldsymbol{\beta}$ indicate particles moving along the image curve in the same or opposite directions.

PROBLEMS

1.1. (a) Show that $\boldsymbol{\alpha}(t) = (\sin 3t \cos t, \sin 3t \sin t, 0)$ is a regular curve.
(b) Find the equation of the tangent line to $\boldsymbol{\alpha}$ at $t = \pi/3$.

1.2. (a) Which of the following are regular curves?
(i) $\boldsymbol{\alpha}(\theta) = (\cos\theta, 1 - \cos\theta - \sin\theta, -\sin\theta)$.
(ii) $\boldsymbol{\beta}(\theta) = (2\sin^2\theta, 2\sin^2\theta\tan\theta, 0)$.
(iii) $\boldsymbol{\gamma}(\theta) = (\cos\theta, \cos^2\theta, \sin\theta)$.
(b) Find the tangent line to each of the above curves at $\theta = \pi/4$.

***1.3.** (a) In Example 1.6, what is the equation of the tangent line at $t = t_0$?
(b) Show that the angle between $(0, 0, 1) = \mathbf{u}$ and $d\boldsymbol{\alpha}/dt$ in Example 1.6 is a constant (i.e., independent of t).

1.4. Show that $f: (-1, 1) \to (-\infty, \infty)$ given by $f(t) = \tan(\pi t/2)$ is a reparametrization.

1.5. Let $g: (0, \infty) \to (0, 1)$ be given by $g(r) = r^2/(r^2 + 1)$. Is this a reparametrization?

1.6. Let $\boldsymbol{\alpha}(\theta) = (e^\theta\cos\theta, e^\theta\sin\theta, 0)$. Prove that the angle between $\boldsymbol{\alpha}$ and $\mathbf{T}$ is constant. (A curve with this property is called a *logarithmic spiral.*)

1.7. Let $\boldsymbol{\alpha}(t)$ be a regular curve. Suppose there is a point $\mathbf{a} \in \mathbf{R}^3$ such that $\boldsymbol{\alpha}(t) - \mathbf{a}$ is orthogonal to $\mathbf{T}(t)$ for all t. Prove that $\boldsymbol{\alpha}(t)$ lies on a sphere. (*Hint:* What should be the center of the sphere?)

***1.8.** Consider the function $\boldsymbol{\alpha}: \mathbf{R} \to \mathbf{R}^3$ by $\boldsymbol{\alpha}(t) = (t^2, t^3, 0)$.
(a) Show that $\boldsymbol{\alpha}$ is C^1 but not regular.
(b) Show that the image of $\boldsymbol{\alpha}$ has a sharp corner by graphing $\boldsymbol{\alpha}$.

***1.9.** Let $\boldsymbol{\alpha}: (a, b) \to \mathbf{R}^3$ be a differentiable curve. Suppose there is a sequence of points $\{t_n\}$ in the interval (a, b) such that the t_n are all distinct, $\lim_{n\to\infty} t_n = t^* \in (a, b)$, and $\boldsymbol{\alpha}(t_n) = \mathbf{x}_0$ for all n (thus "$\boldsymbol{\alpha}$ intersects itself infinitely often at $\mathbf{x}_0$"). Show that $\boldsymbol{\alpha}$ is not regular. (*Hint:* Show that $d\boldsymbol{\alpha}/dt = \mathbf{0}$ at t^*.) This shows that on any finite closed interval, a regular curve can intersect itself only a finite number of times at the same point.

2–2. ARC LENGTH

Sometimes it is useful for technical reasons to consider curves with end points, that is, curves defined on closed intervals:

DEFINITION. A *regular curve segment* is a function $\boldsymbol{\alpha} : [a, b] \longrightarrow \mathbf{R}^3$ together with an open interval (c, d), with $c < a < b < d$, and a regular curve $\boldsymbol{\gamma} : (c, d) \longrightarrow \mathbf{R}^3$ such that $\boldsymbol{\alpha}(t) = \boldsymbol{\gamma}(t)$ for all $t \in [a, b]$.

Thus a curve segment is a curve defined on a closed interval which can be extended to a curve on a slightly larger open interval. In this case it is possible to talk about $d\boldsymbol{\alpha}/dt$ at the end points of the curve segment because we define $(d\boldsymbol{\alpha}/dt)(a)$ to be $(d\boldsymbol{\gamma}/dt)(a)$ and $(d\boldsymbol{\alpha}/dt)(b)$ to be $(d\boldsymbol{\gamma}/dt)(b)$.

Now we shall define the length of a curve segment. The intuitive justification for what we do is as follows. If $\boldsymbol{\alpha}(t)$ is viewed as the path of a particle moving in space, then $|d\boldsymbol{\alpha}/dt|$ is the speed of the particle as a function of time. The integral of speed should be the distance traveled by the particle just as it is in one dimension.

DEFINITION. The *length* of a regular curve segment $\boldsymbol{\alpha} : [a, b] \longrightarrow \mathbf{R}^3$ is

$$\int_a^b \left| \frac{d\boldsymbol{\alpha}}{dt} \right| dt.$$

Note that this is really the familiar formula for the length of a curve in $\mathbf{R}^3$: if $\boldsymbol{\alpha}(t) = (x(t), y(t), z(t))$, then $|d\boldsymbol{\alpha}/dt| = \sqrt{(x')^2 + (y')^2 + (z')^2}$, so that the length of the curve is given by $\int_a^b \sqrt{(x')^2 + (y')^2 + (z')^2}\, dt$.

It makes sense to talk about reparametrizations of curve segments (see Problem 2.7). One would hope that the length of a curve is a geometric property and does not depend on the choice of parametrization. This is the content of the next proposition.

PROPOSITION 2.1. Let $g : [c, d] \longrightarrow [a, b]$ be a reparametrization of a curve segment $\boldsymbol{\alpha} : [a, b] \longrightarrow \mathbf{R}^3$. Then the length of $\boldsymbol{\alpha}$ is equal to the length of $\boldsymbol{\beta} = \boldsymbol{\alpha} \circ g$.

Proof: The length of $\boldsymbol{\beta}$ is

$$\int_c^d \left|\frac{d\boldsymbol{\beta}}{dr}\right| dr = \int_c^d \left|\left(\frac{d\boldsymbol{\alpha}}{dt}\right)\left(\frac{dg}{dr}\right)\right| dr$$
$$= \int_c^d \left|\frac{d\boldsymbol{\alpha}}{dt}\right|\left|\frac{dg}{dr}\right| dr.$$

Case 1: If $dg/dr > 0$, then $|dg/dr| = dg/dr$, $g(c) = a$, $g(d) = b$, and $\int_c^d |d\boldsymbol{\alpha}/dt|\,|dg/dr|\,dr = \int_c^d |d\boldsymbol{\alpha}/dt|\,(dg/dr)\,dr = \int_a^b |d\boldsymbol{\alpha}/dt|\,dt$ by the substitution rule of integral calculus.

Case 2: If $dg/dr < 0$, the proof is similar, using $|dg/dr| = -dg/dr$, $g(c) = b$, and $g(d) = a$.

In both cases we have that the length of $\boldsymbol{\alpha}$ equals the length of $\boldsymbol{\beta}$. ∎

Note that the definition of the length of a curve does not really require $\boldsymbol{\alpha}$ to be regular to make sense—it is sufficient for $\boldsymbol{\alpha}$ to be of class C^1. However, if $\boldsymbol{\alpha}$ is not regular, some segments of the curve may be traversed twice and the formula will count the doubled section twice.

Using the concept of length of a curve, we are able to define an important way to reparametrize a curve. Let $\boldsymbol{\alpha}: (a, b) \longrightarrow \mathbf{R}^3$ be a regular curve and let $t_0 \in (a, b)$. Set $h(t) = \int_{t_0}^t |d\boldsymbol{\alpha}/dt|\,dt$. $s = h(t)$ is called *arc length along* $\boldsymbol{\alpha}$. It actually measures signed arc length along $\boldsymbol{\alpha}$ from $\boldsymbol{\alpha}(t_0)$ with $h(t) < 0$ if $t < t_0$ and $h(t) > 0$ if $t > t_0$.

THEOREM 2.2. h is a one-to-one function mapping (a, b) onto some interval (c, d) and is a reparametrization.

Proof: By the fundamental theorem of calculus, $dh/dt = |d\boldsymbol{\alpha}/dt| > 0$ (since $\boldsymbol{\alpha}$ is regular). Thus h is increasing and so is one-to-one. It is easy to check that if $\boldsymbol{\alpha}$ is of class C^k so is h. Let $g: (c, d) \longrightarrow (a, b)$ be the inverse of h and denote the parameter in (c, d) by s. This means of course that $g(s) = t$ if and only if $h(t) = s$ so that s is the arc length parameter. Because g and h are inverse functions $dg/ds = 1/(dh/dt)$, where the right-hand side is evaluated at $t = g(s)$. This quotient makes sense since $dh/dt \neq 0$. g can be differentiated as often as h can. Thus h is a reparametrization. ∎

As was pointed out in the proof of the above theorem, s is the arc length. By using $g(s)$, any regular curve $\boldsymbol{\alpha}$ can be reparametrized in terms of arc length from a point. Once this has been done we say that the curve has been *parametrized by arc length.* The importance of a curve being parametrized by arc length is carried in the observation that its velocity vector field is its tangent vector field, as may be seen in the following computation.

If $\boldsymbol{\beta}(s)$ is parametrized by arc length, then $s = \int_0^s |d\boldsymbol{\beta}/d\sigma|\,d\sigma$. By the fundamental theorem of calculus, $1 = (d/ds)(\int_0^s |d\boldsymbol{\beta}/d\sigma|\,d\sigma) = |d\boldsymbol{\beta}/d\sigma|$ at

$\sigma = s$; that is, $1 = |d\boldsymbol{\beta}/ds|$. Hence the velocity vector field $d\boldsymbol{\beta}/ds$ is a unit vector field and is thus $\mathbf{T}$. When a curve $\boldsymbol{\beta}$ is parametrized by arc length (or equivalently if its velocity vector field is $\mathbf{T}$) we say that $\boldsymbol{\beta}$ is a *unit speed curve.*

The preceding paragraph and the proof of Theorem 2.2 contain some important facts which we now isolate. We shall assume that 0 is in the domain of $\boldsymbol{\alpha}(t)$ and base arc length at 0.

COROLLARY 2.3. If $\boldsymbol{\alpha}(t)$ is a regular curve and $s = s(t)$ is its arc length, then
(a) $s = s(t) = \int_0^t |d\boldsymbol{\alpha}/dt|\, dt$;
(b) $ds/dt = |d\boldsymbol{\alpha}/dt|$;
(c) $d\boldsymbol{\alpha}/dt = (ds/dt)\mathbf{T}$; and
(d) $\mathbf{T} = d\boldsymbol{\alpha}/ds$.

To take a given regular curve $\boldsymbol{\alpha}$ and reparametrize it by arc length, while always possible in theory, may be very difficult in practice. There are two obstacles to such a program. In the first place, the integral

$$h(t) = \int_{t_0}^{t} \left| \frac{d\boldsymbol{\alpha}}{dt} \right| dt$$

may not be elementary (see Example 2.6) and hence not computable. Secondly, even if $h(t)$ can be determined, it may not be possible to find the inverse function $g(s)$ (see Example 2.7).

EXAMPLE 2.4. Let $\boldsymbol{\alpha}(t) = \mathbf{u} + t\mathbf{v}$ be the straight line of Example 1.1. $d\boldsymbol{\alpha}/dt = \mathbf{v}$, $s = h(t) = \int_0^t |\mathbf{v}|\, dt = t\,|\mathbf{v}|$. Thus $t = g(s) = s/|\mathbf{v}|$. $\boldsymbol{\beta}(s) = \mathbf{u} + s\mathbf{v}/|\mathbf{v}|$ gives the unit speed parametrization of a straight line. Note that the tangent vector field to $\mathbf{x}$ is $\mathbf{T} = \mathbf{v}/|\mathbf{v}|$, and $d\mathbf{T}/ds = \mathbf{0}$.

EXAMPLE 2.5. Let $\boldsymbol{\alpha}(t) = (r \cos t, r \sin t, 0)$ with $r > 0$.

$$\frac{d\boldsymbol{\alpha}}{dt} = (-r \sin t, r \cos t, 0)$$

and $|d\boldsymbol{\alpha}/dt| = r$. $s = h(t) = rt$ and $t = g(s) = s/r$.

$$\boldsymbol{\beta}(s) = \left(r \cos\left(\frac{s}{r}\right), r \sin\left(\frac{s}{r}\right), 0\right)$$

is the unit speed parametrization of a circle of radius r. Note that the tangent vector field of $\boldsymbol{\alpha}$ is

$$\mathbf{T}(s) = \left(-\sin\left(\frac{s}{r}\right), \cos\left(\frac{s}{r}\right), 0\right)$$

and that

$$\frac{d\mathbf{T}}{ds} = \left(-\frac{1}{r}\cos\left(\frac{s}{r}\right), -\frac{1}{r}\sin\left(\frac{s}{r}\right), 0\right)$$

has length $1/r$.

EXAMPLE 2.6. If $\boldsymbol{\alpha}(t)$ is the ellipse $(2 \sin t, \cos t, 0)$, then

$$\frac{d\boldsymbol{\alpha}}{dt} = (2 \cos t, -\sin t, 0).$$

$$\left|\frac{d\boldsymbol{\alpha}}{dt}\right| = \sqrt{4 \cos^2 t + \sin^2 t}$$
$$= \sqrt{4 - 3 \sin^2 t} = 2\sqrt{1 - (\tfrac{3}{4}) \sin^2 t}.$$

But $\sqrt{1 - \frac{3}{4} \sin^2 t}$ does not have an elementary antiderivative and so the integration $h(t) = \int_0^t |d\boldsymbol{\alpha}/dt|\, dt$ cannot be carried out by using the fundamental theorem of calculus. (Definite integrals of this kind are called *elliptic integrals* because they can be interpreted as the arc length of an ellipse. Their values are tabulated in many books of mathematical tables.)

EXAMPLE 2.7. Let $\boldsymbol{\alpha}(t) = (t, t^2/2, 0)$. Then $d\boldsymbol{\alpha}/dt = (1, t, 0)$ and

$$\left|\frac{d\boldsymbol{\alpha}}{dt}\right| = \sqrt{1 + t^2}.$$

Hence

$$s = h(t) = \int_0^t \sqrt{1 + \tau^2}\, d\tau = \tfrac{1}{2}(t\sqrt{1 + t^2} + \ln (t + \sqrt{1 + t^2})).$$

However, it is extremely difficult to find $t = g(s)$ from this equation. Note that $\boldsymbol{\alpha}$ is a parabola, a very simple curve geometrically!

What have we accomplished? Suppose that we want to study regular curves and we are interested only in their geometric shape (that is, we don't care about parametrization). Theorem 2.2 says that we may as well assume that the curve is parametrized by arc length. This will be a very useful technical device in setting up the Frenet-Serret apparatus in the next section.

PROBLEMS

2.1. Find the arc length of the circular helix in Example 1.6 for $0 \leq t \leq 10$.

2.2. Find the arc length of $\boldsymbol{\alpha}(t) = (2 \cosh 3t, -2 \sinh 3t, 6t)$ for $0 \leq t \leq 5$.

2.3. Reparametrize the right circular helix of Example 1.6 by arc length.

2.4. Reparametrize the curve $\boldsymbol{\alpha}(t) = (e^t \cos t, e^t \sin t, e^t)$ by arc length.

2.5. Reparametrize the curve $\boldsymbol{\alpha}(t) = (\cosh t, \sinh t, t)$ by arc length.

2.6. Show that $\boldsymbol{\alpha}(s)$ is a unit speed curve where

$$\boldsymbol{\alpha}(s) = \tfrac{1}{2}(s + \sqrt{s^2 + 1}, (s + \sqrt{s^2 + 1})^{-1}, \sqrt{2} \ln (s + \sqrt{s^2 + 1})).$$

***2.7.** Formulate an appropriate definition of a reparametrization of a curve segment.

2.8. Let $\boldsymbol{\alpha}(t)$ be a regular curve with $|d\boldsymbol{\alpha}/dt| = a$, where a is a fixed positive constant. Show that if s is arc length measured from some point, then $t = (s/a) + c$ for some constant c.

2–3. CURVATURE AND THE FRENET-SERRET APPARATUS

This section is devoted almost entirely to making definitions and developing an intuitive feeling for these definitions. The reader should look again at the last paragraph of the previous section before starting on this section.

DEFINITION. A curve $\boldsymbol{\alpha} : (a, b) \longrightarrow \mathbf{R}^3$ is a *unit speed curve* if $|d\boldsymbol{\alpha}/dt| = 1$.

Note that for a unit speed curve $\boldsymbol{\alpha} = \boldsymbol{\alpha}(t)$, the arc length $s = t - t_0$. We shall assume that t_0 has been chosen to be 0 (so that $s = t$) and will write $\boldsymbol{\alpha}$ as a function of s. Because of this convention we can write unambiguously $\boldsymbol{\alpha}' = \boldsymbol{\alpha}'(s) = \mathbf{T}(s)$ where $\mathbf{T}$ is, as always, the tangent vector field (see Corollary 2.3). For the rest of this section we shall assume that $\boldsymbol{\alpha}(s)$ is a unit speed curve. This assumption amounts to the philosophical statement that we are only interested in the geometric shape of a regular curve since any regular curve can be reparametrized by arc length (Theorem 2.2) and reparametrizing does not change the shape of a curve. In Section 2-6 we will compute the quantities defined below if the curve is not given in a unit speed parametrization.

We now motivate the definition of "curvature" (of a curve). "Curvature" will measure bending and will serve as the central concept of study in this book (and, indeed, in all of differential geometry). The reader probably has some intuitive idea of what "curvature" is. Whatever the definition of "curvature" is, it should satisfy two criteria: (1) the curvature of a straight line (Example 2.4) is zero; and (2) the curvature of a circle (Example 2.5) is the same at each point. In terms of the curvature measuring bending, (1) says that a straight line does not bend at all and (2) says that a circle has constant bending. What is it about a straight line that does not change (i.e., what might we choose to be a measure of bending?)? A glance at Example 2.4 ($\boldsymbol{\alpha}(t) = \mathbf{u} + t\mathbf{v}$) shows that the *tangent vector field of a straight line does not change with the arc length s*. It is $\mathbf{v}/|\mathbf{v}|$, which is independent of s. A glance at Example 2.5 shows that the *tangent vector field of a circle of radius r does change with s but that its derivative has constant length* $1/r$. Because of these considerations we are led to make the following definition.

DEFINITION. The *curvature* of a unit speed curve $\boldsymbol{\alpha}(s)$ is $\kappa(s) = |\mathbf{T}'(s)|$.

From the above discussion, it is clear that $\kappa(s) = 0$ (for all s) if $\boldsymbol{\alpha}$ is a straight line and $\kappa(s) = 1/r$ (for all s) if $\boldsymbol{\alpha}$ is a circle of radius r. This last equality is especially appealing because it says that the smaller the radius is the larger the curvature is (that is, the faster or tighter the circle is bending), which conforms with our intuition.

We may also give a heuristic description of curvature for curves lying in the plane. If $\boldsymbol{\alpha}$ lies in the (x, y) plane, then $\boldsymbol{\alpha}(s)$ takes the form

$$\boldsymbol{\alpha}(s) = (x(s), y(s), 0),$$

hence $\boldsymbol{\alpha}'(s) = \mathbf{T}(s) = (x'(s), y'(s), 0)$. We now let θ be the angle between the horizontal and the tangent vector field to $\boldsymbol{\alpha}$ at s as in Figure 2.5. (*Technical remark:* This angle θ is not really well defined. It is defined only up to a multiple of 2π. We will remedy this problem in Section 3-2. This is the reason we do not separate out what follows as a theorem but merely call it a heuristic description.)

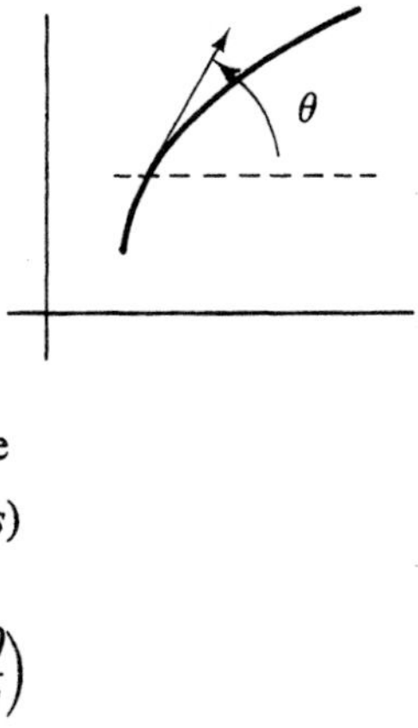

FIGURE 2.5

Because of the description of the angle θ we have

$$x'(s) = \langle \mathbf{T}(s), (1, 0, 0) \rangle = \cos\theta(s)$$

and $\mathbf{T}(s) = (\cos\theta(s), \sin\theta(s), 0)$. Thus

$$\mathbf{T}'(s) = (-\sin\theta(s), \cos\theta(s), 0)\left(\frac{d\theta}{ds}\right)$$

and $\kappa(s) = |\mathbf{T}'(s)| = |d\theta/ds|$. This shows that the *curvature of a plane curve is the rate of change of the angle the tangent vector field makes with the horizontal* (up to sign). This approach to the curvature of plane curves is essentially due to L. Euler (1736).

Having justified the definition of curvature, we shall now develop some machinery to study curvature. This machinery which is called the Frenet-Serret apparatus, is the key to studying the geometry of curves in $\mathbf{R}^3$ and in fact uniquely determines the curve as we will see in Section 2-5.

It is usual in both elementary physics and mathematics to think of a vector as an arrow with a head and a base point (or "point of application"). If we imagine at each point $\boldsymbol{\alpha}(s)$ on the curve the set of all vectors whose base point is $\boldsymbol{\alpha}(s)$, then we obtain at each point $\boldsymbol{\alpha}(s)$ a 3-dimensional vector space. From the point of view of geometry, what is a natural basis for these vector spaces? Certainly if $\mathbf{e}_1 = (1, 0, 0)$, $\mathbf{e}_2 = (0, 1, 0)$, and $\mathbf{e}_3 = (0, 0, 1)$, then

$\{\mathbf{e}_1, \mathbf{e}_2, \mathbf{e}_3\}$ is a basis for these vector spaces (for each s). The problem is that $\{\mathbf{e}_1, \mathbf{e}_2, \mathbf{e}_3\}$ reflects the geometry of $\mathbf{R}^3$ instead of the geometry of the curve and is thus unsatisfactory from a geometric viewpoint. Our line of attack is to take the one "geometric" vector we have (the tangent vector field), find another one (the normal vector field), and use their cross product (the binormal vector field) to obtain an "intrinsically geometric" basis. The following definitions are only valid at those points where $\kappa(s) \neq 0$. Note that $\mathbf{T}'/\kappa$ is a unit vector.

DEFINITION. The *principal normal vector field* to a unit speed curve $\boldsymbol{\alpha}(s)$ is the (unit) vector field $\mathbf{N}(s) = \mathbf{T}'(s)/\kappa(s)$. The *binormal vector field* to $\boldsymbol{\alpha}(s)$ is $\mathbf{B}(s) = \mathbf{T}(s) \times \mathbf{N}(s)$. The *torsion* of $\boldsymbol{\alpha}$ is the real-valued function

$$\tau(s) = -\langle \mathbf{B}'(s), \mathbf{N}(s) \rangle.$$

We will give the geometric meaning of torsion in the next section. (It will measure "how far from lying in a plane" $\boldsymbol{\alpha}$ is.)

DEFINITION.

(a) The *Frenet-Serret apparatus* of the unit speed curve $\boldsymbol{\alpha}(s)$ is

$$\{\kappa(s), \tau(s), \mathbf{T}(s), \mathbf{N}(s), \mathbf{B}(s)\}.$$

(b) If $\boldsymbol{\beta}(t)$ is a regular curve we may write $t = t(s)$ or $s = s(t)$ by Theorem 2.2. Let $\boldsymbol{\alpha}(s) = \boldsymbol{\beta}(t(s))$ be a unit speed reparametrization of $\boldsymbol{\beta}$ and let $\{\kappa(s), \tau(s), \mathbf{T}(s), \mathbf{N}(s), \mathbf{B}(s)\}$ be the Frenet-Serret apparatus of the unit speed curve $\boldsymbol{\alpha}$. The Frenet-Serret apparatus of $\boldsymbol{\beta}(t)$ is

$$\{\kappa(s(t)), \tau(s(t)), \mathbf{T}(s(t)), \mathbf{N}(s(t)), \mathbf{B}(s(t))\}.$$

Because it is so difficult (and often impossible) to find t as a function of s explicitly, we shall need a computational tool for finding the Frenet-Serret apparatus for a non-unit speed curve. This is done in Section 2-6.

EXAMPLE 3.1. Let $\boldsymbol{\alpha}(s) = (r\cos(s/r), r\sin(s/r), 0)$ be a circle of radius $r > 0$. $\mathbf{T}(s) = (-\sin(s/r), \cos(s/r), 0)$ so that $\kappa(s) = |\mathbf{T}'(s)| = 1/r$, as was mentioned earlier in this section. Note that

$$\mathbf{N}(s) = \frac{\mathbf{T}'(s)}{\kappa(s)} = \left(-\cos\left(\frac{s}{r}\right), -\sin\left(\frac{s}{r}\right), 0\right)$$

and $\mathbf{B}(s) = \mathbf{T}(s) \times \mathbf{N}(s) = (0, 0, 1)$. Since $\mathbf{B}' = 0$, we see that $\tau(s) = 0$. We have completed the computation of the Frenet-Serret apparatus in this case. We sketch $\{\mathbf{T}(s), \mathbf{N}(s), \mathbf{B}(s)\}$ at some points in Figure 2.6.

EXAMPLE 3.2. Consider the unit speed circular helix

$$\boldsymbol{\alpha}(s) = (r\cos\omega s, r\sin\omega s, h\omega s),$$

where $\omega = (r^2 + h^2)^{-1/2}$. (See also Example 1.6.)

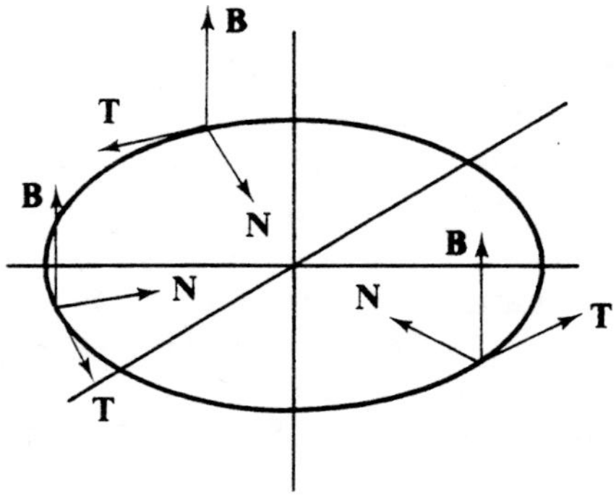

FIGURE 2.6

$$\mathbf{T}(s) = \omega(-r \sin \omega s, r \cos \omega s, h)$$
$$\mathbf{T}'(s) = -\omega^2 r(\cos \omega s, \sin \omega s, 0).$$

Thus $\kappa = \omega^2 r$, which is a constant, yet $\boldsymbol{\alpha}$ is *not* a circle.

$$\mathbf{N} = (-\cos \omega s, -\sin \omega s, 0)$$
$$\mathbf{B} = \mathbf{T} \times \mathbf{N} = \omega(h \sin \omega s, -h \cos \omega s, r)$$
$$\mathbf{B}' = \omega^2 h(\cos \omega s, \sin \omega s, 0).$$

Thus $\tau = -\langle \mathbf{B}' \, \mathbf{N} \rangle = \omega^2 h$. We sketch $\{\mathbf{T}(s), \mathbf{N}(s), \mathbf{B}(s)\}$ at several points in Figure 2.7. Note that the curvature and torsion are both constant in this example.

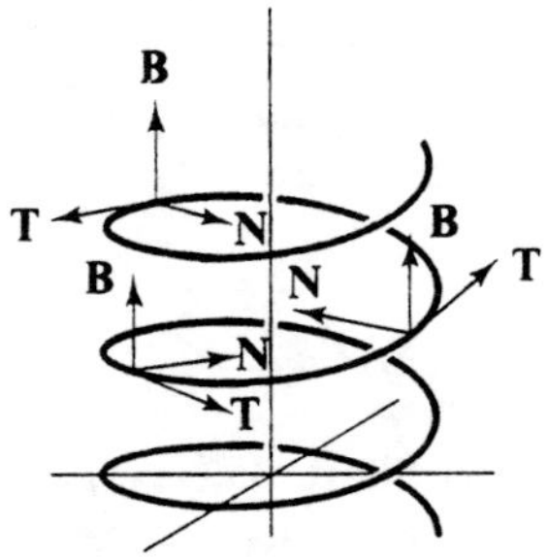

FIGURE 2.7

LEMMA 3.3. Let $\boldsymbol{\alpha}(s)$ be a unit speed curve. Then for every s such that $\kappa(s) \neq 0$, the set $\{\mathbf{T}(s), \mathbf{N}(s), \mathbf{B}(s)\}$ is an orthonormal set.

Proof: $\mathbf{T}(s)$ is a unit vector, so by Lemma 5.1 of Chapter 1, $\langle \mathbf{T}, \mathbf{T}' \rangle = 0$. Since $\mathbf{T}' = \kappa \mathbf{N}$ and $\kappa \neq 0$, we have $\langle \mathbf{T}, \mathbf{N} \rangle = 0$ and so $\mathbf{T}$ and $\mathbf{N}$ are orthogonal. Since $\mathbf{B} = \mathbf{T} \times \mathbf{N}$, Lemma 3.3g of Chapter 1 shows that $\mathbf{B}$ is orthogonal to both $\mathbf{T}$ and $\mathbf{N}$. Since $\mathbf{T}$ and $\mathbf{N}$ are unit vectors, so is $\mathbf{B} = \mathbf{T} \times \mathbf{N}$. ∎

Because of this lemma, at every point on the curve where $\kappa \neq 0$ we have an orthonormal set of vectors $\{\mathbf{T}, \mathbf{N}, \mathbf{B}\}$ that move and twist as we move along the curve. This is the reason that $\{\mathbf{T}, \mathbf{N}, \mathbf{B}\}$ is classically called a "moving frame" or "moving trihedron." Look again at Figures 2.6 and 2.7.

If a curve has isolated points at which κ is zero, then our construction is not valid. However, if κ is zero on an interval, then our intuition demands that the curve be a straight line. This is indeed the case.

PROPOSITION 3.4. Let $\boldsymbol{\alpha}(s)$ be a unit speed curve with $\kappa \equiv 0$ on an interval $[a, b]$. Then the curve segment $\boldsymbol{\alpha}: [a, b] \longrightarrow \mathbf{R}^3$ is a straight line.

Proof: According to Section 1-4 we must produce a vector $\mathbf{v}$ and a point $\mathbf{x}_0$ of $\mathbf{R}^3$ such that $\boldsymbol{\alpha}(s) = s\mathbf{v} + \mathbf{x}_0$. Since $|\mathbf{T}'| = \kappa \equiv 0$, $\mathbf{T}$ is constant. Now for any curve, $\boldsymbol{\alpha}(s) = \int_a^s \mathbf{T}(\sigma)\, d\sigma + \boldsymbol{\alpha}(a)$. Thus, since $\mathbf{T}(\sigma) = \mathbf{T}$ is constant, $\boldsymbol{\alpha}(s) = (s - a)\mathbf{T} + \boldsymbol{\alpha}(a)$. Hence we may let $\mathbf{v} = \mathbf{T}$ and $\mathbf{x}_0 = -a\mathbf{T} + \boldsymbol{\alpha}(a)$. ▌

We note that any unit speed curve can be broken into segments with $\kappa \equiv 0$ on some and $\kappa = 0$ only at the end points of others. By the above proposition, we completely understand the geometry of these segments where $\kappa \equiv 0$. We shall usually consider only the second type of curve and in fact will quite often restrict ourselves to $\boldsymbol{\alpha}: (a, b) \longrightarrow \mathbf{R}^3$ with $\kappa \neq 0$. This is because at this stage we are interested in the *local* behavior of curves, which means the behavior of the curve near a particular point. (If $\kappa(c) \neq 0$, then $\kappa(s) \neq 0$ for all s near c since κ is continuous.)

At an isolated point where $\kappa = 0$ strange things can happen. Problem 4.14 gives an example of a C^∞ curve $\boldsymbol{\alpha}: (-\infty, \infty) \longrightarrow \mathbf{R}^3$ with the image of $(-\infty, 0]$ lying in one plane and the image of $[0, \infty)$ lying in another. Note that κ is zero only at one point ($\boldsymbol{\alpha}(0)$).

PROBLEMS

3.1. Show that $\boldsymbol{\alpha}(s) = (\frac{5}{13}\cos s, \frac{8}{13} - \sin s, -\frac{12}{13}\cos s)$ is unit speed and compute its Frenet-Serret apparatus.

3.2. Show that

$$\boldsymbol{\alpha}(s) = \left(\frac{(1 + s)^{3/2}}{3}, \frac{(1 - s)^{3/2}}{3}, \frac{s}{\sqrt{2}}\right)$$

is a unit speed curve and compute its Frenet-Serret apparatus.

3.3. Show that $\boldsymbol{\alpha}(s) = \frac{1}{2}(\cos^{-1}(s) - s\sqrt{1 - s^2}, 1 - s^2, 0)$ is a unit speed curve and compute its Frenet-Serret apparatus.

3.4. Show that $\boldsymbol{\alpha}(s) = (\sqrt{1 + s^2}, 2s, \ln(s + \sqrt{1 + s^2}))/\sqrt{5}$ is a unit speed curve and compute its Frenet-Serret apparatus.

3.5. Compute the Frenet-Serret apparatus of the curve in Problem 2.6.

*3.6. Let $\alpha(s)$ be a C^k curve in the (x, y) plane. Prove that if $\kappa \neq 0$ then $\tau \equiv 0$. (*Hint:* There are differentiable functions $x(s)$ and $y(s)$ such that $\alpha(s) = (x(s), y(s), 0)$. Show that $\mathbf{B} = \pm(0, 0, 1)$.)

3.7. Let $\alpha(s) = (x(s), y(s), 0)$ be a unit speed curve. Prove that

$$\kappa = |x'y'' - x''y'|.$$

2–4. THE FRENET-SERRET THEOREM AND ITS COROLLARIES

Because the Frenet-Serret apparatus has been defined so geometrically we would expect to get a great deal of information from it. This is indeed the case. After a preliminary lemma, we prove the Frenet-Serret Theorem from which we can derive many geometric corollaries. Even though the lemma below belongs to the realm of linear algebra, we shall prove it in detail here because it is so crucial in the proof of the Frenet-Serret Theorem.

LEMMA 4.1. If $E = \{\mathbf{e}_1, \ldots, \mathbf{e}_n\}$ is an orthonormal set of n elements of an n-dimensional inner product space V, then

(a) E is a basis for V; and

(b) if $\mathbf{v} \in V$, then $\mathbf{v} = \sum_{i=1}^{n} \langle \mathbf{e}_i, \mathbf{v} \rangle \mathbf{e}_i$.

Proof:

(a) Because the number of elements in E is the dimension of V we need only prove that E is linearly independent. Let $c^1, c^2, \ldots, c^n$ be real numbers with $\sum c^i \mathbf{e}_i = \mathbf{0}$. Then $0 = \langle \sum c^i \mathbf{e}_i, \mathbf{e}_j \rangle = \sum c^i \langle \mathbf{e}_i, \mathbf{e}_j \rangle = \sum c^i \delta_{ij} = c^j$ for each j. (Recall that δ_{ij} is the Kronecker delta as defined in Equation (1-1)) of Chapter 1.) Therefore, $\{\mathbf{e}_1, \mathbf{e}_2, \ldots, \mathbf{e}_n\}$ is linearly independent and a basis for V.

(b) Since E is a basis, we know that for each $\mathbf{v} \in V$ there are real numbers v^j such that $\mathbf{v} = \sum v^j \mathbf{e}_j$. Therefore,

$$\langle \mathbf{e}_i, \mathbf{v} \rangle = \langle \mathbf{e}_i, \sum v^j \mathbf{e}_j \rangle = \sum v^j \delta_{ij} = v^i,$$

which proves (b). ▮

The important thing about the above lemma is not that it tells us that $\mathbf{v} \in V$ *can* be expressed as a linear combination of the elements of E, but rather that it tells us *how* to express $\mathbf{v}$ as a linear combination of the elements of E. To appreciate this, consider what you must do if you are given an arbitrary basis $\{\mathbf{u}_1, \mathbf{u}_2, \ldots, \mathbf{u}_n\}$ of V and are asked to write a given vector $\mathbf{v} \in V$ as $\mathbf{v} = \sum a^i \mathbf{u}_i$. You must solve n linear equations in n unknowns, which is, in practice, very difficult. If, however, the basis is an orthonormal one, (V is an inner product space), then it is easy—you need only compute $\langle \mathbf{u}_j, \mathbf{v} \rangle$ for each j.

THEOREM 4.2 (Frenet-Serret). Let $\boldsymbol{\alpha}(s)$ be a unit speed curve with $\kappa \neq 0$ and Frenet-Serret apparatus $\{\kappa, \tau, \mathbf{T}, \mathbf{N}, \mathbf{B}\}$. Then

(a) $\mathbf{T}'(s) = \kappa(s)\mathbf{N}(s)$

(b) $\mathbf{N}'(s) = -\kappa(s)\mathbf{T}(s) + \tau(s)\mathbf{B}(s)$

(c) $\mathbf{B}'(s) = -\tau(s)\mathbf{N}(s)$

for each s.

Proof: Lemma 3.3 asserts that $\{\mathbf{T}(s), \mathbf{N}(s), \mathbf{B}(s)\}$ is an orthonormal set. Therefore, according to Lemma 4.1, we may write any vector $\mathbf{v}$ as

$$\mathbf{v} = \langle \mathbf{T}, \mathbf{v} \rangle \mathbf{T} + \langle \mathbf{N}, \mathbf{v} \rangle \mathbf{N} + \langle \mathbf{B}, \mathbf{v} \rangle \mathbf{B}. \tag{4-1}$$

(a) $\mathbf{T}' = \kappa\mathbf{N}$ is the definition of κ and $\mathbf{N}$.

(b) Since $\mathbf{N}'$ is a vector, we may apply Equation (4-1) with $\mathbf{v} = \mathbf{N}'$. We first compute the coefficient $\langle \mathbf{T}, \mathbf{N}' \rangle$. Differentiating $0 = \langle \mathbf{T}, \mathbf{N} \rangle$, we have $0 = \langle \mathbf{T}', \mathbf{N} \rangle + \langle \mathbf{T}, \mathbf{N}' \rangle = \langle \kappa\mathbf{N}, \mathbf{N} \rangle + \langle \mathbf{T}, \mathbf{N}' \rangle$ so that $\langle \mathbf{T}, \mathbf{N}' \rangle = -\kappa$. Since $\mathbf{N}$ is a unit vector, $\langle \mathbf{N}, \mathbf{N}' \rangle = 0$ and the second coefficient of (4-1) is zero. To compute the third coefficient, $\langle \mathbf{B}, \mathbf{N}' \rangle$, notice that $\langle \mathbf{B}, \mathbf{N} \rangle = 0$ so that by using the definition of τ, $0 = \langle \mathbf{B}', \mathbf{N} \rangle + \langle \mathbf{B}, \mathbf{N}' \rangle = -\tau + \langle \mathbf{B}, \mathbf{N}' \rangle$ and $\langle \mathbf{B}, \mathbf{N}' \rangle = \tau$. Putting these three coefficients in (4-1) with $\mathbf{v} = \mathbf{N}'$ yields Formula (b).

(c) Formula (c) is obtained by using Equation (4-1) with $\mathbf{v} = \mathbf{B}'$. The first coefficient $\langle \mathbf{T}, \mathbf{B}' \rangle$ is zero because $0 = \langle \mathbf{T}, \mathbf{B} \rangle$ implies

$$0 = \langle \mathbf{T}', \mathbf{B} \rangle + \langle \mathbf{T}, \mathbf{B}' \rangle = \langle \kappa\mathbf{N}, \mathbf{B} \rangle + \langle \mathbf{T}, \mathbf{B}' \rangle = \langle \mathbf{T}, \mathbf{B}' \rangle,$$

since $\langle \mathbf{N}, \mathbf{B} \rangle = 0$. The definition of τ gives $\langle \mathbf{N}, \mathbf{B}' \rangle = \langle \mathbf{B}', \mathbf{N} \rangle = -\tau$. Finally, since $\mathbf{B}$ is a unit vector, the last coefficient $\langle \mathbf{B}, \mathbf{B}' \rangle$ is 0. ∎

The reason that we have left spaces in the statement of Theorem 4.2 is that it is easy to remember the form of the equations in matrix format:

$$\begin{pmatrix} \mathbf{T}' \\ \mathbf{N}' \\ \mathbf{B}' \end{pmatrix} = \begin{pmatrix} 0 & \kappa & 0 \\ -\kappa & 0 & \tau \\ 0 & -\tau & 0 \end{pmatrix} \begin{pmatrix} \mathbf{T} \\ \mathbf{N} \\ \mathbf{B} \end{pmatrix}.$$

The fact that the matrix is skew symmetric is very important in more abstract differential geometry.

The equations in the above theorem are naturally called the Frenet-Serret equations. They were independently found by Frenet (1847, published in 1852) and Serret (1851). We show how powerful these equations are by drawing some corollaries and leave other applications for the exercises. The first corollary shows that the vanishing of torsion characterizes plane curves with $\kappa \neq 0$.

COROLLARY 4.3. Let $\boldsymbol{\alpha}(s)$ be a unit speed curve with $\kappa \neq 0$. The following are equivalent:

(a) The image of $\boldsymbol{\alpha}$ lies in a plane (more simply, $\boldsymbol{\alpha}$ is a plane curve).

(b) $\mathbf{B}$ is a constant vector.

(c) $\tau(s) = 0$ for all s.

Proof: The equivalence of (b) and (c) is given by the Frenet-Serret equation $\mathbf{B}' = -\tau\mathbf{N}$. If $\boldsymbol{\alpha}$ is a plane curve, we may assume (by appropriate choice of coordinates in $\mathbf{R}^3$) that $\boldsymbol{\alpha}$ lies in the (x, y) plane. For this case we have already computed the torsion (Problem 3.6) and it is zero, so (a) implies (b).

We now show (b) implies (a). Let $\mathbf{x}_0$ be any point on $\boldsymbol{\alpha}$, say $\mathbf{x}_0 = \boldsymbol{\alpha}(0)$. According to Section 1-4, we must find a vector $\mathbf{v}$ (which does not, of course, depend on s) such that $\langle\boldsymbol{\alpha}(s) - \mathbf{x}_0, \mathbf{v}\rangle$ is identically zero. A glance at the Frenet-Serret equations or a close look at Example 3.1 (when $\mathbf{v} = (0, 0, 1)$) suggests that we should let $\mathbf{v}$ be the binormal vector $\mathbf{B}$.

Since $\mathbf{B}$ is a constant vector field, $\mathbf{B}' = 0$. Thus

$$\langle\boldsymbol{\alpha}(s) - \mathbf{x}_0, \mathbf{B}\rangle' = \langle\boldsymbol{\alpha}'(s), \mathbf{B}\rangle + \langle\boldsymbol{\alpha}(s) - \mathbf{x}_0, \mathbf{B}'\rangle = \langle\mathbf{T}, \mathbf{B}\rangle = 0$$

and $\langle\boldsymbol{\alpha}(s) - \mathbf{x}_0, \mathbf{B}\rangle$ is therefore constant. But $\langle\boldsymbol{\alpha}(s) - \mathbf{x}_0, \mathbf{B}\rangle = 0$ at $s = 0$, so that this constant is zero and $\boldsymbol{\alpha}(s)$ lies in a plane. ▮

Note that the proof actually gives the plane in which $\boldsymbol{\alpha}$ lies. It is the plane through $\mathbf{x}_0$ perpendicular to $\mathbf{B}$. This theorem is actually false if the assumption $\kappa \neq 0$ is omitted. See Problem 4.14.

DEFINITION. The *osculating plane* to a unit speed curve $\boldsymbol{\alpha}$ at the point $\boldsymbol{\alpha}(s)$ is the plane through $\boldsymbol{\alpha}(s)$ perpendicular to $\mathbf{B}$ (and hence spanned by $\mathbf{T}$ and $\mathbf{N}$).

The *normal plane* of $\boldsymbol{\alpha}(s)$ is the plane perpendicular to $\mathbf{T}$.

The *rectifying plane* of $\boldsymbol{\alpha}(s)$ is the plane perpendicular to $\mathbf{N}$.

One of the standard topics in classical curve theory is an investigation into the projections of a given curve onto the above planes. The interested reader may consult, for example, Hicks [1965], Stoker [1969], Laugwitz [1965], Goetz [1970], or Struik [1961].

What is the significance of the osculating plane? If a curve actually lies in a plane ($\tau \equiv 0$), that plane is the osculating plane. More generally, the osculating plane at $\boldsymbol{\alpha}(s)$ is that plane which $\boldsymbol{\alpha}$ is the closest to being in, just as the tangent line at $\boldsymbol{\alpha}(s)$ is the line that $\boldsymbol{\alpha}$ is the closest to being in. See Figure 2.8. (Note that the tangent line lies in the osculating plane.) The curve twists out of the osculating plane, and τ measures this twisting or torsion. As s increases, the curve twists toward the side $\mathbf{B}$ points to if $\tau > 0$ and toward the opposite side if $\tau < 0$. This gives geometric meaning to the sign of τ. Also, since $\mathbf{B}' = -\tau\mathbf{N}$ and $\mathbf{B}$ is the normal to the osculating plane, τ measures how

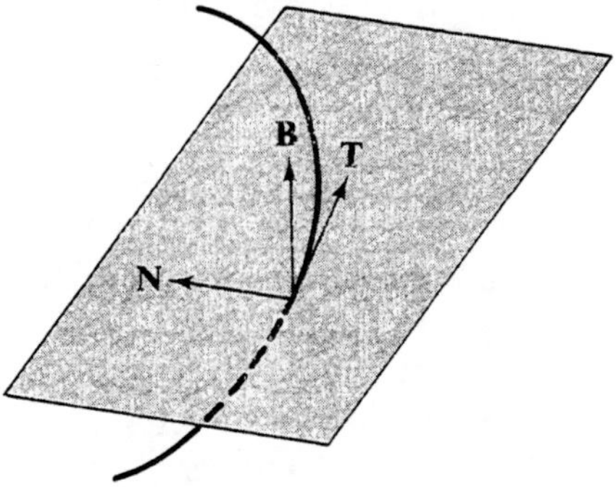

FIGURE 2.8

the osculating plane is turning as s increases. This provides the geometric interpretation of τ which we promised in the last section.

DEFINITION. A (general) *helix* is a regular curve $\boldsymbol{\alpha}$ such that for some fixed unit vector $\mathbf{u}$, $\langle \mathbf{T}, \mathbf{u} \rangle$ is constant. $\mathbf{u}$ is called the *axis* of the helix. Intuitively, the helix grows linearly in the direction of the axis.

EXAMPLE 4.4. The right circular helix of Example 1.6 is a helix with axis (0, 0, 1) according to Problem 1.3.

EXAMPLE 4.5. Any regular plane curve is a helix since $\mathbf{B}$ is constant and may serve as $\mathbf{u}$.

COROLLARY 4.6 (Lancret, 1802). A unit speed curve $\boldsymbol{\alpha}(s)$ with $\kappa \neq 0$ is a helix if and only if there is a constant c such that $\tau = c\kappa$.

Proof: Assume $\boldsymbol{\alpha}$ is a helix. Since $\langle \mathbf{T}, \mathbf{u} \rangle$ is a constant, we may write

$$\langle \mathbf{T}, \mathbf{u} \rangle = \cos \theta$$

where θ is some fixed angle (called the *pitch* of $\boldsymbol{\alpha}$). If θ is an integer multiple of π, then $\mathbf{u} = \mathbf{T}$ or $\mathbf{u} = -\mathbf{T}$. In either case this implies that $\kappa = 0$, which is a contradiction. We may therefore assume that θ is not an integral multiple of π. The following computation shows that $\mathbf{N}$ is perpendicular to $\mathbf{u}$:

$$0 = \langle \mathbf{T}, \mathbf{u} \rangle' = \langle \mathbf{T}', \mathbf{u} \rangle = \kappa \langle \mathbf{N}, \mathbf{u} \rangle.$$

Hence Lemma 4.1 shows that $\mathbf{u} = a\mathbf{T} + b\mathbf{B}$, where $a = \cos\theta$ and $b = \langle \mathbf{B}, \mathbf{u} \rangle$. Since $|\mathbf{u}| = 1$, we may choose the sign of θ so that $b = \sin\theta$. Then

$$\mathbf{0} = \mathbf{u}' = (\cos\theta)\kappa\mathbf{N} - (\sin\theta)\tau\mathbf{N}$$

and $\kappa \cos\theta = \tau \sin\theta$. Since θ is not an integral multiple of π, we have that $\tau = c\kappa$ where $c = \cot\theta$.

We now assume that $\tau = c\kappa$ and show that $\boldsymbol{\alpha}$ is a helix. Motivated by the first half of the proof, we define θ by $\cot\theta = c$ with $0 < \theta < \pi$ and let

$\mathbf{u} = \cos\theta\mathbf{T} + \sin\theta\mathbf{B}$. An application of the Frenet-Serret equations shows that $\mathbf{u}' = \mathbf{0}$, so that $\mathbf{u}$ is constant. Note also that $\langle\mathbf{T}, \mathbf{u}\rangle = \cos\theta$, which is constant, so that $\boldsymbol{\alpha}$ is a helix. ▮

Although Lancret stated this theorem in 1802, the first proof was given by De Saint Venant in 1845.

One of the beautiful things about the above corollary is its constructive nature. If you are given the constant c such that $\tau = c\kappa$, then you can actually *compute* the axis of the helix by first finding its pitch and using the values of $\mathbf{T}$ and $\mathbf{B}$ at a single point.

In Problem 5.4 you will show that if both κ and τ are constant, then $\boldsymbol{\alpha}$ is a circular helix.

We end this section with several propositions which illustrate how the Frenet-Serret Theorem can be applied to derive simple geometric results. Many more such results are contained in the exercises. The reader should note that the general method of proving these results is as follows: (1) express the geometric hypotheses as an algebraic equation using linear algebra; (2) differentiate an appropriate expression (possibly several times), using the Frenet-Serret Theorem and the hypotheses; (3) interpret the result geometrically.

PROPOSITION 4.7. $\boldsymbol{\alpha}(s)$ is a straight line if and only if there is a point $\mathbf{x}_0 \in \mathbf{R}^3$ such that every tangent line to $\boldsymbol{\alpha}$ goes through $\mathbf{x}_0$.

Proof: If $\boldsymbol{\alpha}(s)$ is a straight line, any point on $\boldsymbol{\alpha}(s)$ may be chosen as $\mathbf{x}_0$ since the image of $\boldsymbol{\alpha}(s)$ is the tangent line at each point.

Now suppose that every tangent line to $\boldsymbol{\alpha}$ goes through $\mathbf{x}_0$. Then

$$\boldsymbol{\alpha}(s) - \mathbf{x}_0 = \lambda(s)\mathbf{T}(s)$$

for some function $\lambda(s)$. Either $\kappa = 0$ or $\mathbf{N}(s)$ is defined. In the second case

$$\mathbf{T}(s) = \boldsymbol{\alpha}'(s) = \lambda'(s)\mathbf{T}(s) + \lambda(s)\mathbf{T}'(s) = \lambda'(s)\mathbf{T}(s) + \lambda(s)\kappa(s)\mathbf{N}(s).$$

Hence $(\lambda'(s) - 1)\mathbf{T} + \lambda(s)\kappa(s)\mathbf{N} = \mathbf{0}$. Since $\mathbf{T}$ and $\mathbf{N}$ are linearly independent, $\lambda'(s) \equiv 1$ and $\lambda(s)\kappa(s) \equiv 0$. Thus $\lambda(s) = s + c$, which is not constant, and hence $\kappa(s) \equiv 0$. Then $\boldsymbol{\alpha}$ is a straight line by Proposition 3.4. ▮

PROPOSITION 4.8. Let $\boldsymbol{\alpha}(s)$ be a unit speed curve such that every normal plane to $\boldsymbol{\alpha}(s)$ goes through a given fixed point $\mathbf{x}_0 \in \mathbf{R}^3$. Then the image of $\boldsymbol{\alpha}$ lies on a sphere.

Proof: The normal plane is orthogonal to $\mathbf{T}$, so that $\langle\boldsymbol{\alpha}(s) - \mathbf{x}_0, \mathbf{T}\rangle = 0$. Then $\langle\boldsymbol{\alpha} - \mathbf{x}_0, \boldsymbol{\alpha} - \mathbf{x}_0\rangle' = 2\langle\boldsymbol{\alpha} - \mathbf{x}_0, \mathbf{T}\rangle = 0$ and $\langle\boldsymbol{\alpha} - \mathbf{x}_0, \boldsymbol{\alpha} - \mathbf{x}_0\rangle$ is a constant $a \geq 0$. If $a = 0$, then $\boldsymbol{\alpha}(s) \equiv \mathbf{x}_0$ and $\boldsymbol{\alpha}(s)$ is not regular. Hence $a > 0$ and $\boldsymbol{\alpha}(s)$ lies on a sphere with center $\mathbf{x}_0$ and radius $\sqrt{a}$. ▮

PROPOSITION 4.9. Let $\boldsymbol{\alpha}(s)$ be a unit speed curve with $\kappa \neq 0$. Then $\boldsymbol{\alpha}(s)$ lies in a plane if and only if all osculating planes are parallel.

Proof: If $\boldsymbol{\alpha}(s)$ lies in a plane, that plane is the osculating plane at each point. Hence all osculating planes are parallel.

Suppose all osculating planes are parallel. Then the values of $\mathbf{B}(s)$ at any two points are parallel. Hence $\mathbf{B}(s)$ is constant and $\boldsymbol{\alpha}$ lies in a plane by Proposition 4.3. ∎

PROPOSITION 4.10. Let $\boldsymbol{\alpha}(s)$ be a unit speed curve whose image lies on a sphere of radius r and center $\mathbf{m}$. Then $\kappa \neq 0$. If $\tau \neq 0$, then

$$\boldsymbol{\alpha} - \mathbf{m} = -\rho\mathbf{N} - \rho'\sigma\mathbf{B},$$

where $\rho = 1/\kappa$ and $\sigma = 1/\tau$. Hence $r^2 = \rho^2 + (\rho'\sigma)^2$.

Proof: We have $\langle\boldsymbol{\alpha}(s) - \mathbf{m}, \boldsymbol{\alpha}(s) - \mathbf{m}\rangle = r^2$, so that

$$0 = \langle\boldsymbol{\alpha}(s) - \mathbf{m}, \boldsymbol{\alpha}(s) - \mathbf{m}\rangle' = 2\langle\boldsymbol{\alpha}(s) - \mathbf{m}, \mathbf{T}\rangle.$$

Then

$$\begin{aligned} 0 &= \langle\boldsymbol{\alpha}(s) - \mathbf{m}, \mathbf{T}\rangle' = \langle\mathbf{T}, \mathbf{T}\rangle + \langle\boldsymbol{\alpha}(s) - \mathbf{m}, \mathbf{T}'\rangle \\ &= 1 + \langle\boldsymbol{\alpha}(s) - \mathbf{m}, \kappa\mathbf{N}\rangle, \end{aligned}$$

or $\kappa\langle\boldsymbol{\alpha}(s) - \mathbf{m}, \mathbf{N}\rangle = -1 \neq 0$. Thus $\kappa \neq 0$.

Assume $\tau \neq 0$. $\boldsymbol{\alpha}(s) - \mathbf{m} = a\mathbf{T} + b\mathbf{N} + c\mathbf{B}$, where the coefficients a, b, c may be found by Lemma 4.1.

$$\begin{aligned} a &= \langle\boldsymbol{\alpha}(s) - \mathbf{m}, \mathbf{T}\rangle = 0. \\ b &= \langle\boldsymbol{\alpha}(s) - \mathbf{m}, \mathbf{N}\rangle = -\frac{1}{\kappa} = -\rho. \\ c &= \langle\boldsymbol{\alpha}(s) - \mathbf{m}, \mathbf{B}\rangle. \end{aligned}$$

Since $\langle\boldsymbol{\alpha}(s) - \mathbf{m}, \mathbf{N}\rangle = -\rho$,

$$\begin{aligned} -\rho' &= \langle\boldsymbol{\alpha}(s) - \mathbf{m}, \mathbf{N}\rangle' = \langle\mathbf{T}, \mathbf{N}\rangle + \langle\boldsymbol{\alpha}(s) - \mathbf{m}, -\kappa\mathbf{T} + \tau\mathbf{B}\rangle \\ &= 0 - \kappa\langle\boldsymbol{\alpha}(s) - \mathbf{m}, \mathbf{T}\rangle + \tau\langle\boldsymbol{\alpha}(s) - \mathbf{m}, \mathbf{B}\rangle \\ &= 0 + \tau\langle\boldsymbol{\alpha}(s)) - \mathbf{m}, \mathbf{B}\rangle. \end{aligned}$$

Hence $c = \langle\boldsymbol{\alpha}(s) - \mathbf{m}, \mathbf{B}\rangle = -\rho'/\tau = -\rho'\sigma$. Thus

$$\boldsymbol{\alpha}(s) - \mathbf{m} = -\rho\mathbf{N} - \rho'\sigma\mathbf{B}.$$

Since $\mathbf{N}$ and $\mathbf{B}$ are orthonormal,

$$\begin{aligned} r^2 &= \langle\boldsymbol{\alpha}(s) - \mathbf{m}, \boldsymbol{\alpha}(s) - \mathbf{m}\rangle \\ &= |-\rho\mathbf{N} - \rho'\sigma\mathbf{B}|^2 = \rho^2 + (\rho'\sigma)^2. \quad ∎ \end{aligned}$$

ρ is called the *radius of curvature* and σ the *radius of torsion*.

It is possible to generalize much of the previous material to curves in higher dimensional Euclidean spaces—$\boldsymbol{\alpha} : (a, b) \to \mathbf{R}^n$. The interested reader might consult H. Gluck [1966, 1967]. The Frenet-Serret frame $\{\mathbf{T}, \mathbf{N}, \mathbf{B}\}$ is just one geometrically nice basis for the vector spaces along a curve. There are other possibilities. See R. L. Bishop [1975].

PROBLEMS

General

4.1. Prove that $\kappa\tau = -\langle \mathbf{T}', \mathbf{B}' \rangle$.

4.2. If $\boldsymbol{\alpha}(s)$ is a unit speed curve, prove that $[\boldsymbol{\alpha}', \boldsymbol{\alpha}'', \boldsymbol{\alpha}'''] = \kappa^2\tau$.

4.3. Let $\boldsymbol{\alpha}(s)$ be a C^4 unit speed curve with $\kappa > 0$.
(a) Prove that $[\boldsymbol{\alpha}'', \boldsymbol{\alpha}''', \boldsymbol{\alpha}^{\text{iv}}] = \kappa^5(\tau/\kappa)'$.
(b) Prove that $\boldsymbol{\alpha}$ is a helix if and only if $[\boldsymbol{\alpha}'', \boldsymbol{\alpha}''', \boldsymbol{\alpha}^{\text{iv}}] \equiv 0$.

4.4. Let $\boldsymbol{\alpha}(s)$ be a unit speed curve with $\kappa \neq 0$. Prove

$$\tau = \frac{[\boldsymbol{\alpha}', \boldsymbol{\alpha}'', \boldsymbol{\alpha}''']}{\langle \boldsymbol{\alpha}'', \boldsymbol{\alpha}'' \rangle}.$$

4.5. Describe (if possible) any special geometric shape or form of the curves in Problems 3.1, 3.2, 3.3, and 3.4. If any are helices, find the axis and pitch.

4.6. Find the equation of the normal plane to $\boldsymbol{\alpha}(t) = (e^t, \cos t, 3t^2)$ at $t = 1$. (*Note: t* is not arc length!)

4.7. If $\boldsymbol{\alpha}(s)$ is a unit speed curve with $\kappa \neq 0$, show that the equation of the osculating plane through $\boldsymbol{\alpha}(0)$ is $[\mathbf{x} - \boldsymbol{\alpha}(0), \boldsymbol{\alpha}'(0), \boldsymbol{\alpha}''(0)] \equiv 0$.

4.8. Find the equations of the osculating plane for the curves in Problems 3.1, 3.2, 3.3, and 3.4 at $s = 0$.

4.9. What is the angle between the axis of a helix and the normal to its osculating plane in terms of the pitch?

***4.10.** Prove:
(a) $\mathbf{T} = \mathbf{N} \times \mathbf{B} = -\mathbf{B} \times \mathbf{N}$;
(b) $\mathbf{N} = \mathbf{B} \times \mathbf{T} = -\mathbf{T} \times \mathbf{B}$;
(c) $\mathbf{B} = \mathbf{T} \times \mathbf{N} = -\mathbf{N} \times \mathbf{T}$.

4.11. Let $\boldsymbol{\alpha}(s)$ be a unit speed curve with $\kappa \neq 0$. Find a vector $\mathbf{w}(s)$ such that $\mathbf{T}' = \mathbf{w} \times \mathbf{T}$, $\mathbf{N}' = \mathbf{w} \times \mathbf{N}$, $\mathbf{B}' = \mathbf{w} \times \mathbf{B}$. ($\mathbf{w}$ is called the *Darboux vector* after G. Darboux who investigated curves from a kinematical viewpoint. If a rigid body is thought to move along the curve with unit speed, then the instantaneous motion of this body is described by a translation vector and a rotation vector. The translation is given by the velocity vector $\mathbf{T}$, while the rotation is described by the *Darboux* or *angular velocity vector* $\mathbf{w}$.)

4.12. Prove that $\boldsymbol{\alpha}(s)$ is a straight line if and only if all its tangent lines are parallel.

4.13. Let $\boldsymbol{\alpha}(s)$ be a curve with $\kappa \neq 0$. Prove that $\boldsymbol{\alpha}(s)$ lies in a plane if and only if there is a point $\mathbf{x}_0$ in $\mathbf{R}^3$ such that every osculating plane passes through $\mathbf{x}_0$.

***†4.14.** Let

$$f(t) = \begin{cases} e^{-1/t^2} & \text{if } t \neq 0 \\ 0 & \text{if } t = 0. \end{cases}$$

You may assume f is C^∞ with $f^{(n)}(0) = 0$ for all n. Let $\boldsymbol{\alpha}(t)$ be given by

$$\boldsymbol{\alpha}(t) = \begin{cases} (t, f(t), 0) & \text{if } t < 0 \\ (0, 0, 0) & \text{if } t = 0 \\ (t, 0, f(t)) & \text{if } t > 0. \end{cases}$$

(a) Prove that $\boldsymbol{\alpha}$ is regular and C^∞.
(b) Show that $\kappa = 0$ at $t = 0$. (*Note:* t is not arc length.)
$\boldsymbol{\alpha}$ consists of curves in two different planes joined together at a point where $\kappa = 0$.

Set A—Spherical Images

The next seven problems deal with the notion of tangent, normal, and binormal spherical images. These notions, especially that of the tangent spherical image, are very important in Chapters 3 and 5.

If $\boldsymbol{\alpha}(s)$ is a unit speed curve, then $\mathbf{T}: (a, b) \longrightarrow \mathbf{R}^3$ gives a curve defined by $s \longrightarrow \mathbf{T}(s)$. This curve may not be regular. Since $|\mathbf{T}(s)| = 1$, the image of $\mathbf{T}$ lies on the sphere of radius 1 about $\mathbf{0}$. This curve is called the *tangent spherical image* of $\boldsymbol{\alpha}$. We may also consider the *normal spherical image* $\mathbf{N}$ (defined by $s \longrightarrow \mathbf{N}(s)$) and the *binormal spherical image* (defined by $s \longrightarrow \mathbf{B}(s)$) of $\boldsymbol{\alpha}$.

4.15. Find the tangent, normal, and binormal spherical images of the helix in Example 1.6.

4.16. Let $\boldsymbol{\alpha}(s)$ be a unit speed curve.
(a) Prove that the tangent spherical image of $\boldsymbol{\alpha}$ is a constant curve if and only if $\boldsymbol{\alpha}$ is a straight line.
(b) Prove that the binormal spherical image of $\boldsymbol{\alpha}$ is a constant curve if and only if $\boldsymbol{\alpha}$ is a plane curve.
(c) Prove that the normal spherical image of $\boldsymbol{\alpha}$ is never constant.

***4.17.** Let $\bar{s}$ be arc length along the tangent spherical image of $\boldsymbol{\alpha}$ so that $\bar{s} = \int_0^s |\mathbf{T}'(\sigma)|\, d\sigma$.
(a) Prove that $d\bar{s}/ds = \kappa$.

(b) Find a necessary and sufficient condition for the tangent spherical image of $\boldsymbol{\alpha}$ to be a regular curve.

(c) Let s^* be the arc length along the normal (resp. binormal) spherical image of $\boldsymbol{\alpha}$. Prove that $ds^*/ds = \sqrt{\kappa^2 + \tau^2}$ (resp. $|\tau|$).

4.18. Let $\boldsymbol{\alpha}(s)$ be a unit speed curve with $\kappa > 0$. Let s^* be arc length on the normal spherical image. Prove $\kappa = |ds^*/ds|$ if and only if $\boldsymbol{\alpha}$ is a plane curve.

4.19. Let $\boldsymbol{\alpha}(s)$ be a unit speed curve with $\kappa\tau \neq 0$. Prove that the tangent to the tangent spherical image is parallel to the tangent to the binormal spherical image at corresponding points.

***4.20.** If the tangent spherical image of a unit speed curve $\boldsymbol{\alpha}(s)$ lies in a plane through (0, 0, 0), prove that $\boldsymbol{\alpha}(s)$ is a plane curve.

4.21. Prove that a unit speed curve $\boldsymbol{\alpha}(s)$ is a helix if and only if its tangent spherical image is an arc of a circle.

Problem 6.10 will also deal with the tangent spherical image of a curve.

Set B—Sphere Curves

The next six problems deal with sphere curves, i.e., curves $\boldsymbol{\alpha}(s)$ such that there are constants $\mathbf{m}$ and $r > 0$ with $\langle \boldsymbol{\alpha}(s) - \mathbf{m}, \boldsymbol{\alpha}(s) - \mathbf{m} \rangle = r^2$. The purpose of this set is to give some recent characterizations of sphere curves (Problems 4.26 and 4.27). Bishop [1975] also gives a recent characterization.

4.22. Let $\boldsymbol{\alpha}$ be a regular curve and let $\mathbf{a}$ be a point that belongs to each normal plane of $\boldsymbol{\alpha}$. Prove that $\boldsymbol{\alpha}$ is a sphere curve. (See also Problem 1.7.)

4.23. Show that $\boldsymbol{\alpha}(\theta) = (-\cos 2\theta, -2\cos\theta, \sin 2\theta)$ is a sphere curve by showing $(-1, 0, 0)$ belongs to each normal plane.

4.24. Let $\boldsymbol{\alpha}(s)$ be a unit speed curve with $\kappa \neq 0$, $\tau \neq 0$ and $\rho = 1/\kappa$, $\sigma = 1/\tau$. Assume $\rho^2 + (\rho'\sigma)^2 = \text{constant} = a^2$ where $a > 0$. Prove that the image of $\boldsymbol{\alpha}$ lies on a sphere of radius a. (*Hint:* Show $\boldsymbol{\alpha} + \rho\mathbf{N} + \rho'\sigma\mathbf{B}$ is constant. Call this $\mathbf{m}$. It should be the center of the sphere. This is motivated by Proposition 4.10.) Assume $\rho' \not\equiv 0$ on any interrals.

4.25. Combine the previous result with Proposition 4.10 to prove that if $\boldsymbol{\alpha}(s)$ is a unit speed curve with $\kappa \neq 0$, $\tau \neq 0$, then $\boldsymbol{\alpha}(s)$ lies on a sphere if and only if $\tau/\kappa = (\kappa'/\tau\kappa^2)'$ (or $\tau\rho = -(\rho'/\tau)'$).

4.26. Prove that a unit speed curve $\boldsymbol{\alpha}(s)$ lies on a sphere if and only if $\kappa > 0$ and there exists a differentiable function $f(s)$ with $f\tau = \rho'$, $f' + \tau\rho = 0$. (This result is due to Y-C Wong [1963].)

†**4.27.** Use the previous result to show that $\boldsymbol{\alpha}(s)$ lies on a sphere if and only if there are constants A and B with

$$\kappa\left(A\cos\left(\int_0^s \tau\,ds\right) + B\sin\left(\int_0^s \tau\,ds\right)\right) \equiv 1.$$

(This result is an improvement by Y-C Wong [1972] of a result of S. Breuer and D. Gottlieb [1971].)

Set C—Bertrand Curves

The next five problems deal with a classical topic in the theory of curves, namely Bertrand curves. These curves were first investigated by J. Bertrand [1850]. Similar types of curves are treated in E. Salkowski [1909] and A. Voss [1909]. These problems will not be used in the body of the text.

$\boldsymbol{\alpha}(s)$ and $\boldsymbol{\beta}(s)$ are called *Bertrand curves* if for each s_0, the normal line to $\boldsymbol{\alpha}$ at $s = s_0$ is the same as the normal line to $\boldsymbol{\beta}$ at $s = s_0$. (s *need not* be arc length on both $\boldsymbol{\alpha}$ and $\boldsymbol{\beta}$.) We say that $\boldsymbol{\beta}$ is a *Bertrand mate* for $\boldsymbol{\alpha}$ if $\boldsymbol{\alpha}$ and $\boldsymbol{\beta}$ are Bertrand curves. (Note that this means $\mathbf{N}_\alpha = \pm\mathbf{N}_\beta$.)

4.28. (a) Show that any two circles in the plane with the same center are Bertrand curves.

(b) Let

$$\boldsymbol{\alpha}(s) = \tfrac{1}{2}(\cos^{-1} s - s\sqrt{1 - s^2},\ 1 - s^2,\ 0)$$

and let

$$\boldsymbol{\beta}(s) = \tfrac{1}{2}(\cos^{-1} s - s\sqrt{1 - s^2} - s,\ 1 - s^2 + \sqrt{1 - s^2},\ 0).$$

By Problem 3.3 $\boldsymbol{\alpha}(s)$ is unit speed. $\boldsymbol{\beta}$ is not unit speed. Show that $\boldsymbol{\alpha}$ and $\boldsymbol{\beta}$ are Bertrand curves.

4.29. Prove that the distance between corresponding points of a pair of Bertrand curves is constant. (*Hint:* If $\boldsymbol{\alpha}(s)$ is one of the curves with a unit speed parametrization, then there is a function $\lambda(s)$ such that $\boldsymbol{\alpha}(s) + \lambda(s)\mathbf{N}(s) = \boldsymbol{\beta}(s)$ gives the other curve. Prove λ is constant. $\boldsymbol{\beta}(s)$ may not be a unit speed curve.)

4.30. Prove that the angle between the tangents to two Bertrand curves at corresponding points is constant. (Hint: Show that $\langle \mathbf{T}_\alpha, \mathbf{T}_\beta \rangle$ is constant.)

†**4.31.** Let $\boldsymbol{\alpha}(s)$ be a unit speed curve with $\kappa\tau \neq 0$. Prove there is a curve $\boldsymbol{\beta}(s)$ (s *not* arc length on $\boldsymbol{\beta}$) so that $\boldsymbol{\alpha}$ and $\boldsymbol{\beta}$ are Bertrand curves if and only if there are constants $\lambda \neq 0$ and μ with $1/\lambda = \kappa + \mu\tau$.

4.32. Let $\boldsymbol{\alpha}(s)$ be a unit speed plane curve. Prove there exists a curve $\boldsymbol{\beta}(s)$ so that $\boldsymbol{\alpha}$ and $\boldsymbol{\beta}$ are Bertrand curves.

Problems 4.41, 5.6, and 6.12 will also deal with Bertrand curves.

Set D—Spherical Contact

In Corollary 4.3 and the discussion that followed we saw that the osculating plane at $\boldsymbol{\alpha}(s_0)$ is the plane that $\boldsymbol{\alpha}$ is the closest to being on and $\tau(s_0)$ measured how close $\boldsymbol{\alpha}$ is to actually lying on it. Motivated by the problems on sphere curves, we can ask what is the sphere (osculating sphere) at $\boldsymbol{\alpha}(s_0)$ that $\boldsymbol{\alpha}$ is closest to lying on and how close is $\boldsymbol{\alpha}$ to actually lying on it (spherical contact). Note that if $\boldsymbol{\alpha}$ actually lies on a sphere with center $\mathbf{m}$, then $|\boldsymbol{\alpha}(s) - \mathbf{m}|^2$ is constant (hence all of its derivatives with respect to s are zero) and so we are led to the following definition of contact. Let $\mathbf{m}$ and $r > 0$ be given and let $c(s) = |\boldsymbol{\alpha}(s) - \mathbf{m}|^2$. We say that $\boldsymbol{\alpha}$ has *jth order spherical contact* with the sphere of radius r and center $\mathbf{m}$ at $s = s_0$ if $c(s_0) = r^2$, $c'(s_0) = c''(s_0) = \ldots = c^{(j)}(s_0) = 0$. Note that the larger j is, the closer $c(s)$ is to being a constant function and so the closer $\boldsymbol{\alpha}(s)$ is to lying on a sphere of radius r and center $\mathbf{m}$.

4.33. If $\kappa \neq 0$ compute the first three derivatives of $c(s)$ in terms of $\mathbf{T}$, $\mathbf{N}$, $\mathbf{B}$, κ and τ.

4.34. Prove that $\boldsymbol{\alpha}(s)$ has second order spherical contact at $s = s_0$ if and only if $\mathbf{m} = \boldsymbol{\alpha}(s_0) + (1/\kappa(s_0))\mathbf{N}(s_0) + \lambda\mathbf{B}(s_0)$, where λ is arbitrary. Note that as λ varies with s_0 fixed we get a straight line of possible centers called the *polar axis* of $\boldsymbol{\alpha}$ at s_0. The point

$$\mathbf{m}_c = \boldsymbol{\alpha}(s_0) + \frac{1}{\kappa(s_0)}\mathbf{N}(s_0)$$

is called the *center of curvature* of $\boldsymbol{\alpha}$ at s_0. $\rho(s_0) = 1/\kappa(s_0)$ is the *radius of curvature* at s_0. The circle of radius $\rho(s_0)$ with center $\mathbf{m}_c$ and lying in the osculating plane is called the *osculating circle* of $\boldsymbol{\alpha}$ at s_0. See Figure 2.9.

4.35. If $\tau(s_0) \neq 0$, prove that $\boldsymbol{\alpha}(s)$ has third order spherical contact if and only if $\lambda = -\kappa'/\tau\kappa^2$, where the right-hand side is evaluated at $s = s_0$. The sphere with center

$$\mathbf{m}_s = \boldsymbol{\alpha}(s_0) + \frac{1}{\kappa(s_0)}\mathbf{N}(s_0) - \frac{\kappa'(s_0)}{\tau(s_0)\kappa^2(s_0)}\mathbf{B}(s_0)$$

with third order contact is called the *osculating sphere* of $\boldsymbol{\alpha}$. See Figure 2.9.

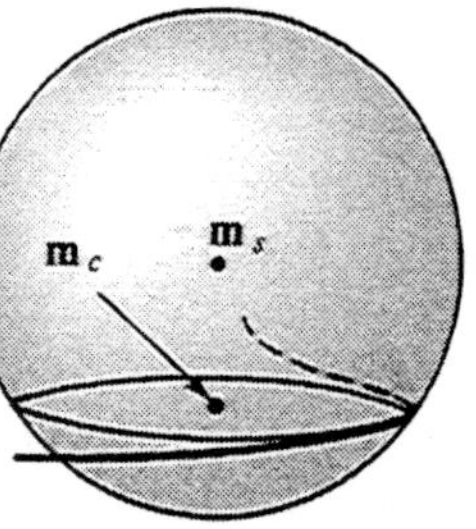

FIGURE 2.9

4.36. Prove that at a point where $\tau(s_0) = 0$ there can be third order spherical contact if and only if $\kappa'(s_0) = 0$. Hence plane curves may not have third order spherical contact anywhere.

4.37. Let $\boldsymbol{\alpha}(s)$ have constant curvature. Prove that the osculating sphere and the osculating circle have the same radius.

4.38. Under what conditions does the osculating sphere have a constant radius, i.e., independent of s_0?

Set E—Involutes and Evolutes

The last six problems deal with another classical topic. Let $\boldsymbol{\alpha}$ and $\boldsymbol{\beta}$ be two regular curves defined on an interval (a, b). $\boldsymbol{\beta}$ is an *involute* of $\boldsymbol{\alpha}$ if $\boldsymbol{\beta}(t_0)$ lies on the tangent line to $\boldsymbol{\alpha}$ at $\boldsymbol{\alpha}(t_0)$ and the tangents to $\boldsymbol{\alpha}$ and $\boldsymbol{\beta}$ at $\boldsymbol{\alpha}(t_0)$ and $\boldsymbol{\beta}(t_0)$ are perpendicular. $\boldsymbol{\beta}$ is an *evolute* of $\boldsymbol{\alpha}$ if $\boldsymbol{\alpha}$ is an involute of $\boldsymbol{\beta}$.

4.39. Suppose that $\boldsymbol{\alpha}(s)$ is a unit speed curve.
(a) If $\boldsymbol{\beta}(s)$ is an involute of $\boldsymbol{\alpha}$ (not necessarily unit speed), prove that $\boldsymbol{\beta}(s) = \boldsymbol{\alpha}(s) + (c - s)\mathbf{T}(s)$, where c is a constant and $\mathbf{T} = \boldsymbol{\alpha}'$.
(b) Under what conditions is $\boldsymbol{\alpha}(s) + (c - s)\mathbf{T}(s)$ a regular curve and hence an involute of $\boldsymbol{\alpha}$?

Note that because of part (a) $|\boldsymbol{\beta} - \boldsymbol{\alpha}|$ is a measure of the arc length of $\boldsymbol{\alpha}$. For this reason $\boldsymbol{\beta}$ can be formed by unwinding a string from the curve $\boldsymbol{\alpha}(s)$. See Figure 2.10 for the case of a plane curve. Because of this, the term *string involute* is sometimes used, especially for plane curves. We shall use this concept in Section 3-5 to give an example of a curve of constant width. The idea of a string involute is due to C. Huygens (1658), who is also known for his work in optics. He discovered involutes while trying to build a more accurate clock. (See Boyer [1968].)

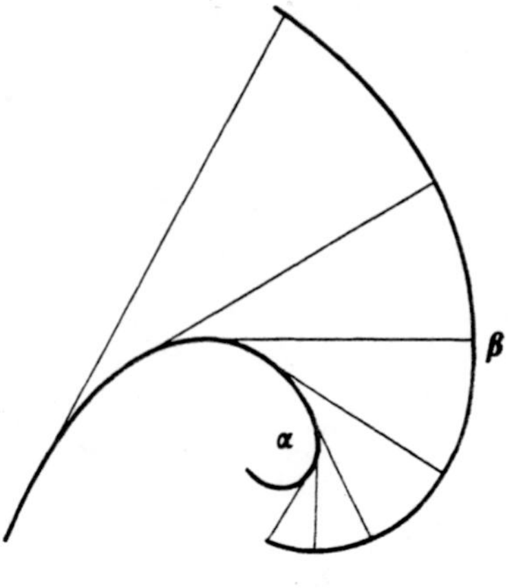

FIGURE 2.10

4.40. (a) Prove that an involute of a plane curve lies in that same plane.
(b) Prove that an involute of a helix is a plane curve. (*Hint:* What should be the normal to this plane?)

4.41. Let $\boldsymbol{\alpha}(s)$ be a unit speed plane curve and let $\boldsymbol{\beta}(s)$ and $\boldsymbol{\gamma}(s)$ be two different involutes of $\boldsymbol{\alpha}$. Prove that $\boldsymbol{\beta}$ and $\boldsymbol{\gamma}$ are Bertrand mates.

4.42. Let $\boldsymbol{\alpha}(s)$ be a unit speed curve with Frenet-Serret apparatus $\{\kappa, \tau, \mathbf{T}, \mathbf{N}, \mathbf{B}\}$ and let $\boldsymbol{\beta}(s)$ be an evolute of $\boldsymbol{\alpha}$.
(a) Prove that $\boldsymbol{\beta}(s) = \boldsymbol{\alpha}(s) + \lambda\mathbf{N} + \mu\mathbf{B}$.
(b) Prove that $\lambda = 1/\kappa$ and $(\lambda' - \tau\mu)/\lambda = (\mu' + \tau\lambda)/\mu$.
(c) Prove that $\tau = (\mu\rho' - \rho\mu')/(\mu^2 + \rho^2)$, where $\rho = 1/\kappa$ as usual.
(d) Prove that $\mu = \rho \cot\left(\int_a^s \tau\, ds + \text{constant}\right)$.

4.43. Prove that an evolute of a plane curve is a helix.

4.44. Let $\boldsymbol{\alpha}(s)$ be a unit speed plane curve.
(a) Prove that the locus of the centers of curvature of $\boldsymbol{\alpha}$ (Problem 4.34) is the unique evolute of $\boldsymbol{\alpha}$ lying in the plane of $\boldsymbol{\alpha}$.
(b) Show that this curve is regular if $\kappa' \neq 0$. (A point where $\kappa' = 0$ is called a *vertex*.)

2–5. THE FUNDAMENTAL EXISTENCE AND UNIQUENESS THEOREM FOR CURVES

After the Frenet-Serret Theorem we drew several corollaries which characterized several types of curves by properties of their curvature and torsion:

$\kappa \equiv 0$	straight line (Proposition 3.4)
$\kappa \neq 0, \quad \tau \equiv 0$	plane curve (Corollary 4.3)
$\dfrac{\kappa}{\tau} = \text{constant}$	helix (Corollary 4.6).

As problems we will have:

$\tau \equiv 0, \quad \kappa = \text{constant} > 0$	circle (Problem 5.3)
$\tau = \text{constant} \neq 0, \quad \kappa = \text{constant} > 0$	circular helix (Problem 5.4).

There were also a large number of exercises where curvature and torsion played a major role in describing what was occurring geometrically.

This is not at all accidental. Theorem 5.2 will tell us that as long as $\kappa \neq 0$, the functions κ and τ completely describe the curve geometrically, except for its position in space. Philosophically, this is the same as saying that two angles and one side completely describe a triangle in plane geometry, or that the radius completely describes a circle. Classically, mathematicians have called $\kappa = \kappa(s)$, $\tau = \tau(s)$ the intrinsic equations of a curve.

The actual proof of Theorem 5.2 will not be used in the sequel and may be omitted. However, the content of the theorem is important and will be used. The proof will depend on the following basic result from the theory of ordinary differential equations which is due, in various formulations, to Picard, Lindelöf, Peano, and Cauchy. See Birkhoff and Rota [1969, Chapter 6] or Coddington and Levinson [1955, Chapter 1]. The dependence of the Fundamental Theorem of Curves on a major theorem of ordinary differential equations shows that in the differential approach to geometry, the heart of geometry is ultimately in differential equations. We will see this again when we study geodesics in Section 4-5 and the Fundamental Theorem of Surfaces in Section 4-10.

THEOREM 5.1 (Picard). Suppose that the $\mathbf{R}^n$-valued function $\mathbf{A}(\mathbf{x}, t)$ is defined and continuous in the closed region $|\mathbf{x} - \mathbf{c}| \leq K$, $|t - a| \leq T$, and satisfies a Lipschitz condition there. Let $M = \sup|\mathbf{A}(\mathbf{x}, t)|$ over this region. Then the differential equation $d\boldsymbol{\alpha}/dt = \mathbf{A}(\boldsymbol{\alpha}, t)$ has a unique solution on the interval $|t - a| \leq \min(T, K/M)$ satisfying $\boldsymbol{\alpha}(a) = \mathbf{c}$. (The technical requirement about the Lipschitz condition is satisfied if $\mathbf{A}$ has bounded partial derivatives with respect to the coordinates of $\mathbf{R}^n$.)

THEOREM 5.2. (Fundamental Theorem of Curves). Any regular curve with $\kappa > 0$ is completely determined, up to position, by its curvature and torsion. More precisely, let (a, b) be an interval about zero, $\bar{\kappa}(s) > 0$ a C^1 function on (a, b), $\bar{\tau}(s)$ a continuous function on (a, b), $\mathbf{x}_0$ a fixed point of $\mathbf{R}^3$, and $\{\mathbf{D}, \mathbf{E}, \mathbf{F}\}$ a fixed right handed orthonormal basis of $\mathbf{R}^3$. Then there exists a unique C^3 regular curve $\boldsymbol{\alpha}: (a, b) \longrightarrow \mathbf{R}^3$ such that:
(a) the parameter is arc length from $\boldsymbol{\alpha}(0)$;
(b) $\boldsymbol{\alpha}(0) = \mathbf{x}_0$, $\mathbf{T}(0) = \mathbf{D}$, $\mathbf{N}(0) = \mathbf{E}$, $\mathbf{B}(0) = \mathbf{F}$; and
(c) $\kappa(s) = \bar{\kappa}(s)$, $\tau(s) = \bar{\tau}(s)$.

Proof: Consider the system of ordinary differential equations

$$\mathbf{u}_j' = \sum_{i=1}^{3} a^i{}_j(s)\mathbf{u}_i, \qquad j = 1, 2, 3,$$

where $(a^i{}_j)$ is the matrix

$$\begin{pmatrix} 0 & \bar{\kappa} & 0 \\ -\bar{\kappa} & 0 & \bar{\tau} \\ 0 & -\bar{\tau} & 0 \end{pmatrix}.$$

Picard's Theorem implies that this system has a unique solution $\mathbf{u}_j(s)$ with $\mathbf{u}_1(0) = \mathbf{D}$, $\mathbf{u}_2(0) = \mathbf{E}$, and $\mathbf{u}_3(0) = \mathbf{F}$. We shall show that the $\mathbf{u}_i$ give the moving trihedron of a regular C^3 space curve $\boldsymbol{\alpha}$ with the required curvature and torsion.

Step 1. The vectors $\mathbf{u}_i$ are orthonormal:

Proof: Let $p_{ij} = \langle \mathbf{u}_i, \mathbf{u}_j \rangle$ so that

$$p_{ij}' = \langle \mathbf{u}_i', \mathbf{u}_j \rangle + \langle \mathbf{u}_i, \mathbf{u}_j' \rangle = \sum a^k{}_i p_{kj} + \sum a^k{}_j p_{ik}.$$

Thus p_{ij} satisfies the initial value problem

$$p_{ij}' = \sum (a^k{}_i p_{kj} + a^k{}_j p_{ik})$$

$$p_{ij}(0) = \delta_{ij} = \begin{cases} 0 & \text{if } i \neq j \\ 1 & \text{if } i = j. \end{cases}$$

By Picard's Theorem, such a system has a unique solution. But

$$\sum (a^k{}_i \delta_{kj} + a^k{}_j \delta_{ik}) = a^j{}_i + a^i{}_j \equiv 0 \equiv \delta_{ij}'.$$

Hence $\delta_{ij} \equiv p_{ij}$ gives a solution, and thus the only solution: $\langle \mathbf{u}_i, \mathbf{u}_j \rangle = \delta_{ij}$ and the $\mathbf{u}_i$ are orthonormal.

Step 2. Let $\boldsymbol{\alpha}(s) = \mathbf{x}_0 + \int_0^s \mathbf{u}_1(\sigma)\, d\sigma$ for $s \in (a, b)$. Then $\boldsymbol{\alpha}(s)$ is C^3, regular, and unit speed:

Proof: $d\boldsymbol{\alpha}/ds = \mathbf{u}_1(s)$ by the fundamental theorem of calculus.

$$\frac{d^2\boldsymbol{\alpha}}{ds^2} = \mathbf{u}_1' = \bar{\kappa}(s)\mathbf{u}_2.$$

Both $\bar{\kappa}$ and $\mathbf{u}_2$ are differentiable.

$$\frac{d^3\boldsymbol{\alpha}}{ds^3} = \bar{\kappa}'\mathbf{u}_2 + \bar{\kappa}\mathbf{u}_2' = \bar{\kappa}'\mathbf{u}_2 + \bar{\kappa}(-\bar{\kappa}\mathbf{u}_1 + \bar{\tau}\mathbf{u}_3).$$

Since $\bar{\kappa}$ is C^1, $\bar{\tau}$ is continuous, and the $\mathbf{u}_i$ are differentiable (hence continuous), $d^3\boldsymbol{\alpha}/ds^3$ is continuous, and $\boldsymbol{\alpha}$ is C^3. $|d\boldsymbol{\alpha}/ds| = |\mathbf{u}_1| = 1$ by Step (1). Hence $\boldsymbol{\alpha}$ is regular and unit speed.

Step 3. $\bar{\kappa} = \kappa$, $\bar{\tau} = \tau$, $\mathbf{u}_1 = \mathbf{T}$, $\mathbf{u}_2 = \mathbf{N}$, $\mathbf{u}_3 = \mathbf{B}$:

Proof: $\boldsymbol{\alpha}' = \mathbf{u}_1$ so $\mathbf{u}_1 \equiv \mathbf{T}$. $\kappa\mathbf{N} = \mathbf{T}' = \mathbf{u}_1' = \bar{\kappa}\mathbf{u}_2$. Since both $\mathbf{N}$ and $\mathbf{u}_2$ are unit vectors and $\bar{\kappa} > 0$, $\bar{\kappa} \equiv \kappa$. Hence $\mathbf{u}_2 \equiv \mathbf{N}$ also. Now $[\mathbf{u}_1, \mathbf{u}_2, \mathbf{u}_3] = \pm 1$ since the vectors are orthonormal. At $s = 0$ we have

$$[\mathbf{u}_1, \mathbf{u}_2, \mathbf{u}_3] = [\mathbf{D}, \mathbf{E}, \mathbf{F}] = +1.$$

Since $[\mathbf{u}_1, \mathbf{u}_2, \mathbf{u}_3]$ is continuous on (a, b), it is always $+1$ and $\{\mathbf{u}_1, \mathbf{u}_2, \mathbf{u}_3\}$ is right handed. Thus $\mathbf{B} = \mathbf{T} \times \mathbf{N} \equiv \mathbf{u}_1 \times \mathbf{u}_2 = \mathbf{u}_3$. Finally,

$$-\tau\mathbf{N} = \mathbf{B}' = \mathbf{u}_3' = -\bar{\tau}\mathbf{u}_2,$$

so $\bar{\tau} \equiv \tau$.

This completes the proof of the existence of a curve $\boldsymbol{\alpha}(s)$ with the required curvature, torsion, starting point and initial moving trihedron. On the other hand, the definition of $\alpha(s)$ was forced if α was to solve the problem. Hence there is a *unique* curve with the required property. ▌

In general, given κ and τ it is very difficult to solve the Frenet-Serret equations and find the curve α. However, it can (almost) be done in the case of a helix.

EXAMPLE 5.3. Let $\alpha(s)$ be a helix with $\kappa > 0$, $\tau = c\kappa$ for some constant c. It will be useful to reparametrize α by a parameter t given by

$$t(s) = \int_0^s \kappa(\sigma)\, d\sigma.$$

Note that this is an allowable change of coordinates since $t' = \kappa > 0$ implies $t(s)$ is one-to-one and both $t(s)$ and $s(t)$ are differentiable. Since $\tau = c\kappa$, the Frenet-Serret equations are

$$\mathbf{T}' = \kappa\mathbf{N}, \qquad \mathbf{N}' = -\kappa\mathbf{T} + c\kappa\mathbf{B}, \qquad \mathbf{B}' = -c\kappa\mathbf{N}.$$

In terms of the parameter t they are

$$\frac{d\mathbf{T}}{dt} = \mathbf{N}, \qquad \frac{d\mathbf{N}}{dt} = -\mathbf{T} + c\mathbf{B}, \qquad \frac{d\mathbf{B}}{dt} = -c\mathbf{N}.$$

Thus $d^2\mathbf{N}/dt^2 = -\mathbf{N} - c^2\mathbf{N} = -\omega^2\mathbf{N}$, where $\omega = \sqrt{1 + c^2}$. Since $\mathbf{N}$ solves this differential equation, we have $\mathbf{N} = \cos \omega t\, \mathbf{a} + \sin \omega t\, \mathbf{b}$ for some fixed vectors $\mathbf{a}$ and $\mathbf{b}$. $d\mathbf{T}/dt = \mathbf{N}$ may be integrated to give $\mathbf{T} = (\sin \omega t\, \mathbf{a} - \cos \omega t\, \mathbf{b} + \mathbf{c})/\omega$. Hence

$$\alpha(s) = \frac{1}{\omega}\left(\int_0^s \sin \omega t(\sigma)\, d\sigma\, \mathbf{a} - \int_0^s \cos \omega t(\sigma)\, d\sigma\, \mathbf{b} + s\mathbf{c} + \mathbf{d}\right).$$

However, the integration constants $\mathbf{a}, \mathbf{b}, \mathbf{c}, \mathbf{d}$ are not arbitrary.

$$\frac{d\mathbf{N}}{dt} = -\omega \sin \omega t\, \mathbf{a} + \omega \cos \omega t\, \mathbf{b}$$

so that

$$0 \equiv \left\langle \mathbf{N}, \frac{d\mathbf{N}}{dt} \right\rangle$$
$$= (-\omega |\mathbf{a}|^2 + \omega |\mathbf{b}|^2) \sin \omega t \cos \omega t + \omega\langle \mathbf{a}, \mathbf{b}\rangle(\cos^2 \omega t - \sin^2 \omega t).$$

At $t = 0$ this equation is $\langle \mathbf{a}, \mathbf{b}\rangle = 0$. Then

$$\tfrac{1}{2}(-\omega |\mathbf{a}|^2 + \omega |\mathbf{b}|^2) \sin 2\omega t \equiv 0$$

and $|\mathbf{a}|^2 = |\mathbf{b}|^2$.

$$1 = |\mathbf{N}|^2 = |\mathbf{a}|^2 \cos^2 \omega t + |\mathbf{b}|^2 \sin^2 \omega t$$

yields $|\mathbf{a}| = |\mathbf{b}| = 1$. Thus $\mathbf{a}$ and $\mathbf{b}$ are orthonormal.

Similarly $0 = \langle \mathbf{T}, \mathbf{N}\rangle$ yields $\langle \mathbf{a}, \mathbf{c}\rangle = 0$ and then $\langle \mathbf{b}, \mathbf{c}\rangle = 0$. $1 = |\mathbf{T}|^2$

gives $|\mathbf{c}| = |c|$ and so $\mathbf{c} = \pm c(\mathbf{a} \times \mathbf{b})$. $d\mathbf{N}/dt = -\mathbf{T} + c\mathbf{B}$ implies that $c\mathbf{B} = d\mathbf{N}/dt + \mathbf{T}$. Hence

$$\begin{aligned}\frac{c}{\omega}(\sin \omega t\, \mathbf{a} - \cos \omega t\, \mathbf{b} + \mathbf{c}) &= c\mathbf{T} = \mathbf{N} \times c\mathbf{B}\\ &= [\cos \omega t\, \mathbf{a} + \sin \omega t\, \mathbf{b}] \times \left[\left(\frac{1}{\omega} - \omega\right)(\sin \omega t\, \mathbf{a} - \cos \omega t\, \mathbf{b}) + \frac{1}{\omega}\mathbf{c}\right]\\ &= \frac{1}{\omega}[\cos \omega t\, (\mathbf{a} \times \mathbf{c}) + \sin \omega t\, (\mathbf{b} \times \mathbf{c}) + c^2(\mathbf{a} \times \mathbf{b})].\end{aligned}$$

Thus $\mathbf{c} = +c(\mathbf{a} \times \mathbf{b})$. In terms of the orthonormal basis $\{\mathbf{a}, \mathbf{b}, \mathbf{a} \times \mathbf{b}\}$ we have

$$\boldsymbol{\alpha}(s) = \frac{1}{\omega}\left(\int_0^s \sin \omega t(\sigma)\, d\sigma, -\int_0^s \cos \omega t(\sigma)\, d\sigma, cs\right) + \mathbf{d}_1, \tag{5-1}$$

where $t(\sigma) = \int_0^\sigma \kappa(s)\, ds$ and $\mathbf{d}_1 = \boldsymbol{\alpha}(0)$.

Note that the above solution requires that you be able to compute some nontrivial integrals, which may not be all that easy to do. See Problems 5.1 and 5.2.

EXAMPLE 5.4. Using Theorem 5.2 we offer an analytic proof of Corollary 4.3. We must show that if $\boldsymbol{\alpha}$ is a unit speed curve with $\tau \equiv 0$, then $\boldsymbol{\alpha}$ lies in a plane. In Example 5.3 we may put $c = 0$. Equation (5-1) yields the curve $\boldsymbol{\beta}(s) = \left(\int_0^s \sin t(\sigma)\, d\sigma, -\int_0^s \cos t(\sigma)\, d\sigma, 0\right)$, where $t(\sigma) = \int_0^\sigma \kappa(s)\, ds$. This is clearly a plane curve, lying in the plane spanned by **a** and **b** and has curvature $\kappa(s)$ and torsion 0. Hence, by Theorem 5.2, it is the same as $\boldsymbol{\alpha}(s)$, up to position. Thus $\boldsymbol{\alpha}(s)$ is also a plane curve.

The preceding proof, although lacking the geometric appeal of the first proof, does contain a valuable insight: a unit speed plane curve may be obtained from its curvature by three integrations. See Problems 5.1 and 5.2.

PROBLEMS

5.1. A plane curve ($\tau \equiv 0$) is a helix. Perform the integration as outlined in Example 5.4 for
(a) $\kappa = \text{constant} > 0$;
(b) $\kappa = 1/(ms + n)$, where m, n are constant.

5.2. Find a unit speed curve $\boldsymbol{\alpha}(s)$ with $\kappa(s) = 1/(1 + s^2)$ and $\tau \equiv 0$.

***5.3.** Prove that the only plane or spherical unit speed curves of constant curvature are circles.

5.4. Let $\boldsymbol{\alpha}(s)$ be a unit speed curve with κ and τ constant so that $\boldsymbol{\alpha}$ is a helix. Prove that $\boldsymbol{\alpha}$ is a circular helix by showing that its projection in a plane perpendicular to $\mathbf{u}$ is a circle. (*Hint:* Equation (5-1).)

5.5. Let $\boldsymbol{\alpha}$ be a helix with axis $\mathbf{u}$. Let Π be a plane perpendicular to $\mathbf{u}$ and let $\boldsymbol{\beta}(s)$ be the projection of $\boldsymbol{\alpha}$ into this plane. ($\boldsymbol{\beta}$ is not unit speed.) Let $\bar{\kappa}$ be the curvature of $\boldsymbol{\beta}$ and prove that $\kappa = \bar{\kappa}\sin^2\theta$, where $\langle \mathbf{u}, \mathbf{T}\rangle = \cos\theta$. (*Hint:* $\mathbf{u} = \mathbf{a} \times \mathbf{b}$.)

5.6. Let $\boldsymbol{\alpha}$ be a C^3 regular curve with $\tau \neq 0$. Prove that $\boldsymbol{\alpha}$ is a circular helix if and only if $\boldsymbol{\alpha}$ has at least two Bertrand mates. (*Hint:* Problem 4.31.)

5.7. Let $\boldsymbol{\alpha}(s)$ be a unit speed curve with $\kappa > 0$ and $\tau > 0$. Let

$$\boldsymbol{\beta}(s) = \int_0^s \mathbf{B}(\sigma)\, d\sigma.$$

(a) Prove that $\boldsymbol{\beta}$ is unit speed.
(b) Show that the Frenet-Serret apparatus $\{\bar{\kappa}, \bar{\tau}, \bar{\mathbf{T}}, \bar{\mathbf{N}}, \bar{\mathbf{B}}\}$ of $\boldsymbol{\beta}$ satisfies $\bar{\kappa} = \tau$, $\bar{\tau} = \kappa$, $\bar{\mathbf{T}} = \mathbf{B}$, $\bar{\mathbf{N}} = -\mathbf{N}$, and $\bar{\mathbf{B}} = \mathbf{T}$.

5.8. Let $\boldsymbol{\alpha}(s)$ be a helix with $\kappa = \tau > 0$. If $\boldsymbol{\beta}(s)$ is defined as in Problem 5.7, show that $\boldsymbol{\alpha}$ and $\boldsymbol{\beta}$ are congruent; that is, they are the same curve up to position in space.

2–6. NON-UNIT SPEED CURVES

In this final section we shall determine the Frenet-Serret apparatus for a curve that is not given a unit speed parametrization. As was pointed out in Section 2-2, in practice it may not be possible to reparametrize by arc length. Thus one must find alternative computational techniques.

Let $\boldsymbol{\beta}(t)$ be a regular curve and let $s(t)$ denote the arc length function. Then $\boldsymbol{\beta}(t) = \boldsymbol{\alpha}(s(t))$, where $\boldsymbol{\alpha}(s)$ is $\boldsymbol{\beta}(t)$ reparametrized by arc length. Note that $ds/dt = |d\boldsymbol{\beta}/dt| > 0$. We wish to determine the Frenet-Serret apparatus in terms of the variable t. We denote derivatives with respect to t by dots: $\dot{\boldsymbol{\beta}} = d\boldsymbol{\beta}/dt$, $\ddot{\boldsymbol{\beta}} = d^2\boldsymbol{\beta}/dt^2$, and $\dddot{\boldsymbol{\beta}} = d^3\boldsymbol{\beta}/dt^3$.

PROPOSITION 6.1. If $\boldsymbol{\beta}(t)$ is a regular curve in $\mathbf{R}^3$, then
(a) $\mathbf{T} = \dot{\boldsymbol{\beta}}/|\dot{\boldsymbol{\beta}}|$;
(b) $\mathbf{B} = \dot{\boldsymbol{\beta}} \times \ddot{\boldsymbol{\beta}}/|\dot{\boldsymbol{\beta}} \times \ddot{\boldsymbol{\beta}}|$;
(c) $\mathbf{N} = \mathbf{B} \times \mathbf{T}$;
(d) $\kappa = |\dot{\boldsymbol{\beta}} \times \ddot{\boldsymbol{\beta}}|/|\dot{\boldsymbol{\beta}}|^3$; and
(e) $\tau = [\dot{\boldsymbol{\beta}}, \ddot{\boldsymbol{\beta}}, \dddot{\boldsymbol{\beta}}]/|\dot{\boldsymbol{\beta}} \times \ddot{\boldsymbol{\beta}}|^2$

Proof: (a) Since $\boldsymbol{\beta}(t) = \boldsymbol{\alpha}(s(t))$, we have $\dot{\boldsymbol{\beta}} = \boldsymbol{\alpha}'\dot{s} = \dot{s}\mathbf{T}$. Because $\dot{s} > 0$, $\dot{s} = |\dot{\boldsymbol{\beta}}|$ and $\mathbf{T} = \dot{\boldsymbol{\beta}}/\dot{s} = \dot{\boldsymbol{\beta}}/|\dot{\boldsymbol{\beta}}|$.

(d) $\ddot{\boldsymbol\beta} = \ddot{s}\mathbf{T} + \dot{s}\dot{\mathbf{T}} = \ddot{s}\mathbf{T} + \dot{s}^2\mathbf{T}' = \ddot{s}\mathbf{T} + \kappa\dot{s}^2\mathbf{N}$. (Since $\dot{\boldsymbol\beta}$ is the velocity, this gives the acceleration in terms of its tangential and normal components.) Hence $\dot{\boldsymbol\beta} \times \ddot{\boldsymbol\beta} = \dot{s}\mathbf{T} \times (\ddot{s}\mathbf{T} + \kappa\dot{s}^2\mathbf{N}) = \kappa\dot{s}^3\mathbf{B}$ and $\kappa\dot{s}^3 = |\dot{\boldsymbol\beta} \times \ddot{\boldsymbol\beta}|$ or $\kappa = |\dot{\boldsymbol\beta} \times \ddot{\boldsymbol\beta}|/\dot{s}^3 = |\dot{\boldsymbol\beta} \times \ddot{\boldsymbol\beta}|/|\dot{\boldsymbol\beta}|^3$.

(b) If $\kappa \neq 0$, then $\mathbf{B} = \dot{\boldsymbol\beta} \times \ddot{\boldsymbol\beta}/\kappa\dot{s}^3 = \dot{\boldsymbol\beta} \times \ddot{\boldsymbol\beta}/|\dot{\boldsymbol\beta} \times \ddot{\boldsymbol\beta}|$.

(c) Since $\{\mathbf{T}, \mathbf{N}, \mathbf{B}\}$ is a right handed orthonormal basis, $\mathbf{N} = \mathbf{B} \times \mathbf{T}$.

(e)

$$\begin{aligned}
\dddot{\boldsymbol\beta} &= \dddot{s}\mathbf{T} + \ddot{s}\dot{\mathbf{T}} + (\kappa\dot{s}^2)^{\boldsymbol\cdot}\mathbf{N} + \kappa\dot{s}^2\dot{\mathbf{N}} \\
&= \dddot{s}\mathbf{T} + \dot{s}\ddot{s}\mathbf{T}' + (\kappa\dot{s}^2)^{\boldsymbol\cdot}\mathbf{N} + \kappa\dot{s}^3\mathbf{N}' \\
&= \dddot{s}\mathbf{T} + \kappa\dot{s}\ddot{s}\mathbf{N} + (\kappa\dot{s}^2)^{\boldsymbol\cdot}\mathbf{N} - \kappa^2\dot{s}^3\mathbf{T} + \kappa\tau\dot{s}^3\mathbf{B} \\
&= (\dddot{s} - \kappa^2\dot{s}^3)\mathbf{T} + (\kappa\dot{s}\ddot{s} + (\kappa\dot{s}^2)^{\boldsymbol\cdot})\mathbf{N} + \kappa\tau\dot{s}^3\mathbf{B}.
\end{aligned}$$

Hence $[\dot{\boldsymbol\beta}, \ddot{\boldsymbol\beta}, \dddot{\boldsymbol\beta}] = \langle\dot{\boldsymbol\beta} \times \ddot{\boldsymbol\beta}, \dddot{\boldsymbol\beta}\rangle = \langle\kappa\dot{s}^3\mathbf{B}, \dddot{\boldsymbol\beta}\rangle = \tau(\kappa\dot{s}^3)^2 = \tau|\dot{\boldsymbol\beta} \times \ddot{\boldsymbol\beta}|^2$. Thus $\tau = [\dot{\boldsymbol\beta}, \ddot{\boldsymbol\beta}, \dddot{\boldsymbol\beta}]/|\dot{\boldsymbol\beta} \times \ddot{\boldsymbol\beta}|^2$. ∎

EXAMPLE 6.2. Let $\boldsymbol\beta(t) = (1 + t^2, t, t^3)$. Then $\dot{\boldsymbol\beta} = (2t, 1, 3t^2)$, $\ddot{\boldsymbol\beta} = (2, 0, 6t)$, and $\dddot{\boldsymbol\beta} = (0, 0, 6)$. Hence $\dot{\boldsymbol\beta} \times \ddot{\boldsymbol\beta} = (6t, -6t^2, -2)$ and $[\dot{\boldsymbol\beta}, \ddot{\boldsymbol\beta}, \dddot{\boldsymbol\beta}] = -12$ so that

$$\kappa = \frac{(36t^2 + 36t^4 + 4)^{1/2}}{(4t^2 + 1 + 9t^4)^{3/2}}$$

$$\tau = \frac{-12}{36t^2 + 36t^4 + 4}$$

$$\mathbf{T} = \frac{(2t, 1, 3t^2)}{(4t^2 + 1 + 9t^4)^{1/2}}$$

$$\mathbf{B} = \frac{(6t, -6t^2, -2)}{(36t^2 + 36t^4 + 4)^{1/2}}$$

and

$$\mathbf{N} = \frac{(-18t^4 + 2, -4t - 18t^3, 6t + 12t^3)}{(4t^2 + 1 + 9t^4)^{1/2}(36t^2 + 36t^4 + 4)^{1/2}}.$$

EXAMPLE 6.3. We show that

$$\boldsymbol\beta(t) = (t, 1 + t^{-1}, t^{-1} - t)$$

for $t > 0$ is a plane curve. $\dot{\boldsymbol\beta} = (1, -t^{-2}, -t^{-2} - 1)$, $\ddot{\boldsymbol\beta} = (0, 2t^{-3}, 2t^{-3})$, and $\dddot{\boldsymbol\beta} = (0, -6t^{-4}, -6t^{-4})$. Clearly $\ddot{\boldsymbol\beta} \times \dddot{\boldsymbol\beta} = 0$ so that $[\dot{\boldsymbol\beta}, \ddot{\boldsymbol\beta}, \dddot{\boldsymbol\beta}] = 0$. By Proposition 6.1, $\tau \equiv 0$ and $\boldsymbol\beta$ is a plane curve, provided $\kappa \neq 0$. $\dot{\boldsymbol\beta} \times \ddot{\boldsymbol\beta} = (2t^{-3}, -2t^{-3}, 2t^{-3}) \neq 0$. Hence, $\kappa \neq 0$.

For a non-unit speed curve $\boldsymbol\beta(t)$, the Frenet-Serret apparatus is a function of t, not s, and does not satisfy the Frenet-Serret equations. However, we do have the following variation.

PROPOSITION 6.4. Let $\boldsymbol{\beta}(t)$ be a regular curve in $\mathbf{R}^3$, and let $v(t) = |\dot{\boldsymbol{\beta}}|$. Then

(a) $\dot{\mathbf{T}} = \kappa v \mathbf{N}$

(b) $\dot{\mathbf{N}} = -\kappa v \mathbf{T} + \tau v \mathbf{B}$

(c) $\dot{\mathbf{B}} = -\tau v \mathbf{N}$.

Proof: Problem 6.5. ∎

PROBLEMS

6.1. Compute the Frenet-Serret apparatus for $\boldsymbol{\beta}(t) = (t, t^2, t^3)$ with $t > 0$.

6.2. Compute the Frenet-Serret apparatus for $\boldsymbol{\beta}(t) = (\cosh t, \sinh t, t)$.

6.3. Compute the Frenet-Serret apparatus for $\boldsymbol{\beta}(t) = (t - \cos t, \sin t, t)$.

6.4. Find the curvature and torsion of $\boldsymbol{\beta}(t) = (e^t \cos t, e^t \sin t, e^t)$.

***6.5.** Prove Proposition 6.4.

6.6. What form do the expressions in Proposition 6.1 take for a curve of constant speed v?

6.7. Let $\boldsymbol{\beta}(t)$ be a regular curve with speed $v = |d\boldsymbol{\beta}/dt|$. Prove that $\kappa = v^{-2}\sqrt{\langle \ddot{\boldsymbol{\beta}}, \ddot{\boldsymbol{\beta}} \rangle - \dot{v}^2}$.

6.8. Prove that $\boldsymbol{\beta}(t)$ is a straight line if and only if $\dot{\boldsymbol{\beta}}$ and $\ddot{\boldsymbol{\beta}}$ are linearly dependent.

†6.9. Let $\boldsymbol{\beta}(t) = (at, bt^2, t^3)$.
(a) Prove that $\boldsymbol{\beta}$ is a helix if and only if $4b^4 = 9a^2$.
(b) What is the axis in this case?

6.10. Let $\boldsymbol{\alpha}(s)$ be a unit speed curve with $\kappa \neq 0$. Let $\bar{\kappa}$ and $\bar{\tau}$ be the curvature and torsion of the tangent spherical image. Prove that $\bar{\kappa} = \sqrt{1 + (\tau/\kappa)^2}$ and $\bar{\tau} = (\tau/\kappa)'/\kappa(1 + (\tau/\kappa)^2)$, where κ and τ are the curvature and torsion of $\boldsymbol{\alpha}$. (See Problem Set 4A.)

6.11. Let $\boldsymbol{\beta}(t)$ be a curve with $\kappa \neq 0$. Prove that $\boldsymbol{\beta}$ is a plane curve if and only if $[\dot{\boldsymbol{\beta}}, \ddot{\boldsymbol{\beta}}, \dddot{\boldsymbol{\beta}}] \equiv 0$.

†6.12. Prove that the product of the torsions of two Bertrand curves at corresponding points is constant. (See Problem Set 4C.)

3

Global Theory of Plane Curves

The geometry of a curve at or near a point in no way restricts or reflects the behavior of the curve at other points "far away" from the given point. In the previous chapter we adopted the microscopic approach (i.e., we looked in very small neighborhoods of points and made assumptions about κ, τ, etc.). In this chapter we shall adopt the macroscopic or global approach and look at the entire curve. The global theorems we shall present are very much in the spirit of much of the modern research in differential geometry. Because we are dealing with global notions, the theorems become harder to prove, but they still do not lose their geometric flavor. To understand the difference between local and global discussions, the reader is urged to look at the curve of Problem 4.14 in Chapter 2 again. About each point of this curve, except one, there is a segment of the curve that lies in a plane. Yet the global behavior is quite strange.

Our first global theorems will deal with plane curves (so that $\tau = 0$). We shall assume that a suitable choice of coordinates has been made in $\mathbf{R}^3$ so that the curve lies in the (x, y) plane. We shall then disregard the z-coordinate and write the curve as if it lies in $\mathbf{R}^2$.

Section 3-1 contains a brief review of line integrals and Green's Theorem, which should be familiar from Calculus III. In Section 3-2 we define the plane curvature of a plane curve. This new notion of curvature has the advantage of being a signed quantity and enables us to globally define a normal vector field to a plane curve. This new notion of curvature coincides with the old except for sign. We also introduce the rotation index of a curve. This measures how many times the tangent vector goes around (the unit circle) as a closed plane curve is traversed once. In Section 3-3 we discuss convex curves and give an alternate description of them in terms of curvature. The Isoperimetric Inequality (which says that the circle is the curve of largest area for a given perimeter) is proved in Section 3-4. In Section 3-5 we study ovals, proving the Four-Vertex Theorem and certain results about curves of constant width. We end the chapter with a brief statement in Section 3-6 about some global theorems for space curves. These results will be proved in Chapter 5.

3–1. LINE INTEGRALS AND GREEN'S THEOREM

We review here the very basic concepts of line integrals for use in this chapter and in Chapters 5 and 6. We shall not use the vector notation since we have no need of it.

Suppose $\boldsymbol{\alpha}(t) = (x(t), y(t))$ is a C^1 parameterization of a geometric curve $\mathcal{C}$ in $\mathbf{R}^2$ with $a \le t \le b$. If f and g are real-valued functions of two variables, then by the line integral $\int_{\mathcal{C}} f\,dx + g\,dy$ we mean the ordinary integral

$$\int_a^b \left[f(x(t), y(t))\frac{dx}{dt} + g(x(t), y(t))\frac{dy}{dt} \right] dt.$$

(Note that there is an obvious generalization to curves in $\mathbf{R}^3$.)

EXAMPLE 1.1 Let $\mathcal{C}$ be the unit circle in $\mathbf{R}^2$ parameterized as

$$\boldsymbol{\alpha}(\theta) = (\cos\theta, \sin\theta), \qquad 0 \le \theta \le 2\pi.$$

Then

$$\begin{aligned}\int_{\mathcal{C}} y\,dx + x\,dy &= \int_0^{2\pi} [-\sin^2\theta + \cos^2\theta]\,d\theta \\ &= \int_0^{2\pi} \cos 2\theta\,d\theta = \frac{\sin 2\theta}{2}\bigg|_0^{2\pi} = 0.\end{aligned}$$

EXAMPLE 1.2. Let $\mathcal{C}$ be as in Example 1.1. Then

$$\int_{\mathcal{C}} x\,dy - y\,dx = \int_0^{2\pi} [\cos^2\theta + \sin^2\theta]\,d\theta = \int_0^{2\pi} d\theta = 2\pi.$$

Sometimes one is given a function to integrate along a curve, such as curvature. $\int_{\mathcal{C}} \kappa\,ds$ may also be thought of as a line integral and is computed in the standard fashion: $\int_{\mathcal{C}} \kappa\,ds = \int_a^b \kappa(s)\,ds$. If one is told to integrate a function along a curve, it is implied that this is with respect to arc length. If the function is given in terms of a different parametrization, a suitable change of coordinates must be made. For example, if it is not practical to reparametrize by arc length (Examples 2.6 and 2.7 of Chapter 2), one would compute $\int_{\mathcal{C}} \kappa\, ds = \int_{\mathcal{C}} \kappa(s(t))(ds/dt)\, dt$, where $\mathcal{C}$ is parametrized by t. Although $s(t)$ is not computable, $ds/dt = |d\boldsymbol{\alpha}/dt|$ is and so is $\kappa(s(t))$ by the results of Section 2-6. Example 1.5 of Chapter 5 carries out this program for an ellipse.

If h is a function defined along a curve, $\int_{\mathcal{C}} dh$ makes sense as the difference in the values of h at the end points:

$$\int_{\mathcal{C}} dh = \int_{\mathcal{C}} \frac{dh}{ds}\, ds = \int_a^b \frac{dh}{ds}\, ds = h(b) - h(a).$$

If $h(s) = p(x(s), y(s))$ for some function $p(x, y)$, there is a second way of computing $\int_{\mathcal{C}} dh$:

$$\int_{\mathcal{C}} dh = \int_{\mathcal{C}} dp = \int_{\mathcal{C}} \frac{\partial p}{\partial x}\, dx + \frac{\partial p}{\partial y}\, dy.$$

Since

$$\frac{\partial p}{\partial x}\frac{dx}{ds} + \frac{\partial p}{\partial y}\frac{dy}{ds} = \frac{dh}{ds}$$

by the chain rule, this latter integral does give the same value for $\int_{\mathcal{C}} dh$.

Note that Example 1.1 was of this latter type with $p(x, y) = xy$: $dp = y\, dx + x\, dy$.

One important tool for computing line integrals is Green's Theorem, which allows us to replace certain line integrals with double integrals. We shall omit the proof of Green's Theorem. The interested reader may refer to an advanced calculus text such as Fulks [1969]. See the next section for the formal concepts of a closed curve and a piecewise C^2 curve.

THEOREM 1.3 (Green). If $\mathcal{C}$ is a closed plane curve made up of C^2 curve segments, which bounds a region $\mathcal{R}$, and which is traversed counterclockwise, then

$$\int_{\mathcal{C}} f\, dx + g\, dy = \iint_{\mathcal{R}} \left(\frac{\partial g}{\partial x} - \frac{\partial f}{\partial y}\right) dx\, dy,$$

for all differentiable functions f and g defined on $\mathcal{R}$.

EXAMPLE 1.4. In Example 1.2 we found $\int_{\mathcal{C}} x\, dy - y\, dx = 2\pi$. On the other hand, setting $f = -y$, $g = x$ in Green's Theorem yields

$$\begin{aligned}\iint_{\mathcal{R}} \left(\frac{\partial g}{\partial x} - \frac{\partial f}{\partial y}\right) dx\, dy &= \iint_{\mathcal{R}} [1 - (-1)]\, dx\, dy \\ &= 2 \iint_{\mathcal{R}} dx\, dy \\ &= 2(\text{area of } \mathcal{R}) = 2\pi.\end{aligned}$$

PROBLEMS

1.1. Compute $\int_{\mathcal{C}} (xy + 1)\,dx + (x^2/y)\,dy$ along each of the following curves:
(a) $\mathcal{C}$ is parametrized as $\boldsymbol{\alpha}(t) = (t, t)$, $0 \leq t \leq 1$;
(b) $\mathcal{C}$ is parametrized as $\boldsymbol{\beta}(t) = (t^2, t^3)$, $0 \leq t \leq 1$.

1.2. Compute $\int_{\mathcal{C}} (x\,dy - y\,dx)/(x^2 + y^2)$ where $\mathcal{C}$ is
(a) the straight line from (0, 1) to (1, 0);
(b) the straight line from (1, 0) to (1, 1); and
(c) the straight line from (1, 1) to (0, 1).

1.3. Verify Green's Theorem for $\int_{\mathcal{C}} (x\,dy - y\,dx)/(x^2 + y^2)$ where $\mathcal{C}$ is the triangle with verticles at (0, 1), (1, 0), and (1, 1). (Both sides of the equation should be 0.)

3–2. THE ROTATION INDEX OF A PLANE CURVE

Suppose that $\boldsymbol{\alpha}: (a, b) \longrightarrow \mathbf{R}^3$ is a plane curve. We can choose coordinates in $\mathbf{R}^3$ so that this plane is given by $z = 0$. We might as well assume that $\boldsymbol{\alpha}: (a, b) \longrightarrow \mathbf{R}^2$, which we do.

Because $\mathbf{R}^2$ has dimension 2 and the tangent vector field of a regular curve is a nonzero vector at each point, we may define the normal to a *plane* curve by using the concept of orientation (Section 1-3). This definition has the advantage of giving a globally defined normal vector field along the curve, as opposed to the normal of Chapter 2, which was only defined when $\kappa \neq 0$. Note that a similar approach for general curves in $\mathbf{R}^3$ would not work because giving one vector in a 3-dimensional vector space does not *uniquely* determine two others to make an oriented basis.

We shall now define the tangent and normal vector fields $\mathbf{t}(s)$, $\mathbf{n}(s)$ and the plane curvature $k(s)$. Lemma 2.2 shows how these concepts compare with the Frenet-Serret apparatus of $\boldsymbol{\alpha}$ viewed as a curve in $\mathbf{R}^3$ (which happens to lie in the (x, y) plane $\mathbf{R}^2$).

DEFINITION. Let $\boldsymbol{\alpha}(s)$ be a unit speed C^2 plane curve. The *tangent vector field*, $\mathbf{t}(s)$, to $\boldsymbol{\alpha}$ is $\mathbf{t}(s) = \boldsymbol{\alpha}'(s)$. The *normal vector field*, $\mathbf{n}(s)$, to $\boldsymbol{\alpha}$ is the unique (unit) vector field $\mathbf{n}(s)$ such that $\{\mathbf{t}(s), \mathbf{n}(s)\}$ gives a right handed orthonormal basis of $\mathbf{R}^2$ for each s. The *plane curvature*, $k(s)$, of $\boldsymbol{\alpha}$ is given by $k(s) = \langle \mathbf{t}'(s), \mathbf{n}(s) \rangle$.

The following lemma gives both an analogue to the first Frenet-Serret equation and a "concrete" realization of $\mathbf{n}(s)$.

LEMMA 2.1. If $\boldsymbol{\alpha}$ is a unit speed plane curve, then

(a) $\mathbf{t}'(s) = k(s)\mathbf{n}(s)$; and

(b) if $\boldsymbol{\alpha}(s)$ is written in the form $\boldsymbol{\alpha}(s) = (x(s), y(s))$ for some C^2 real-valued functions x and y, then

$$\mathbf{t}(s) = (x'(s), y'(s)) \quad \text{and} \quad \mathbf{n}(s) = (-y'(s), x'(s)).$$

Proof: (a) This follows from Lemma 4.1 of Chapter 2 and the fact that $\mathbf{t}$ is a unit vector field.

(b) The formula for $\mathbf{t}$ is obvious. Note that $\mathbf{m}(s) = (-y'(s), x'(s))$ is a unit vector orthogonal to $\mathbf{t}$. Thus we must only show that $\{\mathbf{t}(s), \mathbf{m}(s)\}$ is a right handed basis. Note that (using $\mathbf{e}_1 = (1, 0)$, $\mathbf{e}_2 = (0, 1)$ as usual)

$$\mathbf{t}(s) = x'(s)\mathbf{e}_1 + y'(s)\mathbf{e}_2$$
$$\mathbf{m}(s) = -y'(s)\mathbf{e}_1 + x'(s)\mathbf{e}_2.$$

Hence the matrix $A = (a^i{}_j)$ for the change of basis is

$$A = \begin{pmatrix} x'(s) & -y'(s) \\ y'(s) & x'(s) \end{pmatrix}.$$

This has positive determinant and $\{\mathbf{t}(s), \mathbf{m}(s)\}$ is a right handed basis. Hence $\mathbf{m}(s) = \mathbf{n}(s)$. ∎

LEMMA 2.2. For a unit speed plane curve $\boldsymbol{\alpha}$ we have that

$$\mathbf{t}(s) = \mathbf{T}(s), \qquad \mathbf{n}(s) = \pm\mathbf{N}(s)$$

at all points for which $\mathbf{N}(s)$ is defined, $\kappa(s) = |k(s)|$, $\mathbf{n}(s)$ is differentiable, and $\mathbf{n}'(s) = -k(s)\mathbf{t}(s)$ for all s.

Proof: Problem 2.1. ∎

EXAMPLE 2.3. $\boldsymbol{\alpha}(s) = (r\cos(s/r), r\sin(s/r))$ is a circle of radius r traversed counterclockwise. $\mathbf{t} = (-\sin(s/r), \cos(s/r))$, $\mathbf{n} = (-\cos(s/r), -\sin(s/r))$ and $k = 1/r$.

EXAMPLE 2.4. $\boldsymbol{\alpha}(s) = (r\cos(s/r), -r\sin(s/r))$ is the same circle traversed clockwise. $\mathbf{t} = (-\sin(s/r), -\cos(s/r))$, $\mathbf{n} = (\cos(s/r), -\sin(s/r))$, and $k = -1/r$.

These examples indicate that the sign of k is changed when the direction of traverse is reversed. The sign of k indicates whether $\boldsymbol{\alpha}$ curves in the direction of $\mathbf{n}$ ($k > 0$) or away from $\mathbf{n}$ ($k < 0$).

DEFINITION. A regular curve $\boldsymbol{\beta}(t)$ is *closed* if $\boldsymbol{\beta}$ is periodic, i.e., there is a constant $a > 0$ with $\boldsymbol{\beta}(t) = \boldsymbol{\beta}(t + a)$ for all t. The *period* of $\boldsymbol{\beta}$ is the least such number a. Note that $\mathbf{t}(0) = \mathbf{t}(a)$.

LEMMA 2.5. If $\boldsymbol{\beta}(t)$ is closed with period a and $\boldsymbol{\alpha}(s)$ is $\boldsymbol{\beta}$ reparametrized by arc length, then $\boldsymbol{\alpha}$ is closed with period $L = \int_0^a |d\boldsymbol{\beta}/dt|\, dt$.

Proof:

$$\begin{aligned} s(t + a) &= \int_0^{t+a} \left|\frac{d\boldsymbol{\beta}}{dt}\right| dt \\ &= \int_0^{a} \left|\frac{d\boldsymbol{\beta}}{dt}\right| dt + \int_a^{t+a} \left|\frac{d\boldsymbol{\beta}}{dt}\right| dt \\ &= L + \int_0^{t} \left|\frac{d\boldsymbol{\beta}}{dt}\right| dt \\ &= L + s(t). \end{aligned}$$

Thus $s(t + a) = s(t) + L$ and

$$\boldsymbol{\alpha}(s + L) = \boldsymbol{\alpha}(s(t) + L) = \boldsymbol{\alpha}(s(t + a)) = \boldsymbol{\beta}(t + a) = \boldsymbol{\beta}(t) = \boldsymbol{\alpha}(s(t)) = \boldsymbol{\alpha}(s).$$

Hence $\boldsymbol{\alpha}(s)$ is closed. Since a is the least positive number such that

$$\boldsymbol{\beta}(t + a) = \boldsymbol{\beta}(t) \qquad \text{for all } t,$$

L must be the least positive number such that $\boldsymbol{\alpha}(s + L) = \boldsymbol{\alpha}(s)$ for all s. ∎

Note that the period of a closed unit speed curve is its perimeter (or length).

DEFINITION. A regular curve $\boldsymbol{\beta}(t)$ is *simple* if either $\boldsymbol{\beta}$ is a one-to-one function or if $\boldsymbol{\beta}$ is a closed curve of period a with $\boldsymbol{\beta}(t_1) = \boldsymbol{\beta}(t_2)$ if and only if $t_1 - t_2 = na$ for some integer n.

EXAMPLE 2.6. The unit circle is a simple closed curve of length 2π:

$$\boldsymbol{\alpha}(s) = (\cos s, \sin s).$$

EXAMPLE 2.7. The figure-eight curve is not simple. See Figure 3.1.

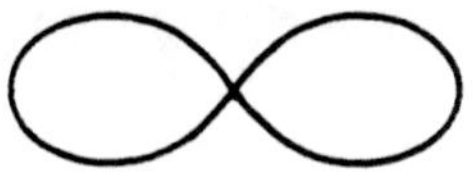

FIGURE 3.1

For a closed curve $\boldsymbol{\alpha}(s)$ it makes sense to ask how much the unit vector $\mathbf{t}$ rotates as the curve is traversed once in the direction of increasing s. This is described by an angle which, because $\mathbf{t}(0) = \mathbf{t}(L)$, must be an integral multiple of 2π. In order to carefully define and find this integer we employ certain normalizations.

Let $\boldsymbol{\alpha}$ be a closed unit speed curve in the plane. We may choose a right handed coordinate system in the plane and the starting point of $\boldsymbol{\alpha}$ so that the image of $\boldsymbol{\alpha}$ lies in the upper half plane and $\boldsymbol{\alpha}(0) = (0, 0)$. In this case $\mathbf{t}(0)$ is horizontal.

At each point P of the image of $\boldsymbol{\alpha}$ define $\bar{\theta}(P)$ to be the angle between the horizontal and $\mathbf{t}$, measured counterclockwise with $0 \leq \bar{\theta} < 2\pi$. $\bar{\theta}$ may not depend continuously on the position of P. However, we can find a continuous function $\theta(s)$ defined for $0 \leq s \leq L$ which also describes the angle. ($\theta(s)$ may be greater than 2π or less than 0.) This is done by breaking $[0, L]$ into small intervals over which $\mathbf{t}$ does not change much and adjusting $\bar{\theta}$ on each interval.

More precisely, points s_i may be chosen with

$$0 = s_0 < s_1 < \ldots < s_n = L$$

such that on each segment $[s_i, s_{i+1}]$ $\mathbf{t}$ does one of four things: (1) points into the upper half plane ($0 < \bar{\theta} < \pi$); (2) points into the left half plane ($\pi/2 < \bar{\theta} < 3\pi/2$); (3) points into the lower half plane ($\pi < \bar{\theta} < 2\pi$); or (4) points into the right half plane ($0 \leq \bar{\theta} \leq \pi/2$ or $3\pi/2 < \bar{\theta} < 2\pi$).

$\bar{\theta}$ is continuous on $[0, s_1]$ so we set $\theta(s) = \bar{\theta}(s)$ for $0 \leq s \leq s_1$. We assume by way of induction that $\theta(s)$ has been defined continuously on $[0, s_k]$. Then on $[s_k, s_{k+1}]$ the angle between the horizontal and $\mathbf{t}$ is well defined up to a multiple of 2π. If the angle is forced to lie in a certain interval of length π (($0, \pi$) for Case (1), ($\pi/2, 3\pi/2$) for Case (2), ($\pi, 2\pi$) for Case (3) and ($3\pi/2, 5\pi/2$) for Case (4)) it is a continuous function of s. By adding an appropriate multiple of 2π it can be made to agree with the known value of $\theta(s)$ at $s = s_k$. In this manner $\theta(s)$ can be continuously defined on $[0, s_{k+1}]$ and eventually on all of $[0, L]$. Note that $\theta(L)$ need not be 0. However, $\theta(L)$ is an integral multiple of 2π.

DEFINITION. The *rotation index* of a closed unit speed plane curve $\boldsymbol{\alpha}(s)$ is the integer $i_{\alpha} = (\theta(L) - \theta(0))/2\pi = \theta(L)/2\pi$.

EXAMPLE 2.8. The curves in Figures 3.2. 3.3, 3.4, and 3.5 have rotation index 1, -1, 0, and 2, respectively. The reader may also want to look ahead to Example 2.10.

Note that

$$\begin{aligned}\mathbf{t}(s) &= \langle \mathbf{t}(s), \mathbf{e}_1 \rangle \mathbf{e}_1 + \langle \mathbf{t}(s), \mathbf{e}_2 \rangle \mathbf{e}_2 \\ &= (\cos \theta(s), \sin \theta(s)).\end{aligned}$$

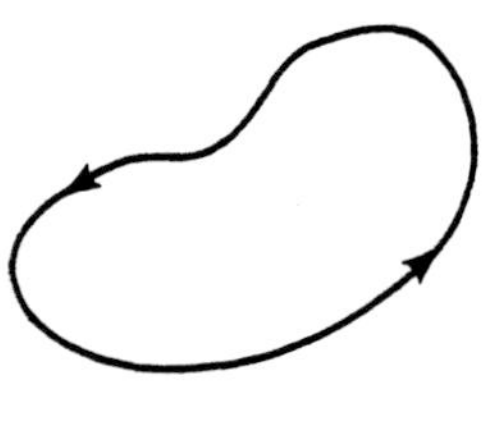

FIGURE 3.2

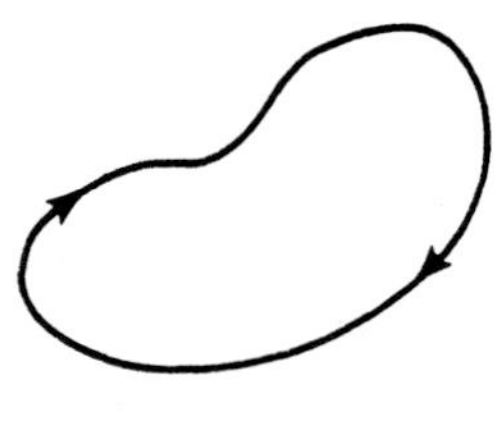

FIGURE 3.3

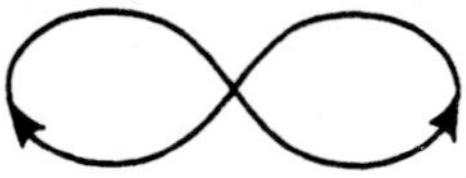

FIGURE 3.4

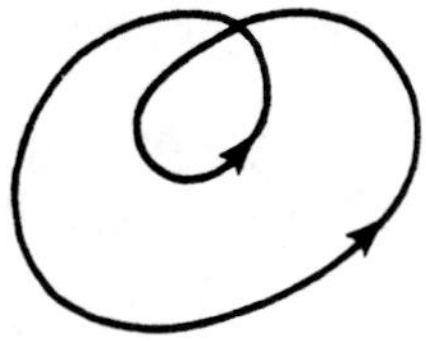

FIGURE 3.5

Since $\mathbf{t}$ is differentiable ($\mathbf{x}$ is C^2), both $\cos(\theta(s))$ and $\sin(\theta(s))$ are differentiable. Since $\theta(s)$ is continuous and both sine and cosine are differentiable, $\theta(s)$ is differentiable. Then $\theta'(s) = k(s)$. (This equality combined with Lemma 2.2 shows that $|\theta'(s)| = \kappa(s)$ for a unit speed plane curve. This was one interpretation of curvature which we promised to make precise in Section 2.3.)

THEOREM 2.9 (Rotation Index Theorem). The rotation index of a simple closed plane curve $\boldsymbol{\alpha}(s)$ is ± 1.

Comment: This theorem seems almost obvious and without need of a formal proof. However, a simple curve can be quite complicated, spiraling a lot, and our intuition based on oval-shaped curves may not hold true. Also, the proof will give us a glimpse of certain techniques.

Proof: Let L be the period of $\boldsymbol{\alpha}(s)$. If $0 \le u < v \le L$, let $\mathbf{a}(u, v)$ be the unit vector from $\boldsymbol{\alpha}(u)$ toward $\boldsymbol{\alpha}(v)$. Let

$$\mathbf{a}(u, u) = \mathbf{t}(u) \qquad (\text{and } \mathbf{a}(0, L) = -\mathbf{t}(0) = -\mathbf{t}(L)).$$

Then $\mathbf{a}$ is a C^2 function in the region Δ of Figure 3.6. As in the case of θ on $[0, L]$, it is possible to define a C^2 function α on Δ, where $\alpha(u, v)$ measures the angle between the horizontal axis and $\mathbf{a}(u, v)$. Note that $\alpha(u, u) = \theta(u)$.

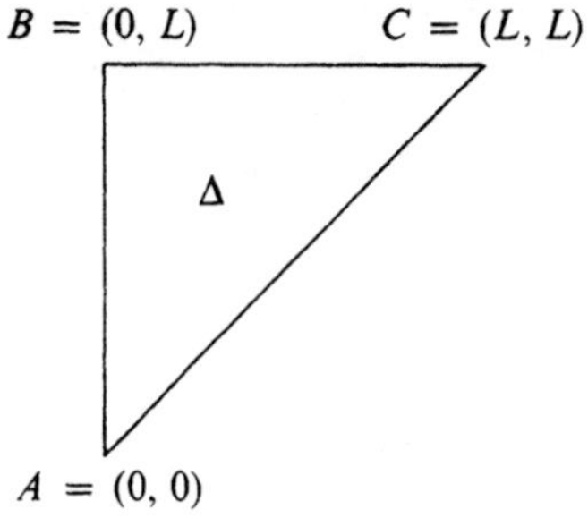

FIGURE 3.6

Now $2\pi i_\alpha = \theta(L) = \theta(L) - \theta(0) = \int_0^L (d\theta/ds)\, ds = \int_\alpha d\theta$. Since

$$\alpha(u, u) = \theta(u),$$

this last line integral equals $\int_{\overline{AC}} d\alpha$.

$$d\alpha = \frac{\partial \alpha}{\partial u}\,du + \frac{\partial \alpha}{\partial v}\,dv,$$

and

$$\frac{\partial^2 \alpha}{\partial u\,\partial v} - \frac{\partial^2 \alpha}{\partial v\,\partial u} \equiv 0 \qquad (\text{since } \alpha \text{ is } C^2).$$

Thus by Green's theorem (1.3), $\int_{\overline{AC}+\overline{CB}+\overline{BA}} d\alpha = 0$. Hence

$$\int_{\overline{AC}} d\alpha = \int_{\overline{AB}} d\alpha + \int_{\overline{BC}} d\alpha.$$

$\int_{\overline{AB}} d\alpha$ is the angle through which $\overrightarrow{OP}$ rotates as P traverses the image of $\boldsymbol{\alpha}$ once in the direction of increasing s. (See Figure 3.7) Since $\boldsymbol{\alpha}$ lies in the upper half plane, $\overrightarrow{OP}$ never points downward and the angle of rotation must be $\pi\epsilon$, with $\epsilon = \pm 1$ according to whether $\boldsymbol{\alpha}$ is traversed counterclockwise or not.

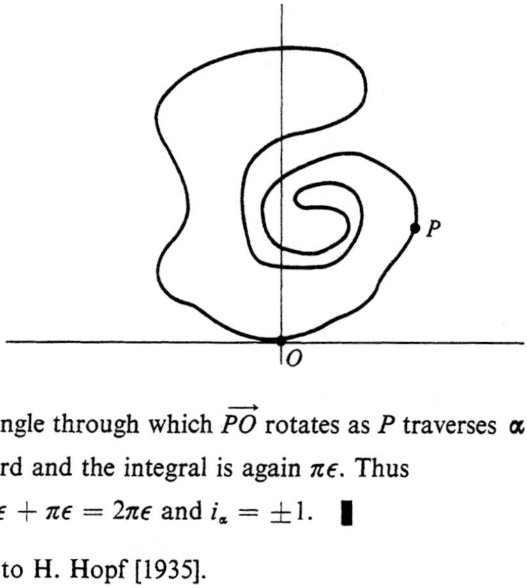

FIGURE 3.7

Likewise, $\int_{\overline{BC}} d\alpha$ is the angle through which $\overrightarrow{PO}$ rotates as P traverses $\boldsymbol{\alpha}$ once. $\overrightarrow{PO}$ never points upward and the integral is again $\pi\epsilon$. Thus

$$2\pi i_{\alpha} = \pi\epsilon + \pi\epsilon = 2\pi\epsilon \text{ and } i_{\alpha} = \pm 1. \quad \blacksquare$$

The above proof is due to H. Hopf [1935].

EXAMPLE 2.10. Consider the simple closed curve in Figure 3.8. The following table gives the values of θ at the labeled points. Hence $i_{\alpha} = 1$ for this curve as $\theta(V) = 2\pi$.

point	A	B	C	D	E	F	G	H	I	J	K
θ	0	$\frac{\pi}{2}$	π	$\frac{3\pi}{2}$	2π	$\frac{5\pi}{2}$	3π	$\frac{7\pi}{2}$	3π	$\frac{5\pi}{2}$	2π

	L	M	N	O	P	Q	R	S	T	U	V
	$\frac{3\pi}{2}$	π	$\frac{\pi}{2}$	0	$\frac{-\pi}{2}$	$\frac{-\pi}{2}$	0	$\frac{\pi}{2}$	π	$\frac{3\pi}{2}$	2π

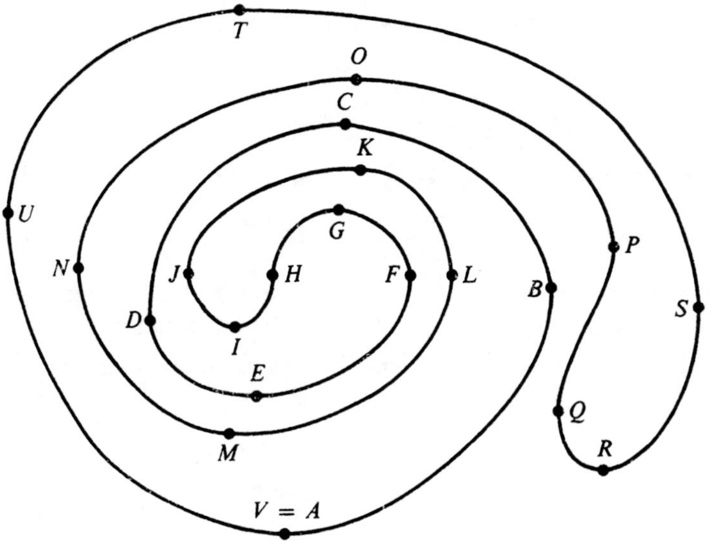

FIGURE 3.8

In the following corollary, and elsewhere, S^1 denotes the *unit circle* in the plane: $S^1 = \{(a, b) \in \mathbf{R}^2 \mid a^2 + b^2 = 1\}$.

COROLLARY 2.11. If $\boldsymbol{\alpha}(s)$ is a simple closed regular plane curve, the *tangent circular image* $\mathbf{t}\colon [0, L] \longrightarrow S^1$ is onto.

Proof: Problem 2.2. ▮

Quite often it is useful to consider curves that are made up of regular curve segments joined together. At these junction points $\mathbf{t}$ may not be defined.

DEFINITION. A *piecewise C^k regular curve* is a continuous function $\boldsymbol{\alpha}$: $[a, b] \longrightarrow R^3$ together with a finite set of points $\{s_i \mid 0 \leq i \leq n\}$ with $a = s_0 < s_1 < \ldots < s_n = b$ such that $\boldsymbol{\alpha}|_{[s_i, s_{i+1}]}$ is a regular C^k curve segment.

At a junction point $\boldsymbol{\alpha}(s_i)$ of a piecewise regular plane curve $\boldsymbol{\alpha}$ let $\mathbf{t}^-(s_i) = \lim_{s \to s_i^-} \mathbf{t}(s)$ and $\mathbf{t}^+(s_i) = \lim_{s \to s_i^+} \mathbf{t}(s)$. The angle from $\mathbf{t}^-(s_i)$ to $\mathbf{t}^+(s_i)$ will be denoted $\Delta\theta_i$. (If $\boldsymbol{\alpha}$ is closed, $\Delta\theta_0 = \Delta\theta_n =$ angle from $\mathbf{t}^-(s_n)$ to $\mathbf{t}^+(s_0)$.) Let $\delta\theta_i$ be the angle through which $\mathbf{t}$ rotates on the segment $\boldsymbol{\alpha}|_{[s_i, s_{i+1}]}$.

DEFINITION. The *rotation index* of a piecewise regular plane closed curve $\boldsymbol{\alpha}$ is

$$i_{\alpha} = \frac{\sum_{i=0}^{n-1} \delta\theta_i + \sum_{i=0}^{n-1} \Delta\theta_i}{2\pi}.$$

Example 2.12. Let $\boldsymbol{\alpha}(t)$ be the triangle with vertices (0, 0), (2, 0), (1, 1).

$$\boldsymbol{\alpha}(t) = \begin{cases} (t, 0) & 0 \leq t \leq 2 \\ (4 - t, t - 2) & 2 \leq t \leq 3 \\ (4 - t, 4 - t) & 3 \leq t \leq 4. \end{cases}$$

Here $s_0 = 0$, $s_1 = 2$, $s_2 = 3$, $s_3 = 4$. Note that t is not arc length. The vectors $\mathbf{t}^-(s_i)$ and $\mathbf{t}^+(s_i)$ are shown in Figure 3.9. Since $\mathbf{t}$ is constant on each segment, $\delta\theta_i = 0$. $\sum_{i=0}^{2} \Delta\theta_i$ is the sum of the *exterior* angles of a triangle, and is thus 2π. Hence the rotation index is 1.

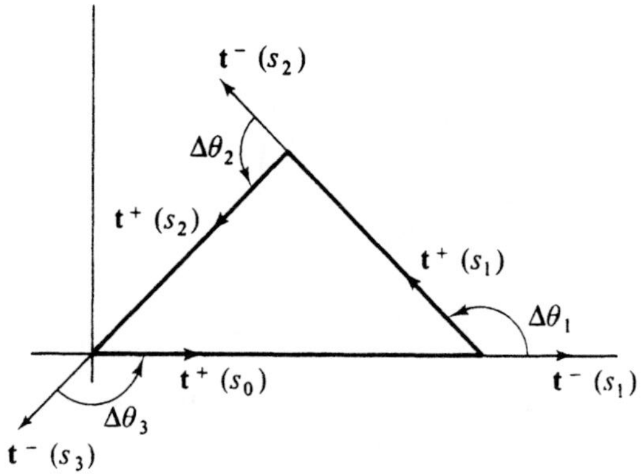

FIGURE 3.9

Note that for any piecewise C^2 curve $\boldsymbol{\alpha}(s)$,

$$\delta\theta_i = \int_{s_i}^{s_{i+1}} k\, ds \quad \text{and} \quad i_{\boldsymbol{\alpha}} = \frac{\int_{\boldsymbol{\alpha}} k\, ds + \sum_{i=0}^{n-1} \Delta\theta_i}{2\pi}.$$

The following proposition is again almost obvious. However, a careful proof is even more involved than that of Theorem 2.9. One way to prove it would be to approximate the piecewise smooth curve by a regular curve. However, development of the necessary ideas would take us too far afield and so we omit the proof. See H. Hopf [1935].

Proposition 2.13. If $\boldsymbol{\alpha}(s)$ is a piecewise regular simple closed plane curve, then $i_{\boldsymbol{\alpha}} = \pm 1$.

There is another "obvious" theorem about simple closed plane curves, called the Jordan Curve Theorem. It states that any simple closed plane curve has an "inside" and an "outside." The interested reader can find a careful proof in Stoker [1969].

PROBLEMS

***2.1.** Prove Lemma 2.2.

***2.2.** Prove Corollary 2.11.

2.3. Let $\mathcal{C}$ be the curve made up of a straight line segment and two circular arcs of radius 2 as in Figure 3.10. Compute the $\delta\theta_i$ and the $\Delta\theta_i$ and verify that the rotation index is 1.

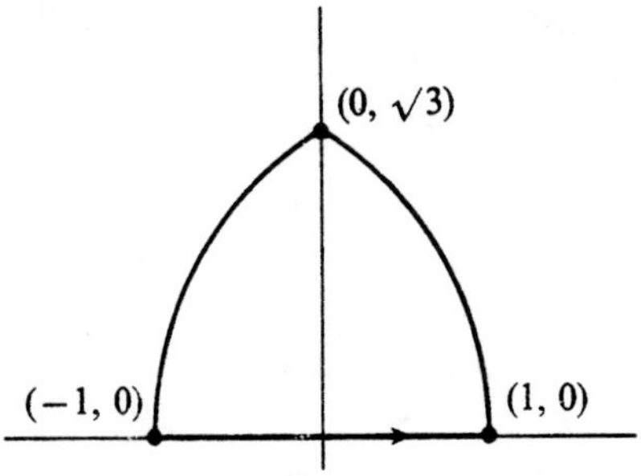

FIGURE 3.10

3–3. CONVEX CURVES

A straight line l divides $\mathbf{R}^2$ into two half planes, H_1 and H_2, such that $H_1 \cup H_2 = \mathbf{R}^2$ and $H_1 \cap H_2 = l$. We say a curve *lies on one side of l* if the image of $\boldsymbol{\alpha}$ is completely contained in one of the half planes H_1 or H_2.

DEFINITION. A regular curve $\boldsymbol{\alpha}$ is *convex* if it lies on one side of each tangent line.

EXAMPLE 3.1. The curves in Figure 3.11 and 3.12 are convex, while those in Figures 3.13 and 3.14 are not convex.

The next theorem will use the concept of monotonicity. A function $f(s)$ is called *monotone increasing* if $s \leq t$ implies $f(s) \leq f(t)$. $f(s)$ is *mono-*

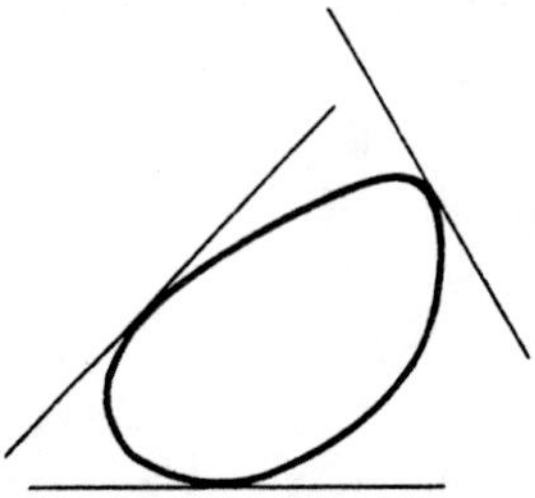

FIGURE 3.11

FIGURE 3.12

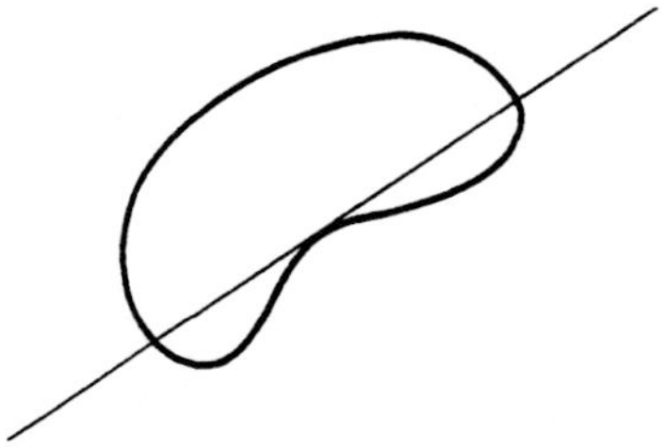

FIGURE 3.13

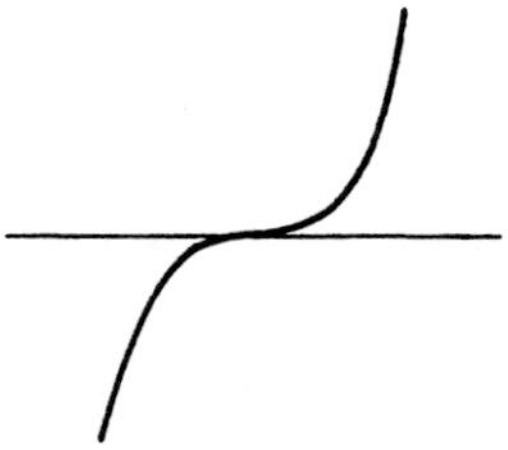

FIGURE 3.14

tone decreasing if $s \leq t$ implies $f(s) \geq f(t)$. f is *monotone* if it is either monotone increasing or monotone decreasing. We note that if f is differentiable, f is monotone if and only if f' has constant sign. ($f' \geq 0$ implies monotone increasing, $f' \leq 0$ implies monotone decreasing. See Problem 3.2.)

THEOREM 3.2. A simple closed regular plane curve $\boldsymbol{\alpha}(s)$ is convex if and only if $k(s)$ has constant sign.

Proof: Since $k(s) = d\theta/ds$, we must show that θ is monotone if and only if $\boldsymbol{\alpha}$ is convex. Suppose θ is monotone. If $\boldsymbol{\alpha}$ is *not* convex there is a point A such that $\boldsymbol{\alpha}$ does not lie on one side of the tangent line l at A. Since $\boldsymbol{\alpha}$ is closed, there are points B and C of $\boldsymbol{\alpha}$ on opposite sides of l which are farthest from l (see Figure 3.15). Note that the tangent lines at A, B, and C must be distinct.

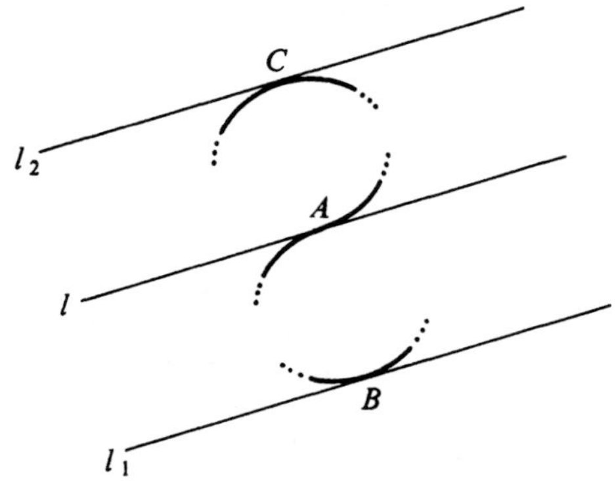

FIGURE 3.15

The tangent lines l_1 and l_2 at B and C must be parallel to l. If this were not the case, one could construct a line through B (or C) parallel to l. Since this line is not tangent to the curve and goes through B (or C), there must be points of the curve on both sides of the line. There would then be points on the curve farther from l than B is.

Two of the three points A, B, C must have tangents $\mathbf{t}$ pointing in the same direction. Thus there exists $s_1 < s_2$ with $\mathbf{t}(s_1) = \mathbf{t}(s_2)$ and $\theta(s_2) = \theta(s_1) + 2\pi n$.

Since θ is monotone, Theorem 2.9 implies $n = -1$, 0, or 1. If $n = 0$, then $\theta(s_1) = \theta(s_2)$ and by monotonicity θ is constant on the interval $[s_1, s_2]$. If $n = \pm 1$, then θ is constant on $[0, s_1]$ and $[s_2, L]$. In either case one of the segments of $\boldsymbol{\alpha}$ between $\boldsymbol{\alpha}(s_1)$ and $\boldsymbol{\alpha}(s_2)$ is a straight line.

Hence the tangent lines at $\boldsymbol{\alpha}(s_1)$ and $\boldsymbol{\alpha}(s_2)$ coincide. But l, l_1, l_2 are distinct. This contradiction implies $\boldsymbol{\alpha}$ is convex.

Now suppose $\boldsymbol{\alpha}$ is convex. If θ is not monotone, there are points $s_1 < s_0 < s_2$ with $\theta(s_1) = \theta(s_2) \neq \theta(s_0)$. We shall prove θ is monotone by showing that if $\theta(s_1) = \theta(s_2)$, then $\theta(s) = \theta(s_1)$ for all s between s_1 and s_2.

If $\theta(s_1) = \theta(s_2)$ for some $0 \leq s_1 < s_2 \leq L$, then $\mathbf{t}(s_1) = \mathbf{t}(s_2)$. By Corollary 2.11, the map $\mathbf{t}\colon [0, L] \longrightarrow S^1$ is onto so there is a point s_3 with $\mathbf{t}(s_3) = -\mathbf{t}(s_1)$. If the tangent lines at s_1, s_2, s_3 are distinct, they are parallel and one is between the others. This can't happen since $\boldsymbol{\alpha}$ is convex. Thus two of the lines coincide and there are points A and B of $\boldsymbol{\alpha}$ lying on the same tangent line l (see Figure 3.16). We next show that the curve $\boldsymbol{\alpha}$ is a straight line between A and B.

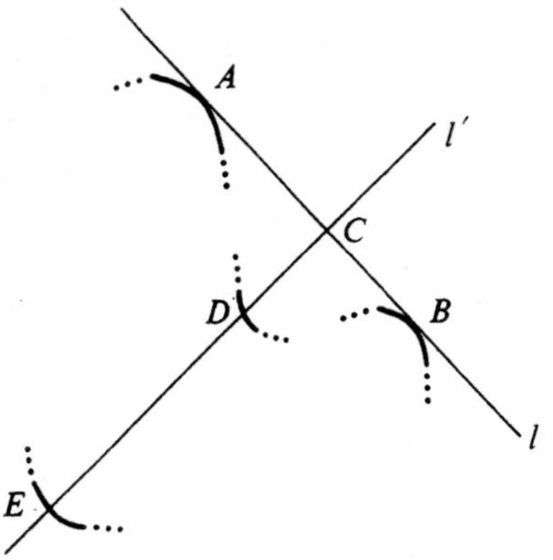

FIGURE 3.16

Suppose some point C of the line segment $\overline{AB}$ is not on $\boldsymbol{\alpha}$. Let l' be the line perpendicular to $\overline{AB}$ at C. l' is not a tangent line since $\boldsymbol{\alpha}$ is convex. Thus l' intersects $\boldsymbol{\alpha}$ in at least two points, D and E, which lie on the same side of l. If D denotes the point closer to C, the tangent line at D has at least one of the points A, B, E on each side, contradicting $\boldsymbol{\alpha}$ being convex.

Hence C is on $\boldsymbol{\alpha}$ and all of $\overline{AB}$ is also. Thus the tangents at A and B have the same direction. Then A, B are $\boldsymbol{\alpha}(s_1)$, $\boldsymbol{\alpha}(s_2)$ and $\boldsymbol{\alpha}|_{[s_1, s_2]}$ is straight. Therefore θ is constant on $[s_1, s_2]$, and θ is monotone on $[0, L]$. ∎

Included in the proof above is the following result.

COROLLARY 3.3. Let $\boldsymbol{\alpha}$ be a simple closed convex curve with horizontal angle θ. If $\theta(s_1) = \theta(s_2)$ with $s_1 < s_2$, then $\boldsymbol{\alpha}$ is a straight line segment on $[s_1, s_2]$.

The assumption that $\alpha(s)$ is simple cannot be relaxed in Theorem 3.2. The curve of Figure 3.5 has θ always increasing (hence $k(s) > 0$), but it is not convex.

COROLLARY 3.4. Let $\alpha(s)$ be a closed convex curve. If a straight line l intersects the image of α in three points, then the entire line segment joining these points lies in the image of α. In particular, if $k \neq 0$, a straight line can intersect α in at most two points.

Proof: Let the three points of intersection be A, B, and F with F between A and B. l must be tangent to α at F or else A and B lie on opposite sides of the tangent line at F, a contradiction. l must also be tangent at A and B or else there would be points of α on both sides of l. But then the proof of Theorem 2.2 shows that the entire straight line segment from A to B is in the image of α. Finally, if $k \neq 0$, α has no straight segments (Corollary 3.3) and l can intersect the image of α in at most two points. ▮

This corollary is true for nonclosed convex curves.

PROBLEMS

3.1. A piecewise C^1 curve may have two tangents at each junction point. Define a piecewise C^1 curve to be *convex* if it lies on one side of each tangent line (and on one side of each tangent line at a junction point). Give an example to show that Theorem 3.2 is false for piecewise C^2 curves.

***3.2.** Prove that a differentiable function f is monotone increasing if $f' \geq 0$ and monotone decreasing if $f' \leq 0$. (*Hint:* Mean Value Theorem.)

3–4. THE ISOPERIMETRIC INEQUALITY

One of the standard problems of early calculus is to find the rectangle (or possibly a triangle) of fixed perimeter bounding the greatest area. The answer, of course, is a square (or equilateral triangle). Quite often the student is told that of all geometric figures with a fixed perimeter, the circle bounds the greatest area. We now give a proof of this fact for regions bounded by regular curves. The proof requires the following result.

LEMMA 4.1. If α is a simple closed plane curve whose image bounds a region $\mathcal{R}$, and which is traversed counterclockwise, then the area of $\mathcal{R}$ is $\int_\alpha x\,dy = -\int_\alpha y\,dx$, where x and y are the coordinates of the plane.

Proof: Problem 4.1. ▮

Theorem (Isoperimetric Inequality). Let $\boldsymbol{\alpha}$ be a simple closed regular plane curve of length (perimeter) L. Let A be the area of the region bounded by $\boldsymbol{\alpha}$. Then $L^2 \geq 4\pi A$ with equality if and only if $\boldsymbol{\alpha}$ is a circle. Thus, of all curves of fixed length L, the circle bounds the greatest area.

Proof: Let l_1, l_2 be two parallel lines tangent to $\boldsymbol{\alpha}$ with $\boldsymbol{\alpha}$ bounded between them. Let $\boldsymbol{\beta}$ be a circle tangent to l_1 and l_2 which does not intersect $\boldsymbol{\alpha}$ and let r be its radius. Choose coordinates x and y for the plane with the origin at the center of the circle and with the y-axis parallel to l_1. Let l_1 be tangent to $\boldsymbol{\alpha}$ at $A = \boldsymbol{\alpha}(0)$ and l_2 at $C = \boldsymbol{\alpha}(s_2)$. See Figure 3.17. The key idea of the proof

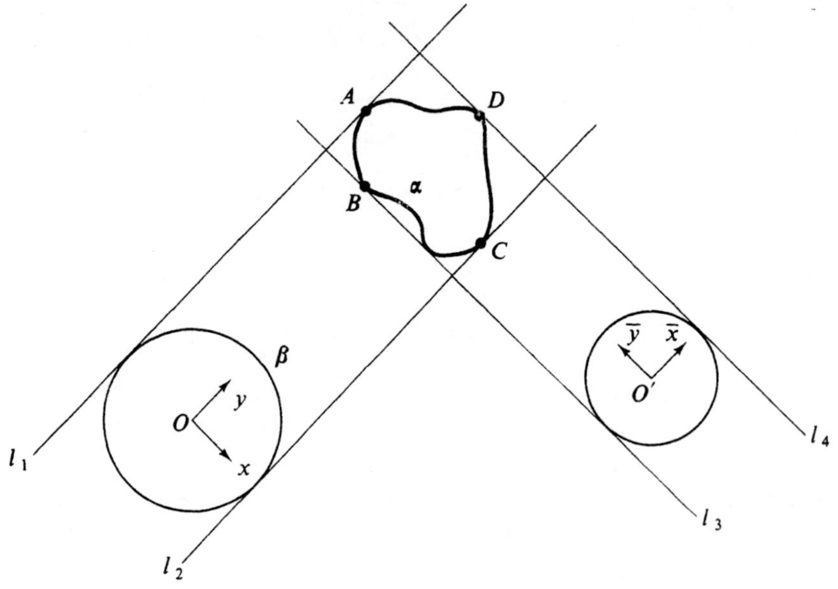

FIGURE 3.17

is to compare the area of the region bounded by $\boldsymbol{\alpha}$ with the area of the circle of radius r. Since both the length and the area enclosed by $\boldsymbol{\alpha}$ are independent of parametrization we may assume that $\boldsymbol{\alpha}$ is a unit speed curve. $\boldsymbol{\alpha}$ may then be written in the (x, y) coordinates as $\boldsymbol{\alpha}(s) = (x(s), y(s))$ where $(x'(s))^2 + (y'(s))^2 = 1$. The curve $\boldsymbol{\beta}$ may be parametrized by $\boldsymbol{\beta}(s) = (z(s), w(s))$ where

(4-1)
$$z(s) = x(s)$$
$$w(s) = \begin{cases} -\sqrt{r^2 - x^2} & 0 \leq s \leq s_2 \\ \sqrt{r^2 - x^2} & s_2 \leq s \leq L. \end{cases}$$

Note that s is arc length on $\boldsymbol{\alpha}$ but not on $\boldsymbol{\beta}$. In fact, with this parametrization $\boldsymbol{\beta}$ may not be regular, but it is C^2.

By Lemma 4.1 the area bounded by $\boldsymbol{\alpha}$ is $A = \int_{\boldsymbol{\alpha}} x\,dy = \int_0^L xy'\,ds$. The area bounded by $\boldsymbol{\beta}$ is $\pi r^2 = -\int_{\boldsymbol{\beta}} y\,dx = -\int_0^L wz'\,ds = -\int_0^L wx'\,ds$. Thus

$$A + \pi r^2 = \int_0^L (xy' - wx')\,ds \leq \int_0^L |xy' - wx'|\,ds$$
$$= \int_0^L |\langle(x', y'), (-w, x)\rangle|\,ds.$$

The Cauchy-Schwarz Inequality (Lemma 1.2 of Chapter 1) together with the fact that $\boldsymbol{\alpha}$ is unit speed tells us that

$$|\langle(x', y'), (-w, x)\rangle| \leq |(x', y')||(-w, x)| = \sqrt{w^2 + x^2} = r$$

(see Equation (4-1)). Thus

$$A + \pi r^2 \leq \int_0^L |\langle(x', y'), (-w, x)\rangle|\,ds \leq \int_0^L r\,ds = rL.$$

We have thus shown that

(4-2) $$A + \pi r^2 \leq rL.$$

If we let $a = A$ and $b = \pi r^2$ and apply Problem 4.2, we obtain

(4-3) $$\sqrt{A\pi r^2} \leq \frac{A + \pi r^2}{2} \leq \frac{rL}{2},$$

where the second inequality comes from (4-2). Hence $A\pi r^2 \leq r^2L^2/4$ or $L^2 \geq 4\pi A$, which is the Isoperimetric Inequality.

Now assume that $L^2 = 4\pi A$. We shall show that $\boldsymbol{\alpha}$ must be a circle. We first show that $x = ry'$. Since $L^2 = 4\pi A$, inequality (4-3) becomes an equality. An application of Problem 4.2 yields $A = \pi r^2$. Inequality (4-2) must also be an equality. Considering the derivation of (4-2) we must have equality where the Cauchy-Schwarz Inequality was used. By Lemma 1.8 of Chapter 1 there is a real number c such that

(4-4) $$(-w, x) = c(x', y').$$

Taking the length of each side of (4-4) yields

$$\sqrt{w^2 + x^2} = |c|\sqrt{(x')^2 + (y')^2} = |c|$$

and so (4-1) implies that $c = \pm r$. On the other hand, (4-4) also says that $c = \langle(x', y'), (-w, x)\rangle$ which (since the first inequality in the derivation of (4-2) must also be an equality) is nonnegative. Hence $c = r$ and (4-4) shows that $x = ry'$.

$A = \pi r^2$ implies that r depends on A and not the choice of l_1. Thus if lines l_3 and l_4 orthogonal to l_1 and l_2 are used, a circle of radius r is tangent to them. If coordinates $\bar{x}$ and $\bar{y}$ with origin at the center of this circle are used (see Figure 3.17), we would derive $\bar{x} = r\bar{y}'$, Since the y-axis is parallel to and in the direction of the $\bar{x}$-axis and the x-axis is parallel to and in the direction of $-\bar{y}$ there are constants d and e such that $\bar{x} = y - d$, $\bar{y} = e - x$. Thus

$$y - d = \bar{x} = r\bar{y}' = -rx'$$

and

$$x^2 + (y - d)^2 = (ry')^2 + (-rx')^2 = r^2((x')^2 + (y')^2) = r^2.$$

Therefore $\boldsymbol{\alpha}$ is a circle of radius r centered at $(0, d)$ in the (x, y) coordinate system. ∎

This proof, which is due to E. Schmidt [1939], and another, which involves Fourier series and is due to Hurwitz (1902), may be found in Chern [1967].

PROBLEMS

***4.1.** Prove Lemma 4.1. (*Hint:* Green's Theorem.)

***4.2.** Let a and b be positive numbers. Prove that $\sqrt{ab} \leq \frac{1}{2}(a + b)$, with equality if and only if $a = b$.

3–5. THE FOUR-VERTEX THEOREM

We shall now consider a certain class of convex curves—those with no straight segments or isolated points where $k = 0$.

DEFINITION. An *oval* is a regular simple closed plane curve with $k > 0$.

Note that an oval is convex since k does not change sign.

DEFINITION. A *vertex* of a regular plane curve is a point where k has a relative maximum or minimum.

The concept of a vertex does not depend on the parametrization of a curve, as you will show in Problem 5.3. Vertices may be found when k is a function of some arbitrary parameter, much as in Section 2-6.

EXAMPLE 5.1. The (non-unit speed) curve $\boldsymbol{\alpha}(t) = (2 \cos t, \sin t)$ is an ellipse. It has exactly four vertices given by $t = 0, \pi/2, \pi, 3\pi/2$. See Problem 5.2.

LEMMA 5.2. If l is a line in $\mathbf{R}^2$, then there are $\mathbf{a}, \mathbf{c} \in \mathbf{R}^2$ with $\mathbf{c} \neq 0$ such that $\mathbf{z} \in l$ if and only if $\langle \mathbf{z} - \mathbf{a}, \mathbf{c} \rangle = 0$.

Proof: Let l be the line given by $\boldsymbol{\alpha}(t) = \mathbf{z}_0 + t\mathbf{v}$ where $\mathbf{0} \neq \mathbf{v} = (v^1, v^2) \in \mathbf{R}^2$. Then $\mathbf{a} = \mathbf{z}_0$ and $\mathbf{c} = (-v^2, v^1)$ give the desired result. ∎

THEOREM 5.3 (Four-Vertex Theorem). An oval $\boldsymbol{\alpha}(s)$ has at least four vertices.

Proof: Since $\boldsymbol{\alpha}$ is C^3, $k' = 0$ at each vertex. If k is constant on any segment, then every point on this segment is a vertex and we are finished. We may therefore assume that $\boldsymbol{\alpha}$ has no circular arcs (and since $k > 0$, no straight line segments either) and that there are distinct vertices A and B where k takes on its global maximum and minimum. Assume also that $A = \boldsymbol{\alpha}(0)$. We now show that the assumption that these are the only vertices leads to a contradiction. Because vertices come in pairs this will prove the theorem.

Let l be the straight line joining A and B. Since we have assumed that there are exactly two vertices, k' is positive on one segment of $\boldsymbol{\alpha}$ and negative on the other. See Figure 3.18.

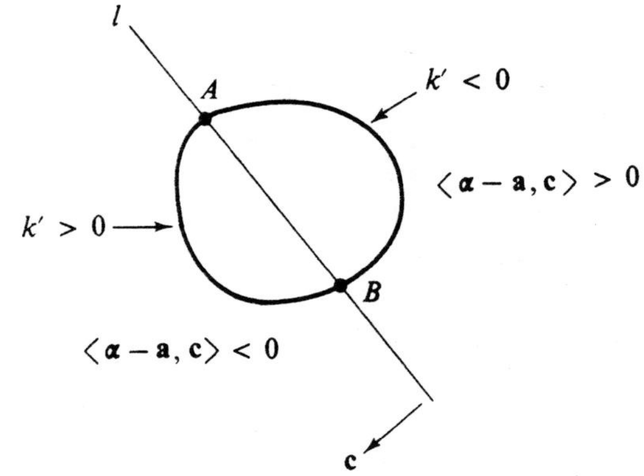

FIGURE 3.18

Lemma 5.2 says that there are constants $\mathbf{a}$ and $\mathbf{c} \neq \mathbf{0}$ such that $\mathbf{z} \in l$ if and only if $\langle \mathbf{z} - \mathbf{a}, \mathbf{c} \rangle = 0$. Because $k > 0$, $\boldsymbol{\alpha}$ must be convex (Theorem 3.2) and so (Corollary 3.4) l intersects $\boldsymbol{\alpha}$ at exactly the two points A and B. Hence $\langle \boldsymbol{\alpha}(s) - \mathbf{a}, \mathbf{c} \rangle$ is positive on one segment of $\boldsymbol{\alpha}$ and negative on the other.

A case by case study shows that $k'(s)\langle \boldsymbol{\alpha}(s) - \mathbf{a}, \mathbf{c} \rangle$ does not change sign on $\boldsymbol{\alpha}$. (It is nonpositive in Figure 3.18.) At some point $k'\langle \boldsymbol{\alpha} - \mathbf{a}, \mathbf{c} \rangle \neq 0$ and so

$$\begin{aligned} 0 \neq \int_0^L k'\langle \boldsymbol{\alpha} - \mathbf{a}, \mathbf{c} \rangle \, ds &= k\langle \boldsymbol{\alpha} - \mathbf{a}, \mathbf{c} \rangle \Big|_0^L - \int_0^L k\langle \boldsymbol{\alpha}', \mathbf{c} \rangle \, ds \\ &= 0 + \int_0^L \langle -k\mathbf{t}, \mathbf{c} \rangle \, ds \\ &= \int_0^L \langle \mathbf{n}', \mathbf{c} \rangle \, ds \\ &= \langle \mathbf{n}, \mathbf{c} \rangle \Big|_0^L = 0. \end{aligned}$$

This contradiction implies there are more than two vertices. Since k' changes sign at a vertex, the number of vertices must be even. Thus there are at least four vertices. ∎

This theorem was first proved by Mukhopadhyaya [1909] and Kneser [1912]. It is, in fact, valid not only for ovals, but also for any simple closed plane curve. (See Kneser [1912] or S. Jackson [1944].) The proof we gave is due to G. Herglotz in a letter to W. Blaschke [1930].

The theorem is false if we omit the hypothesis "closed" (Problem 5.4) or if we omit the hypothesis "simple" as the following example shows.

EXAMPLE 5.4. Consider the nonsimple plane curve given in polar coordinates, (r, ϕ), by $r = 1 - 2\sin\phi$ and whose graph is sketched in Figure 3.19.

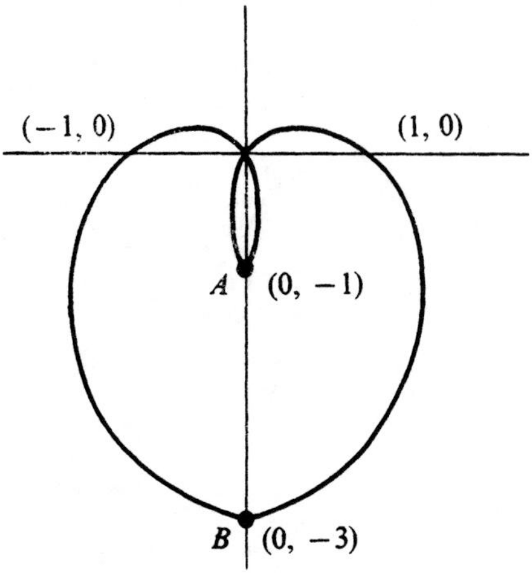

FIGURE 3.19

In rectangular coordinates this is the curve

$$\boldsymbol{\beta}(\phi) = (\cos\phi - 2\sin\phi\cos\phi,\ \sin\phi - 2\sin^2\phi,\ 0).$$

We shall show that $\boldsymbol{\beta}$ has only two vertices. Since the angle θ between the horizontal and $\mathbf{t}$ increases as ϕ increases, $k > 0$ and hence $k = \kappa$. We compute κ via Proposition 6.1 of Chapter 2.

$$\dot{\boldsymbol{\beta}} = (-\sin\phi - 2\cos^2\phi + 2\sin^2\phi,\ \cos\phi - 4\sin\phi\cos\phi,\ 0)$$

$$\ddot{\boldsymbol{\beta}} = (-\cos\phi + 8\sin\phi\cos\phi,\ -\sin\phi - 4\cos^2\phi + 4\sin^2\phi,\ 0)$$

$$\dot{\boldsymbol{\beta}} \times \ddot{\boldsymbol{\beta}} = (0, 0, 9 - 6\sin\phi)$$

$$|\dot{\boldsymbol{\beta}} \times \ddot{\boldsymbol{\beta}}| = 9 - 6\sin\phi$$

$$|\dot{\boldsymbol{\beta}}| = (5 - 4\sin\phi)^{1/2}$$

$$k = \kappa = (9 - 6\sin\phi)(5 - 4\sin\phi)^{-3/2}$$

$$k' = \dot{k}\frac{d\phi}{ds} = (24\cos\phi - 12\cos\phi\sin\phi)(5 - 4\sin\phi)^{-5/2}\left(\frac{d\phi}{ds}\right).$$

We see that $k' = 0$ only when

$$(24\cos\phi - 12\cos\phi\sin\phi) = 12(2 - \sin\phi)\cos\phi$$

is zero. That is, when $\cos\phi = 0$, or $\phi = \pi/2, 3\pi/2$. Thus $\boldsymbol{\beta}$ has only two vertices, which are marked A and B in Figure 3.19.

Let $\boldsymbol{\alpha}$ be an oval and P a point on $\boldsymbol{\alpha}$. By Corollary 2.11, there is a point $\bar{P}$ where the tangent $\mathbf{t}$ is opposite to that at P (i.e., $\mathbf{t}(\bar{P}) = -\mathbf{t}(P)$). The tangent lines at P and $\bar{P}$ are parallel. By the reasoning in the proof of Theorem 3.2, at no other point of $\boldsymbol{\alpha}$ is the tangent line parallel to these two lines. Thus, given a point P on an oval, there is a unique point $\bar{P}$ (called the *opposite point* to P) such that the tangents at P and $\bar{P}$ are parallel and distinct.

DEFINITION. The *width* $w(s)$ of an oval $\boldsymbol{\alpha}$ at $\boldsymbol{\alpha}(s)$ is the distance between the tangent lines at $P = \boldsymbol{\alpha}(s)$ and $\bar{P}$. (See Figure 3.20.)

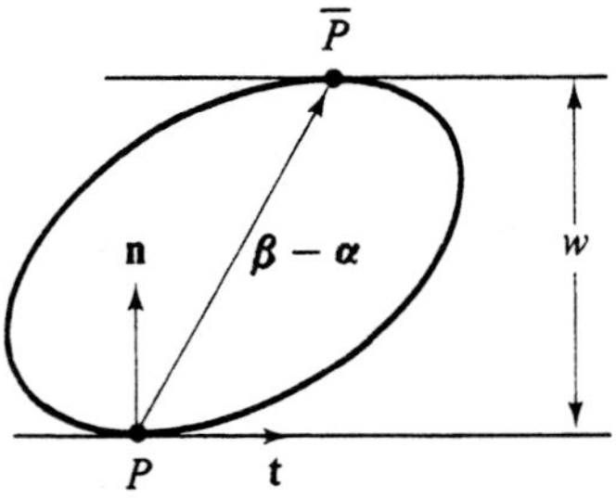

FIGURE 3.20

DEFINITION. An oval has *constant width* if the width at P is independent of the choice of P.

EXAMPLE 5.5. A circle has constant width.

EXAMPLE 5.6. The "piston" for the Wankel engine gives a piecewise differentiable curve of constant width (Figure 3.21). To obtain a differentiable curve of constant width one can take the set of points a fixed distance outside the Wankel piston (Figure 3.22). Note this curve is C^1 but not C^2. (Why?)

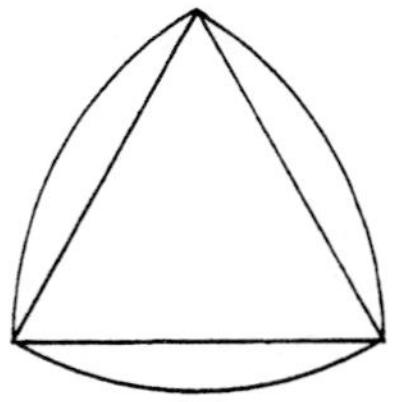

FIGURE 3.21

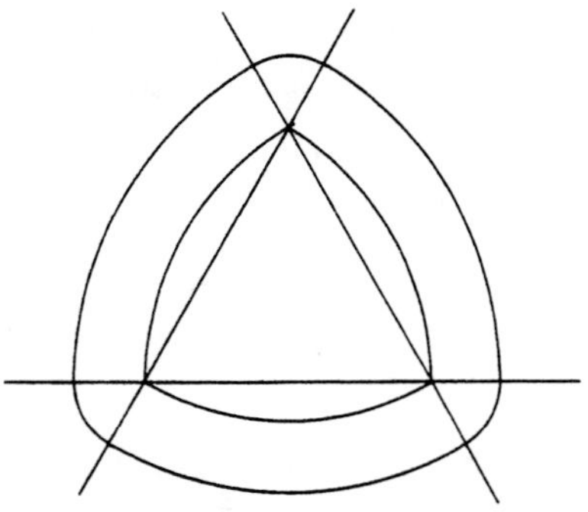

FIGURE 3.22

Problem 5.5. gives another example of a curve of constant width.

The next theorem is surprising because it states that the perimeter of an oval of constant width depends only on the width.

THEOREM 5.7 (Barbier, 1860). If $\boldsymbol{\alpha}$ is an oval of constant width w, then the length of $\boldsymbol{\alpha}$ is πw.

Proof: Let $\boldsymbol{\beta}(s)$ denote the opposite point to $\boldsymbol{\alpha}(s)$. The curve $\boldsymbol{\beta}(s)$ is not a unit speed curve (unless $\boldsymbol{\alpha}$ is a circle). Using Lemma 4.1 of Chapter 2, we have

$$\boldsymbol{\beta}(s) - \boldsymbol{\alpha}(s) = v\mathbf{t}(s) + w\mathbf{n}(s), \tag{5-1}$$

where $v = \langle \boldsymbol{\beta} - \boldsymbol{\alpha}, \mathbf{t} \rangle$ and $w = \langle \boldsymbol{\beta} - \boldsymbol{\alpha}, \mathbf{n} \rangle$. Note that the unit tangent to $\boldsymbol{\beta}(s)$ is $-\mathbf{t}(s)$ and the unit normal is $-\mathbf{n}(s)$.

Let $\bar{s}$ be arc length along $\boldsymbol{\beta}$, so that $d\boldsymbol{\beta}/d\bar{s}$ is the unit tangent, and differentiate Equation (5-1) with respect to $\bar{s}$.

$$\begin{aligned}
-\mathbf{t}(s) &= \frac{d\boldsymbol{\beta}}{d\bar{s}} = \frac{d\boldsymbol{\beta}}{ds}\frac{ds}{d\bar{s}} \\
&= \frac{ds}{d\bar{s}}\frac{d(\boldsymbol{\alpha} + v\mathbf{t} + w\mathbf{n})}{ds} \\
&= \frac{ds}{d\bar{s}}(\mathbf{t} + v'\mathbf{t} + v\mathbf{t}' + w'\mathbf{n} + w\mathbf{n}') \\
&= \frac{ds}{d\bar{s}}(\mathbf{t} + v'\mathbf{t} + vk\mathbf{n} + w'\mathbf{n} - wk\mathbf{t}).
\end{aligned}$$

Comparing coefficients of $\mathbf{t}$ and $\mathbf{n}$, we have $-1 = (ds/d\bar{s})(1 + v' - wk)$ and $0 = (ds/d\bar{s})(vk + w')$ or (provided $ds/d\bar{s} \neq 0$, see Problem 5.7)

$$1 + v' - wk + \frac{d\bar{s}}{ds} = 0 \tag{5-2}$$

and

$$vk + w' = 0. \tag{5-3}$$

Since $\boldsymbol{\alpha}$ has constant width and $k > 0$, Equation (5-3) implies $v = 0$. Then Equation (5-2) yields $1 + (d\bar{s}/ds) = wk$. Let $P = \boldsymbol{\alpha}(0)$ and $\bar{P} = \boldsymbol{\alpha}(s_1)$. Then

$$\int_0^{s_1}\left(1 + \frac{d\bar{s}}{ds}\right) ds = \int_0^{s_1} wk\, ds$$

or

$$\int_0^{s_1} ds + \int_{s=0}^{s=s_1} d\bar{s} = w\int_0^{s_1} k\, ds = w\int_0^{s_1}\frac{d\theta}{ds}\, ds = w\pi.$$

$\int_0^{s_1} ds$ is the arc length of $\boldsymbol{\alpha}$ from P to $\bar{P}$. $\int_{s=0}^{s=s_1} d\bar{s}$ is the arc length of $\boldsymbol{\beta}$ from $\bar{P}$ to P, which is the arc length of $\boldsymbol{\alpha}$ from $\bar{P}$ to P. Hence the length of $\boldsymbol{\alpha}$ is $\int_0^{s_1} ds + \int_{s=0}^{s=s_1} d\bar{s} = w\pi$. ∎

COROLLARY 5.8. If $\boldsymbol{\alpha}$ is an oval of constant width, the straight line joining P to $\bar{P}$ is orthogonal to the tangents at P and $\bar{P}$.

Proof: In the course of the above proof we showed that $v = 0$. This fact coupled with Equation (5-1) implies $\boldsymbol{\beta} = \boldsymbol{\alpha} + w\mathbf{n}$ and so the vector from P to $\bar{P}$ is $w\mathbf{n}$. ∎

PROBLEMS

5.1. If $\boldsymbol{\alpha}(s)$ is an oval, prove that $\mathbf{t}''$ is parallel to $\mathbf{t}$ at at least four points.

5.2. In Example 5.1 show that $k' = 0$ only at the given points. Note that $\boldsymbol{\alpha}$ is not unit speed.

***5.3.** Prove that the concept of a vertex does not depend upon the parametrization.

5.4. Prove that the Four-Vertex Theorem is false if the hypothesis "closed" is omitted by considering the parabola $\boldsymbol{\alpha}(t) = (t, t^2)$.

†5.5. (a) Let $\boldsymbol{\alpha}\colon [0, r] \longrightarrow \mathbf{R}^2$ be a unit speed curve segment and let $\boldsymbol{\beta}(s)$ be the string involute (Problem 4.39 of Chapter 2)

$$\boldsymbol{\beta}(s) = \boldsymbol{\alpha}(s) + (r_0 - s)\mathbf{t}(s)$$

where $r_0 > r$ is some constant. Show that the unit tangent to $\boldsymbol{\beta}$ at $\boldsymbol{\beta}(s)$ is orthogonal to $\mathbf{t}(s)$.

(b) Let A, B, C be three points in the plane and let $\boldsymbol{\alpha}$ be a closed piecewise C^2 curve with junction points at A, B, C and $\mathbf{t}^+ = -\mathbf{t}^-$ at each point. (A, B, C are called *cusps*.) Assume that $\widehat{AB}$ is the longest of the segments of $\boldsymbol{\alpha}$. Let D be a point on the tangent line at A as indicated in Figure 3.23. Define a curve $\boldsymbol{\beta}$ as a string involute of $\boldsymbol{\alpha}$

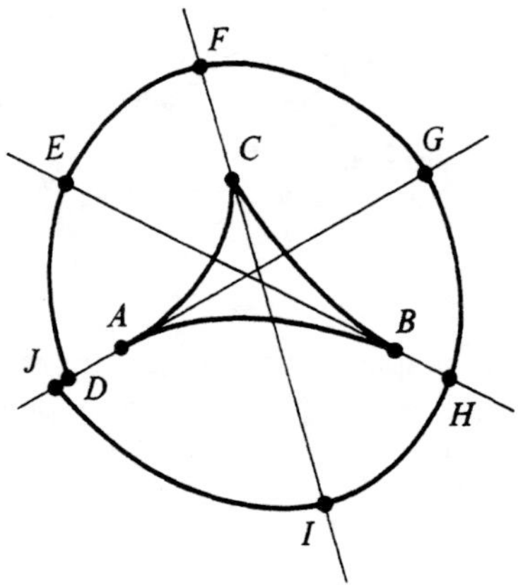

FIGURE 3.23

starting at D: $\widehat{DE}$ is an involute of $\widehat{BA}$, $\widehat{EF}$ is an involute of $\widehat{BC}$, etc. By the first part of the problem, $\widehat{DE}$ and $\widehat{EF}$ meet the line EB at right angles so that $\boldsymbol{\beta}$ is at least C^1. Note that $\overline{EB} = \overline{DA} +$ arc length $\widehat{AB}$. Prove $\overline{JA} = \overline{DA}$ so that $J = D$ and $\boldsymbol{\beta}$ is a closed C^1 curve. Show that $\boldsymbol{\beta}$ has constant width.

5.6. Let $\boldsymbol{\alpha}$ be a plane oval of constant width. Show that the sum of the radii of curvature $(1/k)$ at opposite points is a constant independent of the choice of points.

***5.7.** (a) Let $\boldsymbol{\alpha}(s)$ be a unit speed oval of length L. Let θ be the angle between the horizontal and the tangent $\mathbf{t}(s)$ as in Section 3-2. Prove that $\theta\colon [0, L] \longrightarrow [0, 2\pi]$ is a reparametrization. (*Hint:* $\theta = \int k\, ds$.)
(b) Let $\boldsymbol{\gamma}(\theta)$ be the same oval parametrized by θ so that $\boldsymbol{\alpha}(s) = \boldsymbol{\gamma}(\theta(s))$. Prove that the opposite point to $\boldsymbol{\alpha}(s)$ is $\boldsymbol{\beta}(s) = \boldsymbol{\gamma}(\theta(s) + \pi)$.
(c) Prove that $\boldsymbol{\beta}(s)$ is regular.
(d) Prove that $ds/d\bar{s}$ is nonzero and finite as needed in the proof of Theorem 5.7.

5.8. Assume that $\boldsymbol{\gamma}(\theta)$ is an oval parametrized by θ as in Problem 5.7. Let $w(\theta)$ denote the width of $\boldsymbol{\gamma}$ at $\boldsymbol{\gamma}(\theta)$. Prove that $\int_0^{2\pi} w\, d\theta = 2L$, where $L =$ length of $\boldsymbol{\gamma}$. (*Hint:* Equation (5-2).)

5.9. Let $\boldsymbol{\gamma}$ be an oval parametrized by θ. Prove that

$$\frac{d^2w}{d\theta^2} + w = \frac{1}{k(\theta)} + \frac{1}{k(\theta + \pi)}.$$

(Note that $k(\theta + \pi)$ is the curvature at the point opposite to $\boldsymbol{\gamma}(\theta)$.)

3–6. A PREVIEW

In this section we shall state two theorems about space curves that have the same flavor as those of the preceding sections. However, we have to delay their proofs until Chapter 5 so that we may use certain ideas from the theory of surfaces, namely the concept of curves of shortest length on spheres and integrals over surfaces.

For a closed plane curve $\int_0^L k\, ds$ measured the rotation index:

$$\int_0^L k\, ds = 2\pi i_{\boldsymbol{\alpha}}.$$

For a space curve the analogous object to study is $\int_0^L \kappa\, ds$. For a closed space curve $\int_0^L \kappa\, ds > 0$ as $\kappa \geq 0$, and somewhere $\kappa > 0$. The analogous statement to Theorem 2.3 is

THEOREM 6.1 (Fenchel). If $\boldsymbol{\alpha}$ is a closed unit speed curve of length L, then $\int_0^L \kappa\, ds \geq 2\pi$ with equality if and only if $\boldsymbol{\alpha}$ is a convex plane curve.

A remarkable extension of this theorem was proved independently by I. Fary (1949) and J. Milnor (1950). It depends on the concept of a knot, which will be defined in Chapter 5. For now, think of the special case pictured in Figure 3.24.

FIGURE 3.24

THEOREM 6.2 (Fary-Milnor). If a closed unit speed curve is knotted, then $\int_0^L \kappa \, ds \geq 4\pi$.

This seems plausible when you realize how a knotted curve has to twist about twice as much as an unknotted curve.

4

Local Surface Theory

In the study of curves it was not hard to discern what the geometry was: it was the way in which curves twisted in Euclidean space. This idea leads to the notions of curvature and torsion and a fruitful study of the geometry of curves. We will now discuss the geometry of surfaces. This is a much deeper subject on both a philosophical and mathematical level. There have been many attempts to study geometry including such axiomatic approaches as Hilbert's [1921, 1956]. We shall not attempt to explain this approach, but rather just call the reader's attention to the fact that there are approaches to geometry other than ours. Our viewpoint is that differential geometry provides a unifying thread to geometry (classical plane, spherical, hyperbolic, and projective geometry can all be placed within the framework of differential geometry) and that the approach of Riemann to geometry is the appropriate one. For a more abstract interpretation of Riemann's program in terms of manifolds (as in Chapter 7), see R. Millman and A. Stehney [1973].

Modern geometry was born when Riemann first separated the concept of geometry from the concept of space. His inaugural lecture at Gottingen, *On the Hypotheses which Lie at the Foundation of Geometry* (1854), began: "As is well known, geometry presupposes the concept of space, as well as assuming the basic principles for construction in space. . . . The relationship between these presuppositions is left in the dark." In this chapter we shall first define space (the notion of surface). This will take the first two sections. We are motivated by the idea that a surface should look "locally" like a piece of the Euclidean plane. Then, since our development of surfaces is based upon the study of curves in a surface, we establish the notion of arc length and initiate a study of the curvature of a curve on a surface in Sections 4-3 and 4-4. Having done this, we will discuss in Sections 4-5 and 4-6 what the appropriate notion of "straight line" is in the setting of surfaces. We do this by listing in Section 4-5 the various properties of straight lines in Euclidean space and taking as a definition the idea that the curve does not bend except as the surface bends. We then explore which of the other properties of straight lines in Euclidean space carry over to this setting. Agreeing on a definition of a straight line is, in Riemann's words, agreeing on "constructions in space." The actual terminology for these generalized "straight lines" is *geodesics*. Next we give a development of the concept of curvature of the surface in Sections 4-7 through 4-9. We define the curvature of a surface in two ways. The first way is to break the surface up into (infinitely many) curves and measure the curvature of these curves. The second way is to measure how the normal to the surface changes. (Think how we measured the rate of change of the osculating plane in Chapter 2. We computed $\mathbf{B}'$, that is, the rate of change of its normal.) Section 4-8 has a very extensive set of problems on certain classical topics which tie together several ideas that have been developed so far. The chapter ends with optional material on isometries (length preserving mappings), the Fundamental Theorem of Surfaces, and an investigation of surfaces of constant curvature.

A word is due on the method of differential geometry for attacking these problems. In looking over the first three chapters, the reader should be struck by the usefulness of linear algebra in attacking geometric problems. What, after all, are the Frenet-Serret equations but statements about a (nonobvious) basis for $\mathbf{R}^3$? If we are to follow the approach of earlier chapters here, then we must set up a linear algebraic tool to let us obtain geometric information. This tool will be the tangent space to a surface at a point. The idea is that the linear is significantly easier to work with than is the nonlinear. (For students who have studied differential equations: think how much easier it is to solve linear ordinary differential equations than nonlinear.)

Another point that should be made while we are speaking in an informal way is the difference between local and global. In the study of curves we first introduced only local concepts (in Chapter 2) such as curvature and torsion and obtained information about the local behavior of the curve. In Chapter 3 we then studied the global behavior of a curve (such as the rotation index and convexity). We shall adopt the same strategy in our study of surfaces. We shall first study the local behavior of surfaces, that is, what goes on in a neighborhood of a given point. This is the content of Chapter 4. We then go on to the global behavior of surfaces in Chapter 6. The difference between local and global can be explained as follows: If you are a creature living on the 2-sphere (i.e., the earth) you can only see a very little bit of it at a time, i.e., you can only see "local" things. If, on the other hand, you were an astronaut, then you would be able to see very global things (such as the shortest distance between distant points is the distance along great circles).

This last analogy also brings out another point—the difference between intrinsic and extrinsic. The intrinsic things are the things that an observer who is on the surface sees. The observer doesn't know (for example) *how* his or her world lies in a bigger universe, or in fact that it *does* lie in a bigger universe. He or she only sees concepts that are independent of the embedding of the sphere in space. The astronaut, on the other hand, is out in this space and is cognizant of things such as the normal to the surface while the creature on earth remains totally ignorant of such things. These are the extrinsic concepts. It is sometimes quite difficult to tell exactly which concepts are intrinsic and which are extrinsic. The ninth section of this chapter has as its main theorem the highly nonobvious fact (due to Gauss (1827)) that Gaussian curvature is an intrinsic concept. This last property (being intrinsic) is a very important one and one that we will turn to in the last chapter of this book when we take a very intrinsic (and more abstract!) approach to these ideas.

In the differential geometry of surfaces there is an enormous number of formulas to be derived. We shall emphasize the most important of these formulas by placing a gray background behind each of them.

4–1. BASIC DEFINITIONS AND EXAMPLES

In calculus we are introduced to the concept of a surface through several examples: graphs of functions of two variables, surfaces of revolution, and quadratic surfaces. Here we shall make the concept of a surface more precise and study various geometric properties of surfaces.

There will be an immediate difference between curve theory and surface theory. For a given curve there is a natural geometric parametrization (by arc length) and this parametrization can be used to describe every point on the curve. For a surface there is no natural geometric parametrization. In fact, often it will not be possible to find a parametrization that describes the whole surface, especially in a unique way. As an example, consider the unit sphere $S^2 \subset \mathbf{R}^3$. No matter how we choose a pair of parameters, there will be at least one point that cannot be described by them. Ordinary latitude and longitude fails at the poles ((90° N, 30° E) and (90° N, 60° W) are the same point) and to a lesser extent along the 180° meridian (east or west?). (See Figure 4.1).

DEFINITION. A subset $\mathcal{U}$ of $\mathbf{R}^2$ is *open* if for every point $(a, b) \in \mathcal{U}$ there is a number $\epsilon > 0$ such that $(x, y) \in \mathcal{U}$ whenever

$$(x - a)^2 + (y - b)^2 < \epsilon^2.$$

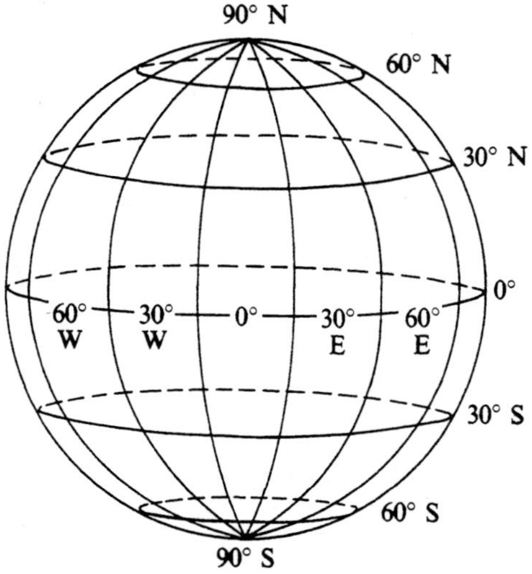

FIGURE 4.1

Thus the statement "$\mathcal{U}$ is open" means that about each point of $\mathcal{U}$ there is a little disk that is contained in $\mathcal{U}$.

EXAMPLE 1.1. $\mathcal{U} = \{(x, y) \in \mathbf{R}^2 \mid x^2 + y^2 < 1\}$ is open.

EXAMPLE 1.2. $\bar{\mathcal{U}} = \{(x, y) \in \mathbf{R}^2 \mid x^2 + y^2 \leq 1\}$ is not open.

EXAMPLE 1.3. $\mathcal{V} = \{(x, y) \in \mathbf{R}^2 \mid x^2 + y^2 > 1\}$ is open.

EXAMPLE 1.4. $\mathcal{W} = \{(x, y) \in \mathbf{R}^2 \mid 1 < x < 2, 3 < y < 5\}$ is open.

EXAMPLE 1.5. $\bar{\mathcal{X}} = \{(x, y) \in \mathbf{R}^2 \mid x \geq 0\}$ is not open.

DEFINITION. A C^k *coordinate patch* (or *simple surface*) is a one-to-one C^k function $\mathbf{x}\colon \mathcal{U} \to \mathbf{R}^3$ for some $k \geq 1$, where $\mathcal{U}$ is an open subset of $\mathbf{R}^2$ with coordinates u^1 and u^2 and $(\partial \mathbf{x}/\partial u^1) \times (\partial \mathbf{x}/\partial u^2) \neq \mathbf{0}$ on $\mathcal{U}$.

Notice how closely this definition resembles that of a (simple) regular C^k curve with the open interval (a, b) replaced by an open set $\mathcal{U}$ and the regularity condition replaced by $(\partial \mathbf{x}/\partial u^1) \times (\partial \mathbf{x}/\partial u^2) \neq \mathbf{0}$. We refer to $(\partial \mathbf{x}/\partial u^1) \times (\partial \mathbf{x}/\partial u^2) \neq \mathbf{0}$ as the regularity condition for surfaces. This makes sense geometrically because a curve $\boldsymbol{\alpha}$ is "one-dimensional" and so the set of vectors tangent to it at one point should be one-dimensional. Since $d\boldsymbol{\alpha}/dt$ is tangent to the curve, insisting that $d\boldsymbol{\alpha}/dt \neq \mathbf{0}$ (i.e., $\{d\boldsymbol{\alpha}/dt\}$ is a linearly independent set) defines the tangent line to $\boldsymbol{\alpha}$ at a point. A surface is "two-dimensional" so that the regularity condition for surfaces should be that we have two linearly independent vectors at each point. According to

Lemma 3.3 of Chapter 1, $(\partial\mathbf{x}/\partial u^1) \times (\partial\mathbf{x}/\partial u^2) \neq \mathbf{0}$ is precisely the condition that $\{\partial\mathbf{x}/\partial u^1, \partial\mathbf{x}/\partial u^2\}$ is a linearly independent set.

As was the case for curves, we shall assume that $k \geq 3$ unless stated otherwise.

EXAMPLE 1.6. Let $f(u^1, u^2)$ be a C^k function of two variables defined in an open set $\mathcal{U}$ of $\mathbf{R}^2$. Set $\mathbf{x}(u^1, u^2) = (u^1, u^2, f(u^1, u^2))$. $\mathbf{x}$ is C^k and one-to-one. Let $f_i = \partial f/\partial u^i$.

$$\begin{aligned}\frac{\partial\mathbf{x}}{\partial u^1} \times \frac{\partial\mathbf{x}}{\partial u^2} &= (1, 0, f_1) \times (0, 1, f_2)\\ &= (-f_1, -f_2, 1) \neq \mathbf{0}.\end{aligned}$$

Thus $\mathbf{x}$ is a C^k simple surface. It is the *graph of a function* and is frequently called a *Monge patch.*

EXAMPLE 1.7. Let $\mathcal{U} = \{(u^1, u^2) \in \mathbf{R}^2 \mid (u^1)^2 + (u^2)^2 < 1\}$ and set

$$\mathbf{x}(u^1, u^2) = (u^1, u^2, \sqrt{1 - (u^1)^2 - (u^2)^2}).$$

This is a special case of Example 1.6. The image is the upper half of the unit sphere $S^2 \subset \mathbf{R}^3$.

EXAMPLE 1.8. Let $\mathcal{V} = \{(v^1, v^2) \in \mathbf{R}^2 \mid (v^1)^2 + (v^2)^2 < 1\}$ and set

$$\mathbf{y}(v^1, v^2) = (v^1, -\sqrt{1 - (v^1)^2 - (v^2)^2}, v^2).$$

This is essentially another case of Example 1.6. The image is the left half of the unit sphere.

EXAMPLE 1.9. Let $\mathcal{W} = \{(w^1, w^2) \in \mathbf{R}^2 \mid -\pi/2 < w^1 < \pi/2, -\pi < w^2 < \pi\}$ and set $\mathbf{z}(w^1, w^2) = (\cos w^1 \cos w^2, \cos w^1 \sin w^2, \sin w^1)$. The image of $\mathbf{z}$ is all of S^2 except the 180° meridian.

$$\begin{aligned}\frac{\partial\mathbf{z}}{\partial w^1} \times \frac{\partial\mathbf{z}}{\partial w^2} &= (-\sin w^1 \cos w^2, -\sin w^1 \sin w^2, \cos w^1)\\ &\quad \times (-\cos w^1 \sin w^2, \cos w^1 \cos w^2, 0)\\ &= (-\cos^2 w^1 \cos w^2, -\cos^2 w^1 \sin w^2, -\sin w^1 \cos w^1).\end{aligned}$$

Since $\cos w^1$ is never zero on $(-\pi/2, \pi/2)$ and $\cos w^2$ is never zero when $\sin w^2$ is, $(\partial\mathbf{z}/\partial w^1) \times (\partial\mathbf{z}/\partial w^2) \neq \mathbf{0}$ on $\mathcal{W}$ and $\mathbf{z}$ is a simple surface since it is easily shown to be one-to-one.

EXAMPLE 1.10. Let $\mathbf{x}\colon \mathbf{R}^2 \to \mathbf{R}^3$ by $\mathbf{x}(u^1, u^2) = ((u^1)^2, (u^2)^2, u^1u^2)$. Then $\partial\mathbf{x}/\partial u^1 = (2u^1, 0, u^2)$, $\partial\mathbf{x}/\partial u^2 = (0, 2u^2, u^1)$, and

$$\frac{\partial\mathbf{x}}{\partial u^1} \times \frac{\partial\mathbf{x}}{\partial u^2} = (-2(u^2)^2, -2(u^1)^2, 4u^1u^2).$$

At the point $(0, 0)$, $(\partial\mathbf{x}/\partial u^1) \times (\partial\mathbf{x}/\partial u^2) = (0, 0, 0)$. Thus $\mathbf{x}$ is not a simple surface because the regularity condition is not satisfied. $\mathbf{x}$ is not one-to-one either since $\mathbf{x}(1, 2) = \mathbf{x}(-1, -2)$. If

$$\mathcal{U} = \{(u^1, u^2) \in \mathbf{R}^2 \mid u^1 > 0, u^2 > 0\},$$

then $\mathbf{x}\colon \mathcal{U} \longrightarrow \mathbf{R}^3$ is a simple surface.

We urge the reader to review the chain rule (Section 1-5) before continuing.

DEFINITION. A C^k *coordinate transformation* is a C^k one-to-one onto function $f\colon \mathcal{V} \longrightarrow \mathcal{U}$ of open sets in $\mathbf{R}^2$ whose inverse $g\colon \mathcal{U} \longrightarrow \mathcal{V}$ is also of class C^k.

Let v^1, v^2 be coordinates in $\mathcal{V}$, u^1, u^2 in $\mathcal{U}$. Set

$$f(v^1, v^2) = (f^1(v^1, v^2), f^2(v^1, v^2)) = (f^1, f^2).$$

Similarly, $g(u^1, u^2) = (g^1(u^1, u^2), g^2(u^1, u^2)) = (g^1, g^2)$. Then since f and g are inverses, we have $f \circ g(u^1, u^2) = (u^1, u^2)$. Thus

$$f^1(g^1(u^1, u^2), g^2(u^1, u^2)) = u^1 \quad \text{and} \quad f^2(g^1(u^1, u^2), g^2(u^1, u^2)) = u^2.$$

If we differentiate using the chain rule, we obtain the four equations:

$$\frac{\partial f^1}{\partial v^1}\frac{\partial g^1}{\partial u^1} + \frac{\partial f^1}{\partial v^2}\frac{\partial g^2}{\partial u^1} = 1, \qquad \frac{\partial f^1}{\partial v^1}\frac{\partial g^1}{\partial u^2} + \frac{\partial f^1}{\partial v^2}\frac{\partial g^2}{\partial u^2} = 0$$

$$\frac{\partial f^2}{\partial v^1}\frac{\partial g^1}{\partial u^1} + \frac{\partial f^2}{\partial v^2}\frac{\partial g^2}{\partial u^1} = 0, \qquad \frac{\partial f^2}{\partial v^1}\frac{\partial g^1}{\partial u^2} + \frac{\partial f^2}{\partial v^2}\frac{\partial g^2}{\partial u^2} = 1.$$

Since $f^i(v^1, v^2) = u^i$ and $g^\alpha(u^1, u^2) = v^\alpha$, we may write the above equations in either of the shortened forms

(1-1) $$\sum_{\alpha=1}^{2} \frac{\partial f^i}{\partial v^\alpha}\frac{\partial g^\alpha}{\partial u^j} = \delta^i{}_j$$

or

(1-2) $$\sum_{\alpha=1}^{2} \frac{\partial u^i}{\partial v^\alpha}\frac{\partial v^\alpha}{\partial u^j} = \delta^i{}_j.$$

(Recall the Kronecker symbol $\delta^i{}_j = 0$ if $i \neq j$ while $\delta^i{}_i = 1$.) Each of the above forms has its advantages. Equation (1-1) explicitly mentions the transformations $f = (f^1, f^2)$ and $g = (g^1, g^2)$, while Equation (1-2) looks more aesthetic in terms of the chain rule. We will use either $(\partial u^i/\partial v^\alpha)$, $(\partial f^i/\partial v^\alpha)$, or $J(f)$ to denote the matrix of partial derivatives of f. It is called the *Jacobian* of f and is named after Carl G. J. Jacobi (1804–1851). Note that $J(f)$ changes from point to point of $\mathcal{V}$, i.e., it is a matrix-valued function defined on $\mathcal{V}$.

Equations (1-1) and (1-2) indicate that $J(f)$ is nonsingular and thus has a nonzero determinant. An important theorem of advanced calculus (the Inverse Function Theorem, see Fulks [1969]) states that if $f\colon \mathcal{V} \longrightarrow \mathcal{U}$ is one-to-one, onto, differentiable, and has nonsingular Jacobian at each point, then the inverse of f is also differentiable and f is a coordinate transformation. We shall not, however, need this result.

These coordinate changes play the role in surface theory that reparametrizations played in curve theory. The idea is that the upper hemisphere is the upper hemisphere whether it is parametrized as in Example 1.7 or Example 1.8 (see Example 1.12 below).

LEMMA 1.11. If $\mathbf{x}\colon \mathcal{U} \longrightarrow \mathbf{R}^3$ is a simple surface and $f\colon \mathcal{V} \longrightarrow \mathcal{U}$ is a coordinate transformation, then $\mathbf{y} = \mathbf{x} \circ f\colon \mathcal{V} \longrightarrow \mathbf{R}^3$ is a simple surface with the same image as $\mathbf{x}$.

Proof: $\mathbf{y}$ is one-to-one since both $\mathbf{x}$ and f are. Its class is the minimum of that of $\mathbf{x}$ and f. Certainly $\mathbf{x}(\mathcal{U}) = \mathbf{y}(\mathcal{V})$. Furthermore the chain rule shows that

$$\frac{\partial \mathbf{y}}{\partial v^\alpha} = \frac{\partial \mathbf{x}}{\partial u^1}\frac{\partial f^1}{\partial v^\alpha} + \frac{\partial \mathbf{x}}{\partial u^2}\frac{\partial f^2}{\partial v^\alpha} = \sum \frac{\partial \mathbf{x}}{\partial u^i}\frac{\partial f^i}{\partial v^\alpha}. \tag{1-3}$$

By a straightforward computation we obtain

$$\frac{\partial \mathbf{y}}{\partial v^1} \times \frac{\partial \mathbf{y}}{\partial v^2} = \det\left(\frac{\partial f^i}{\partial v^\alpha}\right)\frac{\partial \mathbf{x}}{\partial u^1} \times \frac{\partial \mathbf{x}}{\partial u^2} = \det\,(J(f))\frac{\partial \mathbf{x}}{\partial u^1} \times \frac{\partial \mathbf{x}}{\partial u^2}. \tag{1-4}$$

Therefore $(\partial \mathbf{y}/\partial v^1) \times (\partial \mathbf{y}/\partial v^2) \neq \mathbf{0}$ and $\mathbf{y}$ is a simple surface. ∎

EXAMPLE 1.12. Let $\mathcal{U}$ and $\mathcal{V}$ be as in Examples 1.7 and 1.8. Let

$$\tilde{\mathcal{U}} = \{(u^1, u^2) \in \mathcal{U} \mid u^2 < 0\} \quad \text{and} \quad \tilde{\mathcal{V}} = \{(v^1, v^2) \in \mathcal{V} \mid v^2 > 0\}.$$

$\tilde{\mathcal{U}}$ and $\tilde{\mathcal{V}}$ are open. Let $f\colon \tilde{\mathcal{V}} \longrightarrow \tilde{\mathcal{U}}$ be given by

$$f(v^1, v^2) = (v^1, -\sqrt{1 - (v^1)^2 - (v^2)^2}).$$

We show that f is a coordinate transformation. f is one-to-one, onto, and differentiable. Its inverse is $g\colon \tilde{\mathcal{U}} \longrightarrow \tilde{\mathcal{V}}$ given by

$$g(u^1, u^2) = (u^1, \sqrt{1 - (u^1)^2 - (u^2)^2}),$$

which is differentiable. Hence f is a coordinate transformation.

The Jacobian of f is

$$\left(\frac{\partial f^i}{\partial v^\alpha}\right) = \begin{pmatrix} 1 & 0 \\ \dfrac{v^1}{\sqrt{1 - (v^1)^2 - (v^2)^2}} & \dfrac{v^2}{\sqrt{1 - (v^1)^2 - (v^2)^2}} \end{pmatrix}$$

which has nonzero determinant since $v^2 \neq 0$. Note that in Examples 1.7 and 1.8, using $\tilde{\mathcal{U}}$ and $\tilde{\mathcal{V}}$ we have

$$\begin{aligned} \mathbf{x} \circ f(v^1, v^2) &= \mathbf{x}(v^1, -\sqrt{1 - (v^1)^2 - (v^2)^2}) \\ &= (v^1, -\sqrt{1 - (v^1)^2 - (v^2)^2}, \sqrt{(v^2)^2}) \\ &= (v^1, -\sqrt{1 - (v^1)^2 - (v^2)^2}, v^2) \\ &= \mathbf{y}(v^1, v^2). \end{aligned}$$

The discussion for the remainder of this section is absolutely crucial for understanding the rest of this book. We urge the reader to pay especially close attention to these definitions and notations and understand them in the special cases of Examples 1.14 and 1.15.

We shall use the following notation throughout the rest of this book.

NOTATION. If $\mathbf{x}\colon \mathcal{U} \longrightarrow \mathbf{R}^3$ is a simple surface, then

$$\mathbf{x}_1(a, b) = \frac{\partial \mathbf{x}}{\partial u^1}(a, b), \qquad \mathbf{x}_2(a, b) = \frac{\partial \mathbf{x}}{\partial u^2}(a, b)$$

$$\mathbf{x}_1 = \frac{\partial \mathbf{x}}{\partial u^1}, \qquad \mathbf{x}_2 = \frac{\partial \mathbf{x}}{\partial u^2}. \tag{1-5}$$

In this notation Equation (1-3) becomes

$$\mathbf{y}_\alpha = \sum \mathbf{x}_i \frac{\partial u^i}{\partial v^\alpha}.$$

DEFINITION. The *tangent plane* to a simple surface $\mathbf{x}\colon \mathcal{U} \longrightarrow \mathbf{R}^3$ at the point $P = \mathbf{x}(a, b)$ is the plane through P perpendicular to $\mathbf{x}_1(a, b) \times \mathbf{x}_2(a, b)$. The *unit normal* to the surface at P is $\mathbf{n}(a, b) = \mathbf{x}_1 \times \mathbf{x}_2/|\mathbf{x}_1 \times \mathbf{x}_2|$, where the right-hand side is evaluated at (a, b). Note that $\mathbf{n}(a, b)$ exists because $\mathbf{x}_1 \times \mathbf{x}_2 \neq \mathbf{0}$. It is perpendicular to the tangent plane at P.

The set $\{\mathbf{x}_1, \mathbf{x}_2, \mathbf{n}\}$ is linearly independent and hence gives a basis of $\mathbf{R}^3$. In a way it serves the same purpose for surfaces that the Frenet-Serret frame $\{\mathbf{T}, \mathbf{N}, \mathbf{B}\}$ does for a curve. However, as we shall see, it is not an orthonormal basis (in general). Eventually we shall compute the derivatives of $\mathbf{x}_1$, $\mathbf{x}_2$, and $\mathbf{n}$ with respect to the coordinates u^i. The resulting expressions, due to Gauss and Weingarten, will take the place of the Frenet-Serret equations.

The following proposition is immediate from Equation (1-4).

PROPOSITION 1.13. If $\mathbf{x}\colon \mathcal{U} \longrightarrow \mathbf{R}^3$ is a simple surface, $f\colon \mathcal{V} \longrightarrow \mathcal{U}$ is a coordinate transformation, and $\mathbf{y} = \mathbf{x} \circ f$, then
(a) the tangent plane to the simple surface $\mathbf{x}$ at $P = \mathbf{x}(f(a, b))$ is equal to the tangent plane to the simple surface $\mathbf{y}$ at $P = \mathbf{y}(a, b)$;
(b) the normal to the surface $\mathbf{x}$ at P is the same as the normal to the surface $\mathbf{y}$ at P, except possibly it may have the opposite sign.

This proposition says that the tangent plane at P is an intrinsic invariant. This is especially important for the general theory of surfaces as we shall see in the next section. See Problem 1.6.

EXAMPLE 1.14. In Example 1.7 we have

$$\mathbf{x}_1 = \left(1, 0, \frac{-u^1}{\sqrt{1 - (u^1)^2 - (u^2)^2}}\right) \quad \text{and} \quad \mathbf{x}_2 = \left(0, 1, \frac{-u^2}{\sqrt{1 - (u^1)^2 - (u^2)^2}}\right).$$

Therefore $\mathbf{n} = (u^1, u^2, \sqrt{1-(u^1)^2-(u^2)^2})$. At the point $\mathbf{x}(\frac{1}{2}, \frac{1}{2})$ the tangent plane has the equation

$$\frac{1}{2}\left(x-\frac{1}{2}\right)+\frac{1}{2}\left(y-\frac{1}{2}\right)+\frac{\sqrt{2}}{2}\left(z-\frac{\sqrt{2}}{2}\right)=0.$$

See Figure 4.2.

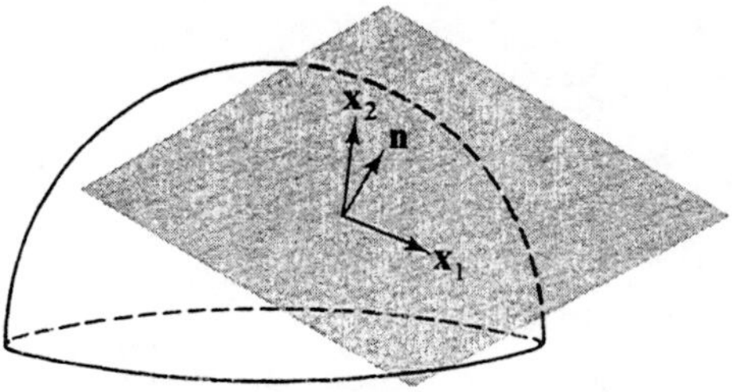

FIGURE 4.2

EXAMPLE 1.15. Consider the Monge patch $\mathbf{x}(u^1, u^2) = (u^1, u^2, u^1u^2)$.

$$\mathbf{x}_1(a, b) = (1, 0, b) \quad \text{and} \quad \mathbf{x}_2(a, b) = (0, 1, a)$$

so that

$$\mathbf{x}_1 \times \mathbf{x}_2 = (-b, -a, 1) \quad \text{and} \quad \mathbf{n}(a, b) = \frac{(-b, -a, 1)}{\sqrt{1+a^2+b^2}}.$$

Note how the normal, and hence the tangent plane, changes as a and b change. At the point $\mathbf{x}(1, 2) = (1, 2, 2)$ the tangent plane has the equation

$$-2(x-1)-(y-2)+(z-2)=0.$$

See Figure 4.3. In this example the tangent plane actually intersects the

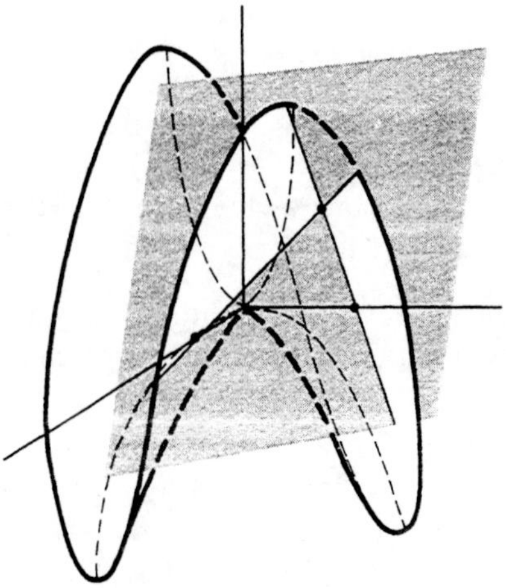

FIGURE 4.3

surface in two straight lines:

$$\boldsymbol{\alpha}(t) = \mathbf{x}(t, 2) = (t, 2, 2t) \quad \text{and} \quad \boldsymbol{\beta}(t) = \mathbf{x}(1, t) = (1, t, t).$$

See Problem 1.8.

DEFINITION. A vector $\mathbf{X}$ is a *tangent vector* to a simple surface $\mathbf{x}\colon \mathcal{U} \longrightarrow \mathbf{R}^3$ at $P = \mathbf{x}(a, b)$ if $\mathbf{X}$ is the velocity vector at P of some curve in $\mathbf{x}(\mathcal{U})$, i.e., if for some $\epsilon > 0$ there is a curve $\boldsymbol{\alpha}\colon (-\epsilon, \epsilon) \longrightarrow \mathbf{x}(\mathcal{U}) \subset \mathbf{R}^3$ with $\boldsymbol{\alpha}(0) = P$, $(d\boldsymbol{\alpha}/dt)(0) = \mathbf{X}$, and $\alpha(t) = \mathbf{x}(\boldsymbol{\alpha}^1(t), \boldsymbol{\alpha}^2(t))$, where the $\boldsymbol{\alpha}^i$ are C^1.

LEMMA 1.16. The set of all tangent vectors to a simple surface $\mathbf{x}\colon \mathcal{U} \longrightarrow \mathbf{R}^3$ at P is a vector space.

Proof: If $\mathbf{X}$ and $\mathbf{Y}$ are tangent vectors at P, then $\mathbf{X} = \dot{\boldsymbol{\alpha}}(0)$ and $\mathbf{Y} = \dot{\boldsymbol{\beta}}(0)$ for some curves $\boldsymbol{\alpha}$ and $\boldsymbol{\beta}$ in $\mathbf{x}(\mathcal{U})$ with $\boldsymbol{\alpha}(0) = \boldsymbol{\beta}(0) = P = \mathbf{x}(a, b)$. There are functions $\alpha^1(t)$, $\alpha^2(t)$, $\beta^1(t)$, $\beta^2(t)$ such that

$$\boldsymbol{\alpha}(t) = \mathbf{x}(\alpha^1(t), \alpha^2(t)) \quad \text{and} \quad \boldsymbol{\beta}(t) = \mathbf{x}(\beta^1(t), \beta^2(t)).$$

(In fact, $x^{-1} \circ \boldsymbol{\alpha}$ defines $(\alpha^1(t), \alpha^2(t))$, where $x^{-1}\colon \mathbf{x}(\mathcal{U}) \longrightarrow \mathcal{U}$ is the inverse of $\mathbf{x}$.) Then

$$\begin{aligned} \mathbf{X} &= \frac{\partial \mathbf{x}}{\partial u^1}\frac{d\alpha^1}{dt}(0) + \frac{\partial \mathbf{x}}{\partial u^2}\frac{d\alpha^2}{dt}(0) \\ \mathbf{Y} &= \frac{\partial \mathbf{x}}{\partial u^1}\frac{d\beta^1}{dt}(0) + \frac{\partial \mathbf{x}}{\partial u^2}\frac{d\beta^2}{dt}(0). \end{aligned} \tag{1-6}$$

Consider $\boldsymbol{\gamma}(t) = \mathbf{x}(\alpha^1(t) + \beta^1(t) - a, \alpha^2(t) + \beta^2(t) - b)$. Since

$$\alpha^1(0) = \beta^1(0) = a \quad \text{and} \quad \alpha^2(0) = \beta^2(0) = b,$$

$\boldsymbol{\gamma}(t)$ make sense and is in $\mathbf{x}(\mathcal{U})$ if $|t|$ is small enough. We claim $\dot{\boldsymbol{\gamma}}(0) = \mathbf{X} + \mathbf{Y}$.

$$\begin{aligned} \frac{d\boldsymbol{\gamma}}{dt} &= \frac{\partial \mathbf{x}}{\partial u^1}\left(\frac{d\alpha^1}{dt} + \frac{d\beta^1}{dt}\right) + \frac{\partial \mathbf{x}}{\partial u^2}\left(\frac{d\alpha^2}{dt} + \frac{d\beta^2}{dt}\right) \\ &= \left(\frac{\partial \mathbf{x}}{\partial u^1}\frac{d\alpha^1}{dt} + \frac{\partial \mathbf{x}}{\partial u^2}\frac{d\alpha^2}{dt}\right) + \left(\frac{\partial \mathbf{x}}{\partial u^1}\frac{d\beta^1}{dt} + \frac{\partial \mathbf{x}}{du^2}\frac{d\beta^2}{dt}\right). \end{aligned}$$

By Equation (1-6) $\dot{\boldsymbol{\gamma}}(0) = \mathbf{X} + \mathbf{Y}$. Hence the sum of two tangent vectors at P is a tangent vector at P ($\boldsymbol{\gamma}(0) = \mathbf{x}(a + a - a, b + b - b) = \mathbf{x}(a, b) = P$).

Assume that $\mathbf{X}$ is a tangent vector at P as above and $r \in \mathbf{R}$, and let $\boldsymbol{\eta}(t) = \boldsymbol{\alpha}(rt)$. For small values of t this makes sense and $\boldsymbol{\eta}(t)$ is a curve in $\mathbf{x}(\mathcal{U})$. $\boldsymbol{\eta}(0) = \boldsymbol{\alpha}(0) = P$. $d\boldsymbol{\eta}/dt = (d\boldsymbol{\alpha}/dt)r$, so that $\dot{\boldsymbol{\eta}}(0) = r\mathbf{X}$. Thus a multiple of a tangent vector is a tangent vector.

All other properties of a vector space are satisfied by the set of tangent vectors to $\mathbf{x}$ at P since the set is a subset of the vector space $\mathbf{R}^3$. Hence the set of all tangent vectors to $\mathbf{x}$ at P forms a vector space. ∎

The next definition is important because it will give examples of tangent vectors to a surface $\mathbf{x}$.

DEFINITION. Let $\mathbf{x}: \mathcal{U} \longrightarrow \mathbf{R}^3$ be a simple surface. The *u^1-curve*, $\boldsymbol{\alpha}$, through $P = \mathbf{x}(a, b)$ is given by $\boldsymbol{\alpha}(u^1) = \mathbf{x}(u^1, b)$. The *$u^2$-curve*, $\boldsymbol{\beta}$, through P is given by $\boldsymbol{\beta}(u^2) = \mathbf{x}(a, u^2)$. A *parametric curve* on $\mathbf{x}$ is one of these curves.

Note that the standard rectangular grid on $\mathcal{U}$ (i.e., the (u^1, u^2) plane) has as its image under $\mathbf{x}$ a grid on $\mathbf{x}(\mathcal{U})$. This grid is sometimes called a *curvilinear coordinate system* on $\mathbf{x}(\mathcal{U})$. The grid lines are the parametric curves. See Figure 4.4.

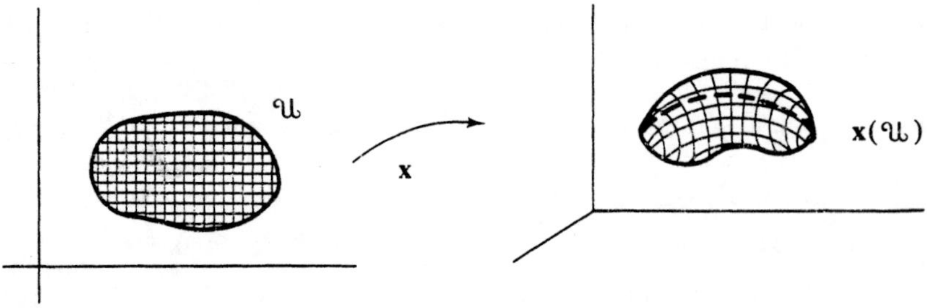

FIGURE 4.4

EXAMPLE 1.17. Consider the upper hemisphere in Example 1.7. The u^1-curve through $P = \mathbf{x}(a, b)$ is $\boldsymbol{\alpha}(u^1) = (u^1, b, \sqrt{1 - (u^1)^2 - b^2})$. The u^2-curve through P is $\boldsymbol{\beta}(u^2) = (a, u^2, \sqrt{1 - a^2 - (u^2)^2})$. See Figure 4.5.

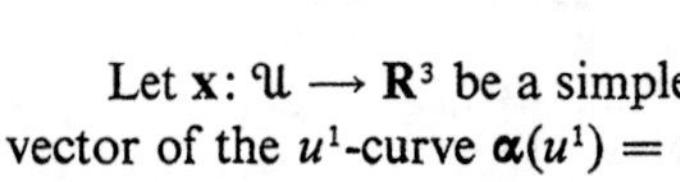

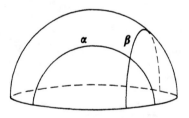

FIGURE 4.5

Let $\mathbf{x}: \mathcal{U} \longrightarrow \mathbf{R}^3$ be a simple surface and $(a, b) \in \mathcal{U}$. What is the velocity vector of the u^1-curve $\boldsymbol{\alpha}(u^1) = \mathbf{x}(u^1, b)$ at $u^1 = a$? It is

$$\frac{d\boldsymbol{\alpha}}{du^1}(a) = \frac{\partial \mathbf{x}}{\partial u^1}(a, b) = \mathbf{x}_1(a, b).$$

Similarly the velocity vector of the u^2-curve $\boldsymbol{\beta}(u^2) = \mathbf{x}(a, u^2)$ at $u^2 = b$ is $\mathbf{x}_2(a, b)$. This means that both $\mathbf{x}_1(a, b)$ and $\mathbf{x}_2(a, b)$ are tangent vectors at $\mathbf{x}(a, b)$.

PROPOSITION 1.18. The set of all tangent vectors to $\mathbf{x}$ at $P = \mathbf{x}(a, b)$ is a vector space of dimension two with basis $\{\mathbf{x}_1(a, b), \mathbf{x}_2(a, b)\}$. Furthermore, viewed as a plane in $\mathbf{R}^3$ through the origin, it is parallel to the tangent plane at P.

Proof: $\mathbf{x}_1(a, b)$ and $\mathbf{x}_2(a, b)$ are tangent vectors at P and are linearly independent. Thus the set of all tangent vectors at P is a vector space (Lemma 1.16) of dimension at least two. It will have dimension two if we can show that $\mathbf{x}_1(a, b)$ and $\mathbf{x}_2(a, b)$ span it.

Let $\mathbf{X}$ be tangent to $\mathbf{x}$ at P and let $\boldsymbol{\gamma}$ be a curve in $\mathbf{x}(\mathcal{U})$ with $\boldsymbol{\gamma}(0) = P$ and $\dot{\boldsymbol{\gamma}}(0) = \mathbf{X}$. There are functions $\gamma^1(t)$, $\gamma^2(t)$ such that $\boldsymbol{\gamma}(t) = \mathbf{x}(\gamma^1(t), \gamma^2(t))$. $((\gamma^1(t), \gamma^2(t))$ is defined by $x^{-1}(\boldsymbol{\gamma}(t))$.) By the chain rule we have

$$\frac{d\boldsymbol{\gamma}}{dt} = \frac{\partial \mathbf{x}}{\partial u^1}\frac{d\gamma^1}{dt} + \frac{\partial \mathbf{x}}{\partial u^2}\frac{d\gamma^2}{dt} = \sum \mathbf{x}_i \frac{d\gamma^i}{dt}. \tag{1-7}$$

Thus

$$\mathbf{X} = X^1\mathbf{x}_1(a, b) + X^2\mathbf{x}_2(a, b), \tag{1-8}$$

where

$$X^i = \dot{\gamma}^i(0).$$

Equation (1-8) says that $\mathbf{x}_1(a, b)$ and $\mathbf{x}_2(a, b)$ span the set of all tangent vectors at P. Thus the dimension of this vector space is two.

The tangent plane at P is perpendicular to $\mathbf{n}(a, b)$ by definition. $\mathbf{n}(a, b)$, which is a multiple of $\mathbf{x}_1(a, b) \times \mathbf{x}_2(a, b)$, is certainly perpendicular to $\mathbf{x}_1(a, b)$ and $\mathbf{x}_2(a, b)$. Hence $\mathbf{n}(a, b)$ is perpendicular to the plane spanned by $\mathbf{x}_1(a, b)$ and $\mathbf{x}_2(a, b)$ and these two planes are parallel. ▌

We quite often identify the vector space of tangent vectors to $\mathbf{x}$ at P with the tangent plane at P by viewing the tail of a tangent vector to be at P.

In the above proof we showed that any vector $\mathbf{X}$ tangent to a surface $\mathbf{x}$ can be written in the form $\mathbf{X} = \sum X^i\mathbf{x}_i$. Suppose $f\colon \mathcal{V} \to \mathcal{U}$ is a coordinate transformation and $\mathbf{y} = \mathbf{x} \circ f$. Then $\mathbf{X} = \sum \bar{X}^\alpha \mathbf{y}_\alpha$. How are the coefficients $\{X^1, X^2\}$ and $\{\bar{X}^1, \bar{X}^2\}$ related? Since $\mathbf{y}_\alpha = \sum \mathbf{x}_i\,(\partial u^i/\partial v^\alpha)$,

$$\sum X^i\mathbf{x}_i = \mathbf{X} = \sum \bar{X}^\alpha\mathbf{y}_\alpha = \sum \bar{X}^\alpha\Big(\frac{\partial u^i}{\partial v^\alpha}\mathbf{x}_i\Big) = \sum \Big(\bar{X}^\alpha \frac{\partial u^i}{\partial v^\alpha}\Big)\mathbf{x}_i.$$

The vectors $\mathbf{x}_1$ and $\mathbf{x}_2$ are independent and so we must have

$$X^i = \sum \bar{X}^\alpha \frac{\partial u^i}{\partial v^\alpha}. \tag{1-9}$$

This formula tells us that if we know the components of the vector $\mathbf{X}$ in the $\mathcal{V}$-coordinate system, we can find the coordinates in the $\mathcal{U}$-coordinate system by multiplying by the Jacobian of $f\colon \mathcal{V} \to \mathcal{U}$:

$$\begin{pmatrix} X^1 \\ X^2 \end{pmatrix} = \begin{pmatrix} \dfrac{\partial u^1}{\partial v^1} & \dfrac{\partial u^1}{\partial v^2} \\[2ex] \dfrac{\partial u^2}{\partial v^1} & \dfrac{\partial u^2}{\partial v^2} \end{pmatrix} \begin{pmatrix} \bar{X}^1 \\ \bar{X}^2 \end{pmatrix}. \tag{1-10}$$

Note that we also have

$$\bar{X}^\alpha = \sum X^i \frac{\partial v^\alpha}{\partial u^i}. \tag{1-11}$$

PROBLEMS

1.1. Let $\mathcal{U} = \{(u^1, u^2) \in \mathbf{R}^2 \mid -\pi < u^1 < \pi,\ -\pi < u^2 < \pi\}$ and define

$$\mathbf{x}(u^1, u^2) = ((2 + \cos u^1)\cos u^2, (2 + \cos u^1)\sin u^2, \sin u^1).$$

(a) Prove that $\mathbf{x}$ is a simple surface. ($\mathbf{x}(\mathcal{U})$ looks like the surface of a donut or innertube.)

(b) Compute $\mathbf{x}_1$, $\mathbf{x}_2$, and $\mathbf{n}$ as functions of u^1 and u^2.

This problem is a special case of the next one, which we shall investigate in detail throughout the chapter.

***1.2.** Consider a curve in the (r, z) plane given by $r = r(t) > 0$, $z = z(t)$. If this curve is rotated about the z-axis, we obtain a *surface of revolution*. (See Figure 4.6.) We may parametrize this surface in the following

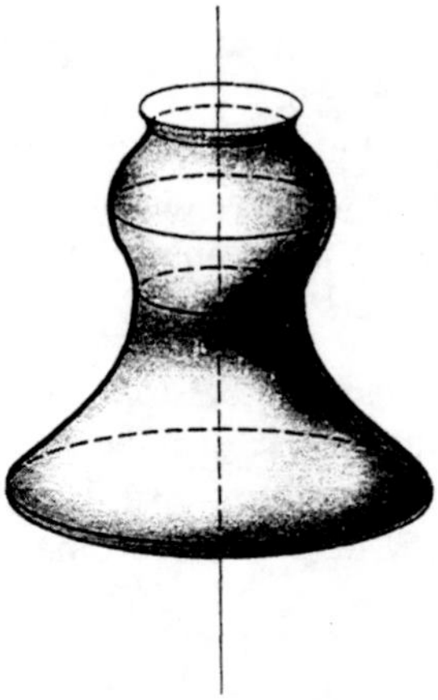

FIGURE 4.6

manner. It is useful to use coordinates t and θ instead of u^1 and u^2, where t measures position on the curve and θ measures how far the curve has been rotated. The surface is given by

$$\mathbf{x}(t, \theta) = (r(t)\cos\theta, r(t)\sin\theta, z(t)).$$

(a) Prove that $\mathbf{x}$ is a simple surface if the original curve

$$\boldsymbol{\alpha}(t) = (r(t), z(t))$$

was regular and one-to-one and if $-\pi < \theta < \pi$ by computing $\mathbf{x}_1 = \partial\mathbf{x}/\partial t$, $\mathbf{x}_2 = \partial\mathbf{x}/\partial\theta$, and $\mathbf{n}$.

(b) Show that Problem 1.1 is a surface of revolution.

(c) Show that Example 1.9 is a surface of revolution.

The t-curves are called *meridians* and the θ-curves are called *circles of latitude*. The z-axis is called the *axis of revolution*.

(d) What are the meridians and circles of latitude of Problem 1.1 and Example 1.9?

1.3. Let $\mathcal{U} = \{(u^1, u^2) \mid 0 < u^1, 0 < u^2 < 2\pi\}$ and

$$\mathbf{x}(u^1, u^2) = (u^1 \cos u^2, u^1 \sin u^2, u^1 + u^2).$$

Show that $\mathbf{x}$ is a simple surface.

1.4. Let $\theta(u^1)$ be a C^k function. Prove

$$\mathbf{x}(u^1, u^2) = (u^2 \cos \theta(u^1), u^2 \sin \theta(u^1), u^1)$$

is a simple surface.

1.5. Let $\mathbf{x}(u^1, u^2) = (u^1 + u^2, u^1 - u^2, u^1u^2)$. Show that $\mathbf{x}$ is a simple surface. Find the normal $\mathbf{n}$ and the equation of the tangent plane at $u^1 = 1, u^2 = 2$.

***1.6.** In Example 1.7 compute the equation of the tangent plane at $u^1 = \frac{1}{2}$, $u^2 = -\frac{1}{2}$. Show that this is the same as the tangent plane in Example 1.8 at $v^1 = \frac{1}{2}$, $v^2 = 1/\sqrt{2}$.

***1.7.** Prove Proposition 1.13.

***1.8.** In Example 1.15 verify that the curves $\boldsymbol{\alpha}$ and $\boldsymbol{\beta}$ do actually lie in the tangent plane.

1.9. Let $\mathbf{x}(\theta, v) = (\cos\theta, \sin\theta, 0) + v(\sin\frac{1}{2}\theta\cos\theta, \sin\frac{1}{2}\theta\sin\theta, \cos\frac{1}{2}\theta)$ with $-\pi < \theta < \pi$, $-\frac{1}{2} < v < \frac{1}{2}$. Compute $\mathbf{n}(\theta, 0)$ and show that

$$\lim_{\theta\to-\pi} \mathbf{n}(\theta, 0) = -\lim_{\theta\to\pi} \mathbf{n}(\theta, 0)$$

while

$$\lim_{\theta\to-\pi} \mathbf{x}(\theta, 0) = \lim_{\theta\to\pi} \mathbf{x}(\theta, 0).$$

See Figure 4.7. This is called the *Möbius band.*

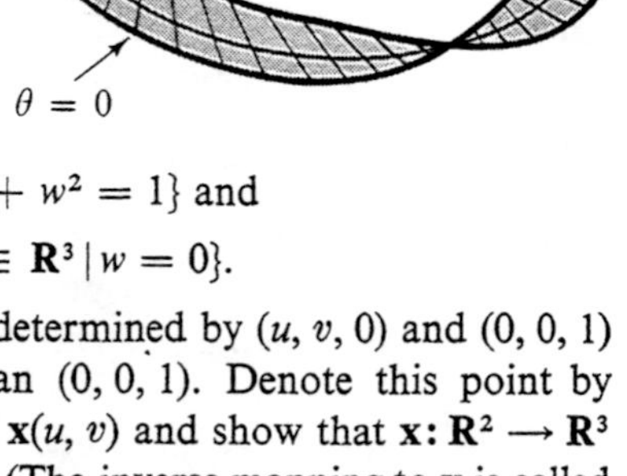

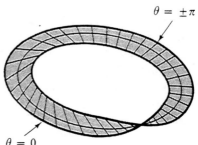

FIGURE 4.7

1.10. Let $S^2 = \{(u, v, w) \in \mathbf{R}^3 \mid u^2 + v^2 + w^2 = 1\}$ and

$$\mathbf{R}^2 = \{(u, v, w) \in \mathbf{R}^3 \mid w = 0\}.$$

If $(u, v, 0)$ belongs to $\mathbf{R}^2$, the line determined by $(u, v, 0)$ and $(0, 0, 1)$ intersects S^2 in a point other than $(0, 0, 1)$. Denote this point by $\mathbf{x}(u, v)$. Compute the actual form of $\mathbf{x}(u, v)$ and show that $\mathbf{x}\colon \mathbf{R}^2 \longrightarrow \mathbf{R}^3$ is a simple surface. See Figure 4.8. (The inverse mapping to $\mathbf{x}$ is called the *stereographic projection.*)

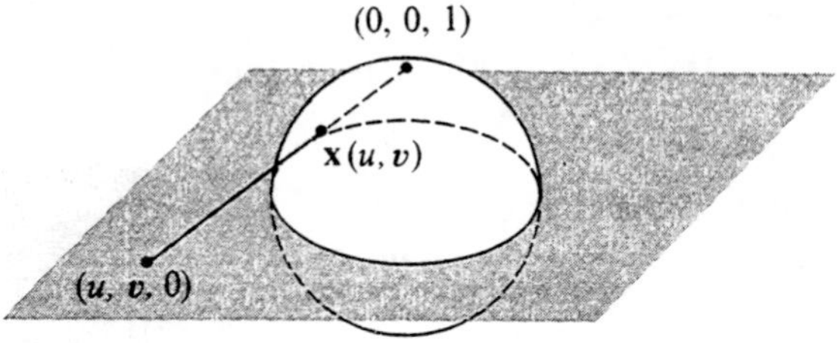

FIGURE 4.8

1.11. Let $\mathbf{x}(r, s) = (r - s, r + s, 2(r^2 + s^2))$, for all $(r, s) \in \mathbf{R}^2$. Show $\mathbf{x}$ is a simple surface. Can you describe the surface geometrically?

1.12. Let $\mathbf{x}(\theta, \phi) = (\sin\phi\cos\theta, 2\sin\phi\sin\theta, 3\cos\phi)$ with $-\pi < \theta < \pi$, $0 < \phi < \pi$. Show $\mathbf{x}$ is a simple surface. What is it?

1.13. Let $\mathbf{x}(u, v) = (\sqrt{1 - u^2}\cos v, \sqrt{1 - u^2}\sin v, u)$ with $-1 < u < 1$ and $-\pi < v < \pi$. Show that $\mathbf{x}$ is a simple surface. What is it?

1.14. Let $\boldsymbol{\alpha}: (a, b) \longrightarrow \mathbf{R}^3$ be a unit speed curve with $\kappa \neq 0$. Let

$$\mathfrak{U} = \{(s, t) \in \mathbf{R}^2 \mid a < s < b, t \neq 0\}.$$

Define $\mathbf{x}: \mathfrak{U} \longrightarrow R^3$ by $\mathbf{x}(s, t) = \boldsymbol{\alpha}(s) + t\boldsymbol{\alpha}'(s)$. Prove that $\mathbf{x}$ is a simple surface, provided $\mathbf{x}$ is one-to-one. $\mathbf{x}$ is called the *tangent developable surface* of $\boldsymbol{\alpha}$.

4–2. SURFACES

The various parametrizations we have given for S^2 (Examples 1.7, 1.8, 1.9, and Problem 1.10) all had one defect in common: they did not describe the entire sphere. However, we shall want to consider geometric sets such as S^2 as surfaces. To do this will require more sophisticated concepts. Basically, a surface will be a collection of simple surfaces that overlap. On the overlap we shall require that the two parametrizations be related by a coordinate transformation. Because we want to emphasize that a surface is made up of patches, we shall now refer to a simple surface exclusively as a coordinate patch. (Both terms mean exactly the same thing.)

DEFINITION. Let M be a subset of $\mathbf{R}^3$ and $\epsilon > 0$. The *ϵ-neighborhood* of $P \in M$ is the set of all points $Q \in M$ such that $d(P, Q) < \epsilon$, where d denotes the usual Euclidean distance in $\mathbf{R}^3$.

Note that the ϵ-neighborhood of P is the intersection of M with the ball in $\mathbf{R}^3$ of radius ϵ and center P. (The student who has studied topology should note that what we are doing is defining the relative topology of M in $\mathbf{R}^3$.)

DEFINITION. If $M \subset \mathbf{R}^3$, a function $g: M \longrightarrow \mathbf{R}^2$ is *continuous* at $P \in M$ if for every open set $\mathcal{U}$ in $\mathbf{R}^2$ with $g(P) \in \mathcal{U}$ there is an ϵ-neighborhood $\mathfrak{N}$ of P with $g(\mathfrak{N}) \subset \mathcal{U}$.

Intuitively, a continuous function is one that sends nearby points to nearby points.

DEFINITION. A coordinate patch $\mathbf{x}: \mathcal{U} \longrightarrow \mathbf{R}^3$ is *proper* if the inverse function $x^{-1}: \mathbf{x}(\mathcal{U}) \longrightarrow \mathcal{U}$ is continuous at each point of $\mathbf{x}(\mathcal{U})$.

EXAMPLE 2.1. The function that sends the open rectangle $(0, 2) \times (0, 1)$ onto the surface in Figure 4.9 is *not* proper. Any ϵ-neighborhood of the point P is mapped in a discontinuous fashion by x^{-1}, if ϵ is sufficiently small.

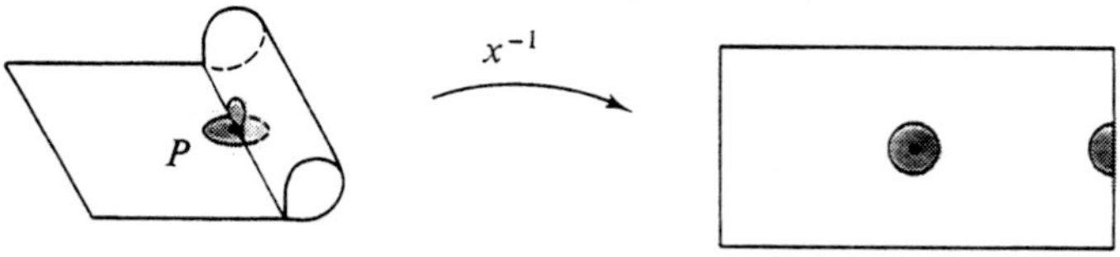

FIGURE 4.9

DEFINITION. A C^k *surface* in $\mathbf{R}^3$ is a subset $M \subset \mathbf{R}^3$ such that for every point $P \in M$ there is a proper C^k coordinate patch whose image is in M and which contains an ϵ-neighborhood of P for some $\epsilon > 0$. Furthermore, if both $\mathbf{x}: \mathcal{U} \longrightarrow \mathbf{R}^3$ and $\mathbf{y}: \mathcal{V} \longrightarrow \mathbf{R}^3$ are such coordinate patches with $\mathcal{U}' = \mathbf{x}(\mathcal{U})$, $\mathcal{V}' = \mathbf{y}(\mathcal{V})$, then $y^{-1} \circ \mathbf{x}: (x^{-1}(\mathcal{U}' \cap \mathcal{V}')) \longrightarrow (y^{-1}(\mathcal{U}' \cap \mathcal{V}'))$ is a C^k coordinate transformation. (See Figure 4.10.)

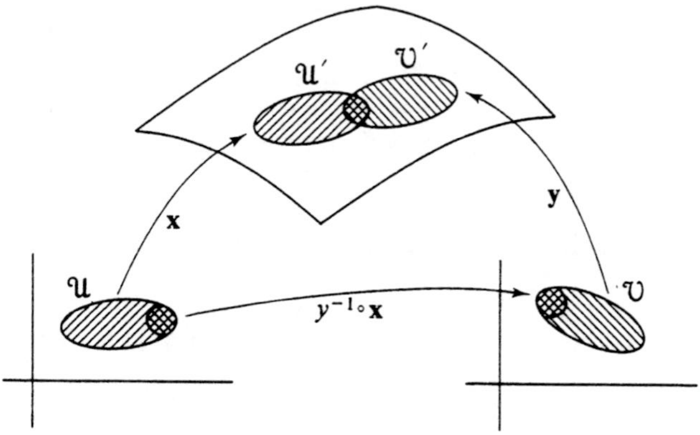

FIGURE 4.10

In practice one usually gives a collection of proper patches such that each $P \in M$ is in at least one of the patches. We shall say the patches *cover M*.

EXAMPLE 2.2. The unit sphere $S^2 = \{(x, y, z) \in \mathbf{R}^3 \,|\, x^2 + y^2 + z^2 = 1\}$ is a surface. We will cover S^2 with six patches. (See also Problem 2.3.) Let $\mathcal{U}^+ = \mathcal{U}^- = \mathcal{V}^+ = \mathcal{V}^- = \mathcal{W}^+ = \mathcal{W}^- = \{(a, b) \in \mathbf{R}^2 \,|\, a^2 + b^2 < 1\}$. The table gives the various patches.

Patch	Definition	Inverse
$\mathbf{x}^+ : \mathcal{U}^+ \longrightarrow \mathbf{R}^3$	$\mathbf{x}^+(u^1, u^2) = (u^1, u^2, \sqrt{1 - (u^1)^2 - (u^2)^2})$	$(x^+)^{-1}(x, y, z) = (x, y)$
$\mathbf{x}^- : \mathcal{U}^- \longrightarrow \mathbf{R}^3$	$\mathbf{x}^-(u^1, u^2) = (u^1, u^2, -\sqrt{1 - (u^1)^2 - (u^2)^2})$	$(x^-)^{-1}(x, y, z) = (x, y)$
$\mathbf{y}^+ : \mathcal{V}^+ \longrightarrow \mathbf{R}^3$	$\mathbf{y}^+(v^1, v^2) = (v^1, \sqrt{1 - (v^1)^2 - (v^2)^2}, v^2)$	$(y^+)^{-1}(x, y, z) = (x, z)$
$\mathbf{y}^- : \mathcal{V}^- \longrightarrow \mathbf{R}^3$	$\mathbf{y}^-(v^1, v^2) = (v^1, -\sqrt{1 - (v^1)^2 - (v^2)^2}, v^2)$	$(y^-)^{-1}(x, y, z) = (x, z)$
$\mathbf{z}^+ : \mathcal{W}^+ \longrightarrow \mathbf{R}^3$	$\mathbf{z}^+(w^1, w^2) = (\sqrt{1 - (w^1)^2 - (w^2)^2}, w^1, w^2)$	$(z^+)^{-1}(x, y, z) = (y, z)$
$\mathbf{z}^- : \mathcal{W}^- \longrightarrow \mathbf{R}^3$	$\mathbf{z}^-(w^1, w^2) = (-\sqrt{1 - (w^1)^2 - (w^2)^2}, w^1, w^2)$	$(z^-)^{-1}(x, y, z) = (y, z)$

The images of the patches are (respectively) the upper, lower, right, left, front, and back hemispheres. Note every point of S^2 belongs to at least one of these hemispheres. We should show that on each overlap the appropriate composite function is a C^k coordinate transformation. We shall only do this for $\mathbf{x}^+$ and $\mathbf{y}^+$. Example 1.12 did it for $\mathbf{x}^+$ and $\mathbf{y}^-$. We leave the other cases to the reader. We omit all $+$ signs on $\mathbf{x}^+$, $\mathcal{U}^+$, etc.

$\mathcal{U}' \cap \mathcal{V}' = \{(x, y, z) \in \mathbf{R}^3 \,|\, y > 0, z > 0\}$ so that

$$x^{-1}(\mathcal{U}' \cap \mathcal{V}') = \{(u^1, u^2) \in \mathcal{U} \,|\, u^2 > 0\}.$$

Now

$$y^{-1} \circ \mathbf{x}(u^1, u^2) = y^{-1}(u^1, u^2, \sqrt{1 - (u^1)^2 - (u^2)^2}) = (u^1, \sqrt{1 - (u^1)^2 - (u^2)^2}),$$

which is certainly a C^k function for any $k \geq 1$ since $(u^1)^2 + (u^2)^2 < 1$. We need also show that $y^{-1} \circ \mathbf{x}$ is a one-to-one function on $x^{-1}(\mathcal{U}' \cap \mathcal{V}')$. If $(y^{-1} \circ \mathbf{x})(a, b) = (y^{-1} \circ \mathbf{x})(c, d)$, then

$$(a, \sqrt{1 - a^2 - b^2}) = (c, \sqrt{1 - c^2 - d^2})$$

so that $a = c$ and $b = \pm d$. However, since both (a, b) and (c, d) belong to $x^{-1}(\mathcal{U}' \cap \mathcal{V}')$, both b and d are positive and $b = d$. Thus $y^{-1} \circ \mathbf{x}$ is one-to-one. The inverse function is $x^{-1} \circ \mathbf{y}$ given by

$$x^{-1} \circ \mathbf{y}(v^1, v^2) = (v^1, \sqrt{1 - (v^1)^2 - (v^2)^2})$$

which is differentiable. Hence $y^{-1} \circ \mathbf{x}$ is a coordinate transformation.

EXAMPLE 2.3. We show that the circular cylinder $S^1 \times (0, 1)$ is a surface. Let $\mathcal{U} = (-3\pi/4, 3\pi/4) \times (0, 1)$ and $\mathbf{x}: \mathcal{U} \longrightarrow \mathbf{R}^3$ by

$$\mathbf{x}(\theta, t) = (\cos\theta, \sin\theta, t).$$

$\mathbf{x}$ is proper. Let $\mathcal{V} = (\pi/4, 7\pi/4) \times (0, 1)$ and $\mathbf{y}: \mathcal{V} \longrightarrow \mathbf{R}^3$ by

$$\mathbf{y}(\phi, s) = (\cos\phi, \sin\phi, s).$$

$\mathbf{y}$ is proper. Every point of $S^1 \times (0, 1)$ is in the image of $\mathbf{x}$ or $\mathbf{y}$. The overlap consists of two separate pieces, a left piece and a right piece. Note that on the right piece both θ and ϕ are between $\pi/4$ and $3\pi/4$. Hence on this piece $y^{-1} \circ \mathbf{x}(\theta, t) = (\theta, t)$, which is certainly a coordinate transformation. (We may also write this as $\phi = \theta$ and $s = t$.) On the left piece θ lies between $-3\pi/4$ and $-\pi/4$ whereas ϕ lies between $5\pi/4$ and $7\pi/4$, so that it is impossible for $\theta = \phi$. In fact, since $\cos\theta = \cos\phi$ and $\sin\theta = \sin\phi$, we have $y^{-1} \circ \mathbf{x}(\theta, t) = (\theta + 2\pi, t)$ (or $\phi = \theta + 2\pi$ and $s = t$). This is also a coordinate transformation. See Figure 4.11.

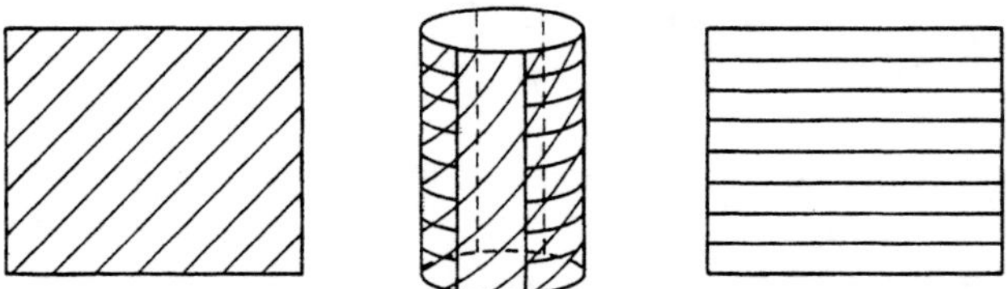

FIGURE 4.11 $\left(-\frac{3\pi}{4}, \frac{3\pi}{4}\right) \times (0, 1)$ $\quad S_1 \times (0, 1)$ $\quad \left(\frac{\pi}{4}, \frac{7\pi}{4}\right) \times (0, 1)$

Example 2.4. A surface of revolution (Problem 1.2) can be made into a surface in a similar manner. See Problem 2.1.

In advanced calculus there is a difficult theorem, which we shall not prove, called the Implicit Function Theorem. (See Fulks [1969].) It states that if $f: \mathbf{R}^3 \to \mathbf{R}$ is a differentiable function such that $(f_x, f_y, f_z) \neq \mathbf{0}$ at all points of $M = \{(x, y, z) \mid f(x, y, z) = 0\}$, then M is a surface. In fact, if $f_z \neq 0$ at a point P of M, there is a Monge patch in M that contains P.

Example 2.5. Let $f(x, y, z) = x^2 + y^2 + z^2 - 1$. Then the set of points where $f = 0$ is the surface S^2.

Example 2.6.

$$f(x, y, z) = \frac{x^2}{a^2} + \frac{y^2}{b^2} + \frac{z^2}{c^2} - 1$$

gives a surface called an *ellipsoid.*

Example 2.7. $f(x, y, z) = x^2 - y^2 - z$ gives a surface called a *hyperbolic paraboloid.*

Example 2.8.

$$f(x, y, z) = \frac{x^2}{a^2} + \frac{y^2}{b^2} - \frac{z^2}{c^2} - 1$$

gives a surface called a *hyperboloid of one sheet.*

EXAMPLE 2.9.

$$f(x, y, z) = \frac{x^2}{a^2} - \frac{y^2}{b^2} - \frac{z^2}{c^2} - 1$$

gives a surface called a *hyperboloid of two sheets.*

PROBLEMS

2.1. The coordinate patch of Problem 1.2 does not cover the entire surface of revolution—it omits points that would correspond to $\theta = \pm\pi$. Define a second coordinate patch with $0 < \phi < 2\pi$, check the overlap condition, and thus show that a surface of revolution is a surface as in Example 2.3.

2.2. Problem 1.1 gave one coordinate patch for a *torus* (or inner tube; see Figure 4.12). Find three more patches making the entire torus a surface (including the points where $u^1 = \pi$ and those where $u^2 = \pi$).

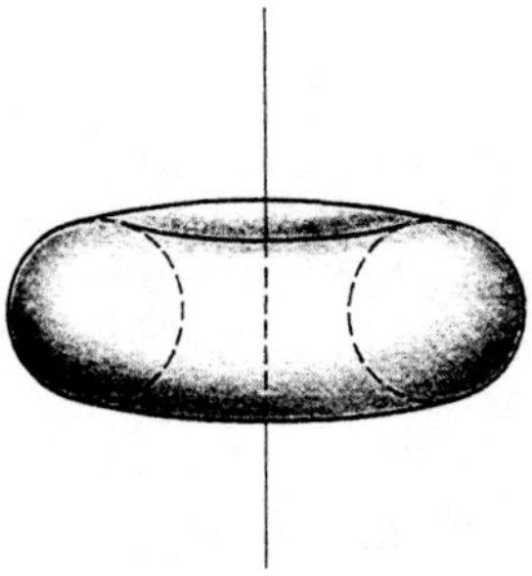

FIGURE 4.12

2.3. Note that every point of S^2 is either in the image of the patch in Example 1.7 or the patch in Problem 1.10. Show that these two patches make S^2 into a surface by computing the coordinate transformation on the overlap.

2.4. Describe some possible parametrizations of the ellipsoid

$$\frac{x^2}{a^2} + \frac{y^2}{b^2} + \frac{z^2}{c^2} = 1.$$

2.5. Give another coordinate patch for the surface in Problem 1.9 so that the curve $\theta = \pm\pi$ is included, thus making that example into a surface. Note that Problem 1.9 says that it is impossible to make a choice of unit normal at each point of this surface in a continuous fashion.

4-3. THE FIRST FUNDAMENTAL FORM AND ARC LENGTH

In this chapter we are concerned with local properties of surfaces. We must understand them before we can consider global properties. For that reason we shall not use the terminology of the last section very much now. Normally we shall assume we are in a coordinate patch, which is equivalent to looking at just one small portion of a surface at a time.

One problem will arise by staying in one coordinate patch. Any definition that seems to depend upon the coordinates will have to be checked as to its form in another coordinate patch in order to determine if the concept is really geometric (i.e., independent of coordinate patch). We have already done this in the case of the tangent plane at a point. (Proposition 1.13.)

Let M be a surface in $\mathbf{R}^3$ and $P \in M$. If $\mathbf{X}$ and $\mathbf{Y}$ are two vectors tangent to M at P, it makes sense to compute $\langle \mathbf{X}, \mathbf{Y} \rangle$. In terms of a proper coordinate patch $\mathbf{x}: \mathcal{U} \longrightarrow \mathbf{R}^3$ about P we can write $\mathbf{X} = \sum X^i \mathbf{x}_i$ and $\mathbf{Y} = \sum Y^j \mathbf{x}_j$. Then, since the inner product is a bilinear function,

$$\langle \mathbf{X}, \mathbf{Y} \rangle = \sum X^i Y^j \langle \mathbf{x}_i, \mathbf{x}_j \rangle = \sum X^i Y^j g_{ij},$$

where we make the definition

$$g_{ij}(u^1, u^2) = \langle \mathbf{x}_i(u^1, u^2), \mathbf{x}_j(u^1, u^2) \rangle \quad \text{or} \quad g_{ij} = \langle \mathbf{x}_i, \mathbf{x}_j \rangle. \tag{3-1}$$

The functions g_{ij} define a symmetric matrix at each point in the image of $\mathbf{x}$. It is important to note that g_{ij} is a function defined on $\mathcal{U}$ and depends critically on which coordinate patch we have picked. See Example 3.1. The functions g_{ij} are sometimes referred to as the *metric coefficients*, the *coefficients of the metric tensor*, or the *coefficients of the Riemannian metric*. Their definition is due to K. Gauss (1827) and, in a more abstract formulation, to G. F. B. Riemann (1854).

DEFINITION. The *tangent space* of a surface M at $P \in M$ is the set T_PM of all vectors tangent to M at P.

Propositions 1.13 and 1.18 show that this is the same as the tangent plane at P to any of the coordinate patches whose image contains P. The standard inner product $\langle \ , \ \rangle$ of $\mathbf{R}^3$ can be restricted to this vector space, where it is still an inner product. The matrix (g_{ij}) is the representation of the restricted inner product with respect to the basis $\{\mathbf{x}_1, \mathbf{x}_2\}$. Thus (g_{ij}) is a nonsingular positive definite matrix. (Recall that a two by two matrix (a_{ij}) is positive definite if and only if $a_{12} = a_{21}$, $a_{11} > 0$, and $\det (a_{ij}) > 0$.)

The rule which assigns to any two tangent vectors $\mathbf{X}, \mathbf{Y} \in T_PM$ their inner product is called the *first fundamental form* of the surface. The terminology "form" refers to the fact that an inner product is a bilinear form in the sense of linear algebra.

EXAMPLE 3.1. Consider the coordinate patch of Example 1.7:

$$\mathbf{x}(u^1, u^2) = (u^1, u^2, \sqrt{1 - (u^1)^2 - (u^2)^2})$$
$$\mathbf{x}_1 = \left(1, 0, \frac{-u^1}{\sqrt{1 - (u^1)^2 - (u^2)^2}}\right)$$
$$\mathbf{x}_2 = \left(0, 1, \frac{-u^2}{\sqrt{1 - (u^1)^2 - (u^2)^2}}\right).$$

Then

$$g_{11} = \frac{1 - (u^2)^2}{1 - (u^1)^2 - (u^2)^2}$$
$$g_{12} = \frac{u^1u^2}{1 - (u^1)^2 - (u^2)^2} = g_{21}$$
$$g_{22} = \frac{1 - (u^1)^2}{1 - (u^1)^2 - (u^2)^2}.$$

Thus

$$(g_{ij}) = \frac{1}{1 - (u^1)^2 - (u^2)^2}\begin{pmatrix} 1 - (u^2)^2 & u^1u^2 \\ u^1u^2 & 1 - (u^1)^2 \end{pmatrix}.$$

Note that this matrix is positive definite since $(u^1)^2 + (u^2)^2 < 1$. We also note that $\det (g_{ij}) = 1/(1 - (u^1)^2 - (u^2)^2) > 0$.

In the coordinate patch of Example 1.8 you can compute that the corresponding matrix is

$$(\bar{g}_{\alpha\beta}) = \frac{1}{1 - (v^1)^2 - (v^2)^2}\begin{pmatrix} 1 - (v^2)^2 & v^1v^2 \\ v^1v^2 & 1 - (v^1)^2 \end{pmatrix}.$$

Note that this matrix is *not* found by replacing u^1 and u^2 by v^1 and $-\sqrt{1 - (v^1)^2 - (v^2)^2}$ as given by $f\colon \tilde{\mathcal{V}} \to \tilde{\mathcal{U}}$. This is because the two matrices represent the inner product with respect to different bases. Later in this section we shall determine the actual relation between (g_{ij}) and $(\bar{g}_{\alpha\beta})$.

EXAMPLE 3.2. Consider the Monge patch $\mathbf{x}(u^1, u^2) = (u^1, u^2, f(u^1, u^2))$ of Example 1.6. $\mathbf{x}_1 = (1, 0, f_1)$ and $\mathbf{x}_2 = (0, 1, f_2)$. $g_{11} = 1 + (f_1)^2$, $g_{12} = g_{21} = f_1 f_2$, and $g_{22} = 1 + (f_2)^2$. Also,

$$\det (g_{ij}) = 1 + (f_1)^2 + (f_2)^2 > 0.$$

EXAMPLE 3.3. Consider the surface $\mathbf{x}(u, v) = (u^2, uv, v^2)$, where $u > 0$ and $v > 0$. Then $\mathbf{x}_1 = (2u, v, 0)$ and $\mathbf{x}_2 = (0, u, 2v)$. $g_{11} = 4u^2 + v^2$,

$g_{12} = g_{21} = uv$, and $g_{22} = u^2 + 4v^2$. Note that

$$\det(g_{ij}) = 4u^4 + 16u^2v^2 + 4v^4 > 0.$$

The term "metric" used with the functions g_{ij} refers to the fact that they play an important role in measuring. They obviously help to measure lengths and angles in T_PM. We shall see below that they appear in formulas for the lengths of curves and (later) for areas of regions.

Let $\boldsymbol{\alpha}(t)$ be a regular curve whose image is contained in that of a coordinate patch $\mathbf{x}\colon \mathcal{U} \longrightarrow \mathbf{R}^3$ so that $\boldsymbol{\alpha}(t) = \mathbf{x}(\alpha^1(t), \alpha^2(t))$ for some real-valued functions $\alpha^1(t)$ and $\alpha^2(t)$ defined by $(\alpha^1(t), \alpha^2(t)) = x^{-1} \circ \boldsymbol{\alpha}(t)$. If we wish to know the length of the segment $\boldsymbol{\alpha}\colon [a, b] \longrightarrow \mathbf{R}^3$, we must compute $\int_a^b |d\boldsymbol{\alpha}/dt|\, dt$. But $d\boldsymbol{\alpha}/dt = \sum \mathbf{x}_i\, d\alpha^i/dt$ so that

$$\begin{aligned} \left|\frac{d\boldsymbol{\alpha}}{dt}\right| &= \sqrt{\left\langle \frac{d\boldsymbol{\alpha}}{dt}, \frac{d\boldsymbol{\alpha}}{dt} \right\rangle} \\ &= \sqrt{\left\langle \sum \mathbf{x}_i \frac{d\alpha^i}{dt}, \sum \mathbf{x}_j \frac{d\alpha^j}{dt} \right\rangle} \\ &= \sqrt{\sum g_{ij} \frac{d\alpha^i}{dt}\frac{d\alpha^j}{dt}}. \end{aligned}$$

Hence the length of the segment is $\int_a^b \sqrt{\sum g_{ij}(d\alpha^i/dt)(d\alpha^j/dt)}\, dt$. Thus if a curve segment $\boldsymbol{\alpha}$ is defined in terms of local coordinates by α^1 and α^2, it is easy (at least theoretically) to compute its length with the use of the metric coefficients g_{ij}.

NOTATION.

(3-2) $$g = \det(g_{ij})$$

(3-3) $$g^{kl} = (k, l) \text{ entry of the inverse matrix of } (g_{ij}).$$

LEMMA 3.4. For a coordinate patch $\mathbf{x}\colon \mathcal{U} \longrightarrow \mathbf{R}^3$

(a) $g = |\mathbf{x}_1 \times \mathbf{x}_2|^2$;

(b) $g^{11} = g_{22}/g$, $g^{12} = g^{21} = -g_{12}/g$, $g^{22} = g_{11}/g$; and

(c) for all i and j, $\sum_{k=1}^{2} g_{ik}g^{kj} = \delta_i^{\,j}$.

Proof:

(a) Let θ be the angle between $\mathbf{x}_1$ and $\mathbf{x}_2$ so that

$$\begin{aligned} |\mathbf{x}_1 \times \mathbf{x}_2|^2 &= |\mathbf{x}_1|^2 |\mathbf{x}_2|^2 \sin^2\theta = |\mathbf{x}_1|^2 |\mathbf{x}_2|^2 (1 - \cos^2\theta) \\ &= |\mathbf{x}_1|^2 |\mathbf{x}_2|^2 \left(1 - \frac{\langle \mathbf{x}_1, \mathbf{x}_2\rangle^2}{|\mathbf{x}_1|^2 |\mathbf{x}_2|^2}\right) \\ &= |\mathbf{x}_1|^2 |\mathbf{x}_2|^2 - \langle \mathbf{x}_1, \mathbf{x}_2\rangle^2 \\ &= g_{11}g_{22} - g_{12}g_{21} = g. \end{aligned}$$

(b) A straightforward calculation shows that

$$\begin{pmatrix} \dfrac{g_{22}}{g} & -\dfrac{g_{12}}{g} \\ -\dfrac{g_{12}}{g} & \dfrac{g_{11}}{g} \end{pmatrix} \quad \text{and} \quad \begin{pmatrix} g_{11} & g_{12} \\ g_{21} & g_{22} \end{pmatrix}$$

are inverse matrices.

(c) $\sum g_{ik}g^{kj}$ is the (i, j) entry of the product $(g_{ij})(g^{kl})$, which is the identity matrix $(\delta_i{}^j)$. ▮

EXAMPLE 3.5. In Examples 1.7 and 3.1 we have $g = 1/(1 - (u^1)^2 - (u^2)^2)$ so that $g^{11} = 1 - (u^1)^2$, $g^{12} = g^{21} = -u^1u^2$, and $g^{22} = 1 - (u^2)^2$.

Suppose now that there is a coordinate transformation $f\colon \mathcal{V} \longrightarrow \mathcal{U}$. How are the metric coefficients g_{ij}, their determinant g, and the inverse coefficients g^{kl} in the $\mathcal{U}$-coordinate system related to those in the $\mathcal{V}$ system for a surface $\mathbf{x}\colon \mathcal{U} \longrightarrow \mathbf{R}^3$?

Let

$$\mathbf{y} = \mathbf{x} \circ f, \quad \mathbf{y}_\alpha = \frac{\partial \mathbf{y}}{\partial v^\alpha}, \quad \bar{g}_{\alpha\beta} = \langle \mathbf{y}_\alpha, \mathbf{y}_\beta \rangle,$$

$$\bar{g} = \det(\bar{g}_{\alpha\beta}), \quad \text{and} \quad (\bar{g}^{\gamma\delta}) = (\bar{g}_{\alpha\beta})^{-1}.$$

Just as $\mathbf{y}_\alpha = \sum \mathbf{x}_i\,(\partial u^i/\partial v^\alpha)$ (Equation (1-3)), we have $\mathbf{x}_i = \sum \mathbf{y}_\alpha\,(\partial v^\alpha/\partial u^i)$ so that

$$g_{ij} = \langle \mathbf{x}_i, \mathbf{x}_j \rangle = \sum \langle \mathbf{y}_\alpha, \mathbf{y}_\beta \rangle \frac{\partial v^\alpha}{\partial u^i}\frac{\partial v^\beta}{\partial u^j} = \sum \bar{g}_{\alpha\beta}\frac{\partial v^\alpha}{\partial u^i}\frac{\partial v^\beta}{\partial v^j}.$$

Thus

(3-4) $$g_{ij} = \sum \bar{g}_{\alpha\beta}\frac{\partial v^\alpha}{\partial u^i}\frac{\partial v^\beta}{\partial u^j}.$$

As a matrix equation this is

(3-5) $$\begin{pmatrix} g_{11} & g_{12} \\ g_{21} & g_{22} \end{pmatrix} = \begin{pmatrix} \dfrac{\partial v^1}{\partial u^1} & \dfrac{\partial v^2}{\partial u^1} \\ \dfrac{\partial v^1}{\partial u^2} & \dfrac{\partial v^2}{\partial u^2} \end{pmatrix} \begin{pmatrix} \bar{g}_{11} & \bar{g}_{12} \\ \bar{g}_{21} & \bar{g}_{22} \end{pmatrix} \begin{pmatrix} \dfrac{\partial v^1}{\partial u^1} & \dfrac{\partial v^1}{\partial u^2} \\ \dfrac{\partial v^2}{\partial u^1} & \dfrac{\partial v^2}{\partial u^2} \end{pmatrix}$$

If J is the Jacobian of $f\colon \mathcal{V} \longrightarrow \mathcal{U}$, Equation (3-5) is

(3-6) $$(g_{ij}) = (J^{-1})^t(\bar{g}_{\alpha\beta})J^{-1}$$

since

$$\begin{pmatrix} \dfrac{\partial v^1}{\partial u^1} & \dfrac{\partial v^1}{\partial u^2} \\ \dfrac{\partial v^2}{\partial u^1} & \dfrac{\partial v^2}{\partial u^2} \end{pmatrix}$$

is the Jacobian of $g: \mathcal{U} \longrightarrow \mathcal{V}$ and the inverse of the Jacobian of $f: \mathcal{V} \longrightarrow \mathcal{U}$. (See Equation (1-2) of Chapter 1 and the preceding discussion.)

$$\begin{aligned} g &= \det (g_{ij}) = \det ((J^{-1})^t(\bar{g}_{\alpha\beta})(J^{-1})) \\ &= \det (J^{-1}) \det (\bar{g}_{\alpha\beta}) \det (J^{-1}) = \bar{g}(\det (J^{-1}))^2. \end{aligned}$$

Thus

$$g = \bar{g}\left(\det \frac{\partial v^\alpha}{\partial u^i}\right)^2. \tag{3-7}$$

Finally

$$(g^{kl}) = (g_{ij})^{-1} = [(J^{-1})^t(\bar{g}_{\alpha\beta})(J^{-1})]^{-1} = J(\bar{g}_{\alpha\beta})^{-1}J^t = J(\bar{g}^{\gamma\delta})J^t.$$

Thus

$$g^{kl} = \sum \bar{g}^{\gamma\delta} \frac{\partial u^k}{\partial v^\gamma} \frac{\partial u^l}{\partial v^\delta}. \tag{3-8}$$

For future reference we gather together the various formulas associated with a coordinate transformation $f: \mathcal{V} \longrightarrow \mathcal{U}$ for a coordinate patch $\mathbf{x}: \mathcal{U} \longrightarrow \mathbf{R}^3$.

$$X^i = \sum \bar{X}^\alpha \frac{\partial u^i}{\partial v^\alpha}, \tag{3-9}$$

where $\mathbf{X} = \sum X^i\mathbf{x}_i = \sum \bar{X}^\alpha \mathbf{y}_\alpha$ is a tangent vector to the surface,

$$g_{ij} = \sum \bar{g}_{\alpha\beta} \frac{\partial v^\alpha}{\partial u^i} \frac{\partial v^\beta}{\partial u^j} \tag{3-10}$$

$$g = \bar{g}\left(\det \left(\frac{\partial v^\alpha}{\partial u^i}\right)\right)^2 \tag{3-11}$$

and

$$g^{kl} = \sum \bar{g}^{\gamma\delta} \frac{\partial u^k}{\partial v^\gamma} \frac{\partial u^l}{\partial v^\delta}. \tag{3-12}$$

Interchanging the roles of $\mathcal{U}$ and $\mathcal{V}$, we obtain

$$\bar{X}^\alpha = \sum X^i \frac{\partial v^\alpha}{\partial u^i} \tag{3-9'}$$

$$\bar{g}_{\alpha\beta} = \sum g_{ij} \frac{\partial u^i}{\partial v^\alpha} \frac{\partial u^j}{\partial v^\beta} \tag{3-10'}$$

$$\bar{g} = g\left(\det \left(\frac{\partial u^i}{\partial v^\alpha}\right)\right)^2 \tag{3-11'}$$

and

$$\bar{g}^{\gamma\delta} = \sum g^{kl} \frac{\partial v^\gamma}{\partial u^k} \frac{\partial v^\delta}{\partial u^l}. \tag{3-12'}$$

In Equations (3-9), (3-10), (3-11), and (3-12) there is one fundamental difference. While all four tell us how to write an expression in the $\mathcal{U}$-coordinates if we know how to write the analogous expression in the $\mathcal{V}$ system, some involve the Jacobian of $f: \mathcal{V} \longrightarrow \mathcal{U}$ ((3-9) and (3-12)) and some involve the Jacobian of $g: \mathcal{U} \longrightarrow \mathcal{V}$ ((3-10) and (3-11)). Classically, transformation laws (3-9) and (3-12) are called contravariant transformations, while (3-10) is called a covariant transformation. Transformation (3-11) is neither covariant nor contravariant because it does not involve matrix multiplication. We shall not belabor the difference between covariant and contravariant. However, from time to time we may mention whether a given quantity is covariant, contravariant, or neither. In the abstract approach of Chapter 7 it is easy to distinguish between covariant and contravariant.

Our notation (due to Ricci, Levi-Civita, and Einstein, circa 1900) is designed to aid in remembering such formulas. Whenever we are summing an expression, such as $\sum X^i \mathbf{x}_i$, it is with an upper index versus a lower index. In any transformation equation (like (3-9) or (3-10)) the dummy summation indices are upper versus lower. All other indices appear on both sides of the equation in the same position (upper or lower). We shall use Latin indices (i, j, k) in the $\mathcal{U}$-coordinate system and Greek indices (α, β, γ) in the $\mathcal{V}$ system. With these rules in mind it is easy to recall Equations (3-9), (3-10), (3-11), (3-12). In some texts the summation sign is omitted completely and one must remember to sum whenever there is a repeated index.

EXAMPLE 3.6. Consider the sphere S^2 and, in particular, the overlap of the coordinate patches in Examples 1.7 and 1.8. To make the following computation a little easier to read, we set $r = u^1$, $s = u^2$, $a = v^1$, and $b = v^2$,

$$f(a, b) = (a, -\sqrt{1 - a^2 - b^2}) = (r, s)$$

and

$$g(r, s) = (r, \sqrt{1 - r^2 - s^2}) = (a, b).$$

The matrix $(J(f))^{-1} = J(g)$ is

$$\begin{pmatrix} 1 & 0 \\ \dfrac{-r}{\sqrt{1 - r^2 - s^2}} & \dfrac{-s}{\sqrt{1 - r^2 - s^2}} \end{pmatrix}.$$

$$(\bar{g}_{\alpha\beta}) = \begin{pmatrix} \dfrac{1 - b^2}{1 - a^2 - b^2} & \dfrac{ab}{1 - a^2 - b^2} \\ \dfrac{ab}{1 - a^2 - b^2} & \dfrac{1 - a^2}{1 - a^2 - b^2} \end{pmatrix}$$

$$= \begin{pmatrix} \dfrac{r^2 + s^2}{s^2} & \dfrac{r\sqrt{1 - r^2 - s^2}}{s^2} \\ \dfrac{r\sqrt{1 - r^2 - s^2}}{s^2} & \dfrac{1 - r^2}{s^2} \end{pmatrix}.$$

Thus

$$(J^{-1})^t(\bar{g}_{\alpha\beta})(J^{-1}) = \begin{pmatrix} 1 & \dfrac{-r}{\sqrt{1-r^2-s^2}} \\ 0 & \dfrac{-s}{\sqrt{1-r^2-s^2}} \end{pmatrix}$$

$$\cdot \begin{pmatrix} \dfrac{r^2+s^2}{s^2} & \dfrac{r\sqrt{1-r^2-s^2}}{s^2} \\ \dfrac{r\sqrt{1-r^2-s^2}}{s^2} & \dfrac{1-r^2}{s^2} \end{pmatrix}$$

$$\cdot \begin{pmatrix} 1 & 0 \\ \dfrac{-r}{\sqrt{1-r^2-s^2}} & \dfrac{-s}{\sqrt{1-r^2-s^2}} \end{pmatrix}$$

$$= \begin{pmatrix} \dfrac{1-s^2}{1-r^2-s^2} & \dfrac{rs}{1-r^2-s^2} \\ \dfrac{rs}{1-r^2-s^2} & \dfrac{1-r^2}{1+r^2-s^2} \end{pmatrix} = (g_{ij}).$$

DIGRESSION (This material may be omitted.)

If $\boldsymbol{\alpha}(t)$ is a curve in a simple surface $\mathbf{x}\colon \mathcal{U} \longrightarrow \mathbf{R}^3$, and if s is arc length on $\boldsymbol{\alpha}$, then

$$\frac{ds}{dt} = \sqrt{\sum g_{ij}\frac{d\alpha^i}{dt}\frac{d\alpha^j}{dt}} \quad \text{or} \quad \left(\frac{ds}{dt}\right)^2 = \sum g_{ij}\frac{d\alpha^i}{dt}\frac{d\alpha^j}{dt}.$$

A classical geometer would drop all reference to the parameter t (and in fact the curve itself) and write this equation as

$$ds^2 = \sum g_{ij}\, du^i\, du^j \tag{3-13}$$

and call ds^2 the first fundamental form. To a classical geometer the symbols ds, du^i, du^j had meaning as infinitesimals. To a modern geometer they have meaning as linear functionals, as we shall describe below.

Equation (3-10) can also be derived from (3-13). We must have

$$ds^2 = \sum \bar{g}_{\alpha\beta}\, dv^\alpha\, dv^\beta.$$

$dv^\alpha = \sum (\partial v^\alpha/\partial u^i)\, du^i$, so that

$$\sum \bar{g}_{\alpha\beta}\, dv^\alpha\, dv^\beta = \sum \bar{g}_{\alpha\beta}\frac{\partial v^\alpha}{\partial u^i}\frac{\partial v^\beta}{\partial u^j}\, du^i\, du^j.$$

If we accept the fact that the symbols $du^i\, du^j$ are independent, then

$$ds^2 = \sum g_{ij}\, du^i\, du^j$$

must yield

$$g_{ij} = \sum \bar{g}_{\alpha\beta}\frac{\partial v^\alpha}{\partial u^i}\frac{\partial v^\beta}{\partial u^j}.$$

Classically the expressions du^1 and du^2 expressed the difference in the coordinates of two infinitesimally close points. The modern viewpoint is to view them as linear functions on T_PM. Before we indicate how this is done, let us review some more linear algebra.

DEFINITION. A *linear functional* on a real vector space V is a linear function $\rho: V \longrightarrow \mathbf{R}$.

EXAMPLE 3.7. Let $V = \mathbf{R}^3$ and define ρ by $\rho(x, y, z) = x - 2y + z$.

EXAMPLE 3.8. Let V be any real vector space with an inner product. Let $\mathbf{p} \in V$ and define $\rho: V \longrightarrow \mathbf{R}$ by $\rho(\mathbf{v}) = \langle \mathbf{p}, \mathbf{v} \rangle$.

The set of all linear functionals defined on V forms a vector space under the usual concepts of function addition and multiplication by real numbers: $(r\phi + \psi)(\mathbf{v}) = r(\phi(\mathbf{v})) + \psi(\mathbf{v})$. This vector space is called the *dual space* of V, and is denoted V^*.

THEOREM 3.9. If V is a real vector space of dimension n, then so is V^*.

Proof: By the previous observation we need only show that V^* has dimension n. Let $\{\mathbf{u}_1, \mathbf{u}_2, \ldots, \mathbf{u}_n\}$ be a basis of V. Define a linear functional $\phi^i: V \longrightarrow \mathbf{R}$ by $\phi^i(\mathbf{u}_j) = \delta^i{}_j$. Then $\phi^i(\sum a^j\mathbf{u}_j) = \sum a^j\phi^i(\mathbf{u}_j) = \sum a^j\delta^i{}_j = a^i$. We claim $\{\phi^1, \ldots, \phi^n\}$ is a basis of V^*. First we show that the ϕ^i are independent. Suppose there are real numbers a_i with $\sum a_i\phi^i$ being the zero function. Then $0 = (\sum a_i\phi^i)(\mathbf{u}_j) = \sum a_i(\phi^i(\mathbf{u}_j)) = \sum a_i\delta^i{}_j = a_j$. Hence the ϕ^i are independent. Now we need to show that they span V^*. Let $\rho \in V^*$ and let $c_i = \rho(\mathbf{u}_i)$. Then a straightfoward calculation shows that $\sum c_i\phi^i(\mathbf{v}) = \rho(\mathbf{v})$ for any $\mathbf{v} \in V$ so that $\rho = \sum c_i\phi^i$. Hence $\{\phi^1, \ldots, \phi^n\}$ is a basis and V^* has dimension n. ∎

The basis $\{\phi^1, \phi^2, \ldots, \phi^n\}$ found in the previous proof is called the *dual basis* of V^* with respect to the basis $\{\mathbf{u}_1, \mathbf{u}_2, \ldots, \mathbf{u}_n\}$. Suppose one has another basis $\{\mathbf{v}_1, \mathbf{v}_2, \ldots, \mathbf{v}_n\}$ of V with $\mathbf{u}_i = \sum a^\alpha{}_i\mathbf{v}_\alpha$. Let $\{\psi^1, \psi^2, \ldots, \psi^n\}$ be the basis of V^* dual to $\{\mathbf{v}_1, \mathbf{v}_2, \ldots, \mathbf{v}_n\}$. Then $\psi^\alpha = \sum b^\alpha{}_i\phi^i$ for some matrix $(b^\alpha{}_i)$. What is $b^\alpha{}_j$?

$$\begin{aligned} b^\alpha{}_j &= \sum b^\alpha{}_i\delta^i{}_j = \sum b^\alpha{}_i\phi^i(\mathbf{u}_j) = (\sum b^\alpha{}_i\phi^i)(\mathbf{u}_j) = \psi^\alpha(\mathbf{u}_j) \\ &= \psi^\alpha(\sum a^\beta{}_j\mathbf{v}_\beta) = \sum a^\beta{}_j\psi^\alpha(\mathbf{v}_\beta) = \sum a^\beta{}_j\delta^\alpha{}_\beta = a^\alpha{}_j. \end{aligned}$$

Thus $(b^\alpha{}_j) = (a^\alpha{}_j)$. If $\mathbf{u}_i = \sum a^\alpha{}_i\mathbf{v}_\alpha$, then

$$\psi^\alpha = \sum a^\alpha{}_i\phi^i. \tag{3-14}$$

If $(c^j{}_\beta)$ denotes the inverse of $(a^\alpha{}_i)$, then

$$\phi^j = \sum c^j{}_\beta\psi^\beta. \tag{3-15}$$

Now we shall indicate how the differentials du^1, du^2 may be viewed as linear functionals on T_PM. If $\mathbf{X} = \sum X^i\mathbf{x}_i \in T_PM$, then $du^j(\mathbf{X})$ is defined to be X^j. This is certainly a linear functional, and $\{du^1, du^2\}$ is the dual basis (of $T_P M^*$) with respect to $\{\mathbf{x}_1, \mathbf{x}_2\}$. This is also consistent with the manner in which dual bases transform:

$$\mathbf{x}_i = \sum \frac{\partial v^\alpha}{\partial u^i}\mathbf{y}_\alpha \quad \text{while} \quad du^i = \sum \frac{\partial u^i}{\partial v^\alpha}\, dv^\alpha.$$

Note that $(\partial u^i/\partial v^\alpha)$ is the inverse of $(\partial v^\alpha/\partial u^i)$ as required by Equation (3-15).

It thus makes sense to call $ds^2 = \sum g_{ij}\, du^i\, du^j$ the first fundamental form because it is a bilinear form:

$$\begin{aligned} ds^2(\mathbf{X}, \mathbf{Y}) &= \sum g_{ij}\, du^i\, du^j(\mathbf{X}, \mathbf{Y}) = \sum g_{ij}\, du^i(\mathbf{X})\, du^j(\mathbf{Y}) \\ &= \sum g_{ij} X^i Y^j = \langle \mathbf{X}, \mathbf{Y} \rangle. \end{aligned}$$

The matrix (g^{ij}) can be viewed as representing an inner product on $T_P M^*$, namely $\langle du^i, du^j \rangle = g^{ij}$. This is natural in the following sense. $\mathbf{X}^{(i)} = \sum g^{il}\mathbf{x}_l$ has the property that $\langle \mathbf{X}^{(i)}, \mathbf{Y} \rangle = du^i(\mathbf{Y})$. $g^{ij} = \langle \mathbf{X}^{(i)}, \mathbf{X}^{(j)} \rangle$. That is, the inner product of du^i and du^j is defined to equal that of the vectors that represent du^i and du^j via the inner product in $T_P M$.

PROBLEMS

***3.1.** Show that the metric coefficients for a surface of revolution (Problem 1.2) are given by the matrix $\begin{pmatrix} \dot r^2 + \dot z^2 & 0 \\ 0 & r^2 \end{pmatrix}$.

3.2. Compute the metric coefficients for the parametrization of S^2 in Problem 1.10.

3.3. Compute the inverse of metric coefficients in Examples 3.2 and 3.3.

3.4. A coordinate patch $\mathbf{x}\colon \mathcal{U} \longrightarrow \mathbf{R}^3$ is called an *orthogonal net* if the u^1-curves meet the u^2-curves at right angles, that is, $\mathbf{x}_1$ is perpendicular to $\mathbf{x}_2$ everywhere. Prove that the meridians and circles of latitude of a surface of revolution form an orthogonal net.

***3.5.** For a coordinate patch $\mathbf{x}\colon \mathcal{U} \longrightarrow \mathbf{R}^3$ show that u^1 is arc length on the u^1-curves if and only if $g_{11} \equiv 1$.

3.6. Let x and y be cartesian coordinates of the plane while r and θ are polar coordinates. Show that $x = r\cos\theta$, $y = r\sin\theta$ is a C^1 coordinate transformation for $r > 0$. Show that the metric coefficient matrix for the plane with respect to the x, y coordinates is $\begin{pmatrix} 1 & 0 \\ 0 & 1 \end{pmatrix}$. Determine the metric coefficients for polar coordinates.

3.7. Consider the coordinate patch $x(r, \theta) = (r\cos\theta, r\sin\theta, r)$ where $r > 0$. Find the coefficients of the metric tensor (g_{ij}). Now consider the curve $r(t) = e^{t(\cot\beta)/2}$, $\theta(t) = t/\sqrt{2}$ in this surface, where $0 \le t \le \pi$ and β is a constant. Find the length of the curve. Show that β is the angle between this curve and the line $\theta = $ constant.

4-4. NORMAL CURVATURE, GEODESIC CURVATURE, AND GAUSS'S FORMULAS

Let $\boldsymbol{\gamma}(s)$ be a unit speed curve whose image lies on a surface $M \subset \mathbf{R}^3$. $\boldsymbol{\gamma}$ has Frenet-Serret apparatus $\{\kappa, \tau, \mathbf{T}, \mathbf{N}, \mathbf{B}\}$. The reader is warned not to confuse $\mathbf{N}$, the normal vector field to $\boldsymbol{\gamma}$ as defined in Chapter 2, with $\mathbf{n}$, the normal vector to a coordinate patch in M. They can very well point in different directions (and usually do!).

EXAMPLE 4.1. Let $M = S^2$ and let $\boldsymbol{\gamma}(s)$ be the curve

$$\boldsymbol{\gamma}(s) = \left(\frac{\sqrt{2}}{2}\cos\sqrt{2}\,s, \frac{\sqrt{2}}{2}\sin\sqrt{2}\,s, \frac{\sqrt{2}}{2}\right).$$

Then

$$\boldsymbol{\gamma}' = \mathbf{T} = (-\sin\sqrt{2}\,s, \cos\sqrt{2}\,s, 0)$$
$$\mathbf{T}' = (-\sqrt{2}\cos\sqrt{2}\,s, -\sqrt{2}\sin\sqrt{2}\,s, 0)$$
$$\mathbf{N} = (-\cos\sqrt{2}\,s, -\sin\sqrt{2}\,s, 0).$$

At $s = \pi/2\sqrt{2}$, $\mathbf{N} = (0, -1, 0)$. Using Example 1.14 we find

$$\mathbf{n} = \left(0, \frac{\sqrt{2}}{2}, \frac{\sqrt{2}}{2}\right) \quad \text{at} \quad \boldsymbol{\gamma}\left(\frac{\pi}{2\sqrt{2}}\right) = \left(0, \frac{\sqrt{2}}{2}, \frac{\sqrt{2}}{2}\right) = \mathbf{x}\left(0, \frac{\sqrt{2}}{2}\right).$$

See Figure 4.13.

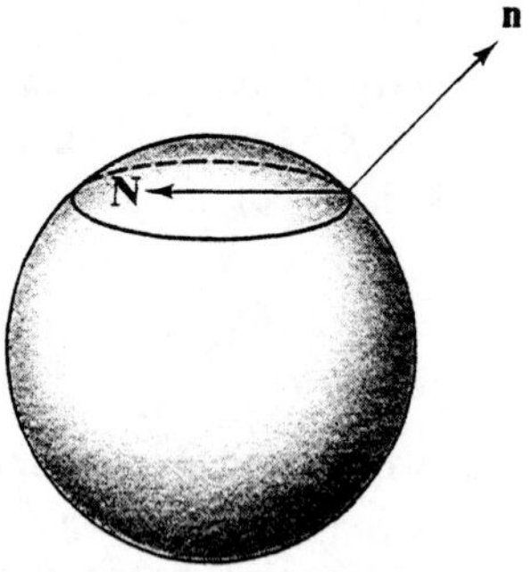

FIGURE 4.13

NOTATION. If $\mathbf{x}\colon \mathcal{U} \longrightarrow \mathbf{R}^3$ is a simple surface and $\boldsymbol{\gamma}(s)$ is a unit speed curve in the image of $\mathbf{x}$, then

$$\mathbf{x}_{ij}(a, b) = \frac{\partial^2 \mathbf{x}}{\partial u^j\,\partial u^i}(a, b) \tag{4-1}$$

$$\mathbf{x}_{ij} = \frac{\partial^2 \mathbf{x}}{\partial u^j\,\partial u^i} \tag{4-2}$$

$$\mathbf{S} = \mathbf{n} \times \mathbf{T}. \tag{4-3}$$

We shall call $\mathbf{S}$ the *intrinsic normal* of $\boldsymbol{\gamma}$. It is well defined on a surface M up to sign, just as $\mathbf{n}$ is. The two choices of sign correspond to different orientations. Do not let the terminology mislead you: $\mathbf{S}$ is normal to the curve but tangent to the surface. Note that $\mathbf{x}_{ij} = \mathbf{x}_{ji}$ under the assumption that $\mathbf{x}$ is C^2.

For the curve $\boldsymbol{\gamma}(s)$ whose image is contained in the coordinate patch $\mathbf{x}\colon \mathcal{U} \longrightarrow \mathbf{R}^3$, we write, as usual, $\boldsymbol{\gamma}(s) = \mathbf{x}(\gamma^1(s), \gamma^2(s))$. By Equation (1-7) we have $\mathbf{T} = \boldsymbol{\gamma}' = \sum (\partial\mathbf{x}/\partial u^i)(\gamma^i)' = \sum \mathbf{x}_i(\gamma^i)'$ or

$$\boldsymbol{\gamma}'(s) = \sum \mathbf{x}_i(\gamma^1(s), \gamma^2(s))(\gamma^i(s))'. \tag{4-4}$$

We wish to differentiate this equation with respect to s. What is the derivative of $\mathbf{x}_i(\gamma^1(s), \gamma^2(s))$? An application of the chain rule yields

$$\begin{aligned}\frac{d\mathbf{x}_i(\gamma^1(s), \gamma^2(s))}{ds} &= \frac{\partial \mathbf{x}_i}{\partial u^1}\frac{d\gamma^1}{ds} + \frac{\partial \mathbf{x}_i}{\partial u^2}\frac{d\gamma^2}{ds} \\ &= \mathbf{x}_{i1}\frac{d\gamma^1}{ds} + \mathbf{x}_{i2}\frac{d\gamma^2}{ds} = \sum \mathbf{x}_{ij}\frac{d\gamma^j}{ds}.\end{aligned}$$

This equation and (4-4) imply

$$\boldsymbol{\gamma}''(s) = \sum_{i,j} \mathbf{x}_{ij}(\gamma^i)'(\gamma^j)' + \sum_i \mathbf{x}_i(\gamma^i)''. \tag{4-5}$$

If $P \in M$, we may let $N_PM = \{r\mathbf{n} \mid r \in \mathbf{R}\}$. N_PM is the set of all vectors perpendicular to M at P and is called the *normal space* of M at P. Clearly $\mathbf{R}^3 = T_PM + N_PM$, and thus any vector can be decomposed as a sum of a vector tangent to M at P and a vector normal to M at P. In particular, this may be done for the vector $\boldsymbol{\gamma}''(s)$:

$$\boldsymbol{\gamma}''(s) = \mathbf{X}(s) + \mathbf{V}(s),$$

where $\mathbf{X}(s)$ is tangent to M and $\mathbf{V}(s)$ is normal to M.

Since $\mathbf{T}(s)$ is tangent to M, $\langle \mathbf{V}, \mathbf{T}\rangle = 0$. $\langle \boldsymbol{\gamma}'', \mathbf{T}\rangle = 0$ and thus $\langle \mathbf{X}(s), \mathbf{T}\rangle = 0$ also. But $\langle \mathbf{X}(s), \mathbf{n}\rangle = 0$. Hence $\mathbf{X}(s)$ is perpendicular to both $\mathbf{n}$ and $\mathbf{T}$ and is thus a multiple of $\mathbf{S} = \mathbf{n} \times \mathbf{T}$. We may then define two functions $\kappa_n(s)$ and $\kappa_g(s)$ by $\kappa_n(s) = \langle \boldsymbol{\gamma}''(s), \mathbf{n}(\gamma^1(s), \gamma^2(s))\rangle$ and $\kappa_g(s) = \langle \boldsymbol{\gamma}''(s), \mathbf{S}(s)\rangle$ so that

$$\kappa(s)\mathbf{N}(s) = \mathbf{T}'(s) = \boldsymbol{\gamma}''(s) = \kappa_n(s)\mathbf{n}(s) + \kappa_g(s)\mathbf{S}(s). \tag{4-6}$$

DEFINITION. The *normal curvature* of a unit speed curve $\boldsymbol{\gamma}$ is the normal component of $\boldsymbol{\gamma}''$ (i.e., is κ_n). The *geodesic curvature* of $\boldsymbol{\gamma}$ is the component of $\boldsymbol{\gamma}''$ in the direction of $\mathbf{S} = \mathbf{n} \times \mathbf{T}$ (i.e., is κ_g).

In Section 4-7 we will see that the normal curvature helps measure how M is curving in $\mathbf{R}^3$. In Section 4-5 we will see that the geodesic curvature measures how $\boldsymbol{\gamma}$ is curving in M.

Note that although $\kappa \geq 0$, κ_n and κ_g can be negative. In fact, a change of the sign of $\mathbf{n}$ changes the signs of κ_n and κ_g. Note also that κ, κ_n, and κ_g are related by

$$\kappa^2 = \kappa_n^2 + \kappa_g^2. \tag{4-7}$$

(See Problem 4.1.)

DEFINITION. The *coefficients of the second fundamental form* of a simple surface $\mathbf{x}\colon \mathcal{U} \longrightarrow \mathbf{R}^3$ are the functions L_{ij} defined on $\mathcal{U}$ by $L_{ij} = \langle \mathbf{x}_{ij}, \mathbf{n} \rangle$. The *Christoffel symbols* are the functions $\Gamma_{ij}{}^k$ $(1 \leq i, j, k \leq 2)$ defined on $\mathcal{U}$ by $\Gamma_{ij}{}^k = \sum_{l=1}^{2} \langle \mathbf{x}_{ij}, \mathbf{x}_l \rangle g^{lk}$.

Since $\mathbf{x}_{ij} = \mathbf{x}_{ji}$, we have $L_{ij} = L_{ji}$ and $\Gamma_{ij}{}^k = \Gamma_{ji}{}^k$. Proposition 4.2 will show that the L_{ij} measure the normal component of $\mathbf{x}_{ij}$ while the $\Gamma_{ij}{}^k$ measure the tangential components. The reason that the L_{ij} are called the coefficients of the second fundamental form is that the assignment $\mathrm{II}(\mathbf{X}, \mathbf{Y}) = \sum_{i,j} L_{ij} X^i Y^j$ is a symmetric bilinear form (but not necessarily positive definite) on $T_P M$ just as the first fundamental form

$$\mathrm{I}(\mathbf{X}, \mathbf{Y}) = \langle \mathbf{X}, \mathbf{Y} \rangle = \sum g_{ij} X^i Y^j$$

is. We will study the second fundamental form and its relation to κ_n in Section 4-7.

PROPOSITION 4.2. Let $\mathbf{x}\colon \mathcal{U} \longrightarrow \mathbf{R}^3$ be a simple surface. Then

(a) (Gauss's formulas)

$$\mathbf{x}_{ij} = L_{ij}\mathbf{n} + \sum_k \Gamma_{ij}{}^k \mathbf{x}_k \tag{4-8}$$

(b) for any unit speed curve, $\boldsymbol{\gamma}(s) = \mathbf{x}(\gamma^1(s), \gamma^2(s))$

$$\kappa_n = \sum_{i,j} L_{ij}(\gamma^i)'(\gamma^j)' \tag{4-9}$$

and

$$\kappa_g \mathbf{S} = \sum_k [(\gamma^k)'' + \sum_{i,j} \Gamma_{ij}{}^k (\gamma^i)'(\gamma^j)'] \mathbf{x}_k. \tag{4-10}$$

Proof:

(a) Since $\mathbf{n}, \mathbf{x}_1, \mathbf{x}_2$ are linearly independent, and hence a basis of $\mathbf{R}^3$, we may express any arbitrary vector as a linear combination of these three vectors. In particular, there are functions, a_{ij} and $b_{ij}{}^m$, defined on $\mathcal{U}$ such that $\mathbf{x}_{ij} = a_{ij}\mathbf{n} + \sum b_{ij}{}^m \mathbf{x}_m$. Thus $L_{ij} = \langle \mathbf{x}_{ij}, \mathbf{n} \rangle = a_{ij}$ since $\langle \mathbf{x}_m, \mathbf{n} \rangle = 0$. Also

$\langle \mathbf{x}_{ij}, \mathbf{x}_l \rangle = \sum_m b_{ij}{}^m g_{ml}$ so that $\langle \mathbf{x}_{ij}, \mathbf{x}_l \rangle g^{lk} = \sum_m b_{ij}{}^m g_{ml} g^{lk}$. If we sum this equation over l and apply Lemma 3.4, we obtain

$$\Gamma_{ij}{}^k = \sum_l \langle \mathbf{x}_{ij}, \mathbf{x}_l \rangle g^{lk} = \sum_{l,m} b_{ij}{}^m g_{ml} g^{lk} = \sum_m b_{ij}{}^m \delta_m{}^k = b_{ij}{}^k.$$

Thus $a_{ij} = L_{ij}$ and $b_{ij}{}^k = \Gamma_{ij}{}^k$, which proves Gauss's formulas.

(b) Substitute Equation (4-8) into Equation (4-5) to get

$$\gamma'' = \sum \mathbf{x}_k (\gamma^k)'' + \sum \mathbf{x}_k \Gamma_{ij}{}^k (\gamma^i)'(\gamma^j)' + \sum \mathbf{n} L_{ij} (\gamma^i)'(\gamma^j)'$$

or

$$\gamma'' = \sum [(\gamma^k)'' + \sum \Gamma_{ij}{}^k (\gamma^i)'(\gamma^j)'] \mathbf{x}_k + (\sum L_{ij} (\gamma^i)'(\gamma^j)') \mathbf{n}.$$

Since $\mathbf{x}_1$ and $\mathbf{x}_2$ are tangent to the surface, a comparison of this equation with Equation (4-6) completes the proof. ∎

From Equation (4-8) we obtain $\langle \mathbf{x}_{ij}, \mathbf{x}_l \rangle = \sum_k \Gamma_{ij}{}^k g_{kl}$. This quantity is classically written $\Gamma_{ij|l}$ and is called a *Christoffel symbol of the first kind*, while $\Gamma_{ij}{}^k$ is called a *Christoffel symbol of the second kind*. We have no use for $\Gamma_{ij|l}$ and will use the word Christoffel symbol to refer to $\Gamma_{ij}{}^k$. These symbols were first used by G. B. Christoffel (1829–1900) who used $\left\{ \begin{matrix} ij \\ k \end{matrix} \right\}$ for $\Gamma_{ij}{}^k$.

The remainder of this section is devoted to the calculations of the Christoffel symbols in terms of the metric coefficients and several applications of these calculations.

PROPOSITION 4.3. For a coordinate patch $\mathbf{x}: \mathfrak{U} \longrightarrow \mathbf{R}^3$ with metric coefficients g_{ij},

$$\Gamma_{ij}{}^l = \frac{1}{2} \sum_{k=1}^{2} g^{kl} \left(\frac{\partial g_{ik}}{\partial u^j} - \frac{\partial g_{ij}}{\partial u^k} + \frac{\partial g_{kj}}{\partial u^i} \right). \tag{4-11}$$

Proof: This proof (which is due to Gauss) makes use of a classical technique called "cyclic permutation of indices."

$$\frac{\partial g_{ij}}{\partial u^k} = \frac{\partial}{\partial u^k} \langle \mathbf{x}_i, \mathbf{x}_j \rangle = \langle \mathbf{x}_{ik}, \mathbf{x}_j \rangle + \langle \mathbf{x}_i, \mathbf{x}_{jk} \rangle \tag{4-12}$$

$$\frac{\partial g_{ik}}{\partial u^j} = \langle \mathbf{x}_{ij}, \mathbf{x}_k \rangle + \langle \mathbf{x}_i, \mathbf{x}_{kj} \rangle \tag{4-13}$$

$$\frac{\partial g_{jk}}{\partial u^i} = \langle \mathbf{x}_{ji}, \mathbf{x}_k \rangle + \langle \mathbf{x}_j, \mathbf{x}_{ki} \rangle. \tag{4-14}$$

Since we assume that $\mathbf{x}$ is of class C^3, $\mathbf{x}_{ij} = \mathbf{x}_{ji}$. Hence, we may combine (4-12), (4-13), and (4-14) to obtain

$$\frac{1}{2} \left(\frac{\partial g_{ik}}{\partial u^j} - \frac{\partial g_{ij}}{\partial u^k} + \frac{\partial g_{jk}}{\partial u^i} \right) = \langle \mathbf{x}_{ij}, \mathbf{x}_k \rangle. \tag{4-15}$$

If we multiply this result by g^{kl} and sum over k, we obtain

$$\frac{1}{2}\sum g^{kl}\left(\frac{\partial g_{ik}}{\partial u^j} - \frac{\partial g_{ij}}{\partial u^k} + \frac{\partial g_{kj}}{\partial u^i}\right) = \sum \langle \mathbf{x}_{ij}, \mathbf{x}_k \rangle g^{kl} = \Gamma_{ij}{}^l. \quad \blacksquare$$

Note that this formula shows that $\Gamma_{ij}{}^k$ is completely determined by the metric coefficients. We say that the $\Gamma_{ij}{}^k$ are *intrinsic*. This means that they can be determined by measurements within the surface. More generally, an intrinsic concept is one which depends only on (g_{ij}). If we were two-dimensional beings living on a surface, we would be able to compute lengths and angles and thus the metric coefficients and anything that depends on them. The intrinsic concepts would be the only geometric concepts we would have. We would know nothing about a normal vector. (An amusing interpretation of this is in E. A. Abbott's *Flatland*, a story about a two-dimensional creature in a two-dimensional world with insight into the third dimension.)

PROPOSITION 4.4. The geodesic curvature of a surface curve is intrinsic.

Proof: Let $\epsilon_{ij} = [\mathbf{n}, \mathbf{x}_i, \mathbf{x}_j]$ (see Section 1-3 for triple bracket notation) so that $\epsilon_{11} = \epsilon_{22} = 0$ and, by Lemma 3.4, $\epsilon_{12} = -\epsilon_{21} = \sqrt{g}$. Since ϵ_{ij} is either zero or depends only on g, ϵ_{ij} is intrinsic. Now

$$\kappa_g = \langle \kappa_g \mathbf{S}, \mathbf{S} \rangle = [\kappa_g \mathbf{S}, \mathbf{n}, \mathbf{T}].$$

Equation (4-10) shows that

$$\begin{aligned}\kappa_g &= \sum (\gamma^{k\prime\prime} + \sum \Gamma_{ij}{}^k \gamma^{i\prime}\gamma^{j\prime})[\mathbf{x}_k, \mathbf{n}, \sum \mathbf{x}_l \gamma^{l\prime}] \\ &= \sum (\gamma^{k\prime\prime} + \sum \Gamma_{ij}{}^i \gamma^{i\prime}\gamma^{j\prime})\gamma^{l\prime}[\mathbf{n}, \mathbf{x}_l, \mathbf{x}_k]\end{aligned}$$

or

(4-16) $$\kappa_g = \sum (\gamma^{k\prime\prime} + \sum \Gamma_{ij}{}^k \gamma^{i\prime}\gamma^{j\prime})\gamma^{l\prime}\epsilon_{lk},$$

which is intrinsic. $\blacksquare$

Although (4-16) gives a formula for κ_g, sometimes it is useful to have an extrinsic formula.

(4-17) $$\begin{aligned}\kappa_g &= \langle \mathbf{T}', \mathbf{S} \rangle = [\mathbf{T}', \mathbf{n}, \mathbf{T}] = [\mathbf{n}, \mathbf{T}, \mathbf{T}'] = \langle \mathbf{n}, \mathbf{T} \times \mathbf{T}' \rangle \\ &= \langle \mathbf{n}, \mathbf{T} \times \kappa \mathbf{N} \rangle = \kappa \langle \mathbf{n}, \mathbf{B} \rangle = \kappa \cos \alpha,\end{aligned}$$

where α is the angle between the unit normal $\mathbf{n}$ of the surface, and the binormal $\mathbf{B}$ of the curve. Note that $\kappa_g = \kappa \cos \alpha$ makes sense even when $\kappa = 0$ and neither $\mathbf{B}$ nor α is defined because $\kappa_g = 0$ when $\kappa = 0$.

EXAMPLE 4.5. Let $\mathbf{x}(u^1, u^2) = (u^1, u^2, f(u^1, u^2))$ be a Monge patch. We compute some of the coefficients of the second fundamental form and then some of the Christoffel symbols both extrinsically and intrinsically. $\mathbf{x}_1 = (1, 0, f_1)$ and $\mathbf{x}_2 = (0, 1, f_2)$, $\mathbf{n} = (-f_1, -f_2, 1)/\sqrt{1 + (f_1)^2 + (f_2)^2}$,

$\mathbf{x}_{11} = (0, 0, f_{11})$, $\mathbf{x}_{12} = \mathbf{x}_{21} = (0, 0, f_{12})$ and $\mathbf{x}_{22} = (0, 0, f_{22})$. By Gauss's formulas (4-8), $\mathbf{x}_{11} = L_{11}\mathbf{n} + \Gamma_{11}{}^1\mathbf{x}_1 + \Gamma_{11}{}^2\mathbf{x}_2$. Thus

$$L_{11} = \langle \mathbf{x}_{11}, \mathbf{n} \rangle = \frac{f_{11}}{\sqrt{1 + (f_1)^2 + (f_2)^2}}.$$

Considering the first coordinate of $\mathbf{x}_{11}$, we note

$$0 = \frac{L_{11}(-f_1)}{\sqrt{1 + (f_1)^2 + (f_2)^2}} + \Gamma_{11}{}^1.$$

Thus we obtain $\Gamma_{11}{}^1 = f_1 f_{11}/(1 + (f_1)^2 + (f_2)^2)$. A similar calculation yields $\Gamma_{11}{}^2 = f_2 f_{11}/(1 + (f_1)^2 + (f_2)^2)$.

Alternatively we could compute $\Gamma_{11}{}^2$ intrinsically by using Equation (4-11).

$$g^{12} = \frac{-f_1 f_2}{1 + (f_1)^2 + (f_2)^2} \quad \text{and} \quad g^{22} = \frac{1 + (f_1)^2}{1 + (f_1)^2 + (f_2)^2}.$$

$$\frac{\partial g_{11}}{\partial u^1} = \frac{\partial(1 + (f_1)^2)}{\partial u^1} = 2f_1 f_{11},$$

$$\frac{\partial g_{12}}{\partial u^1} = \frac{\partial(f_1 f_2)}{\partial u^1} = f_{11} f_2 + f_1 f_{21},$$

and

$$\frac{\partial g_{11}}{\partial u^2} = 2f_1 f_{12}.$$

$$\begin{aligned}
\Gamma_{11}{}^2 &= \frac{1}{2} \sum g^{k2}\left(\frac{\partial g_{1k}}{\partial u^1} - \frac{\partial g_{11}}{\partial u^k} + \frac{\partial g_{k1}}{\partial u^1}\right) \\
&= \frac{1}{2}\left[\frac{(-f_1 f_2)(2f_1 f_{11} - 2f_1 f_{11} + 2f_1 f_{11})}{(1 + (f_1)^2 + (f_2)^2)} \right. \\
&\quad \left. + \frac{(1 + (f_1)^2)(f_{11} f_2 + f_1 f_{21} - 2f_1 f_{12} + f_{11} f_2 + f_1 f_{21})}{(1 + (f_1)^2 + (f_2)^2)}\right] \\
&= \frac{[-2(f_1)^2 f_2 f_{11} + 2f_{11} f_2 + 2(f_1)^2 f_{11} f_2]}{2(1 + (f_1)^2 + (f_2)^2)} \\
&= \frac{f_{11} f_2}{(1 + (f_1)^2 + (f_2)^2)}.
\end{aligned}$$

Notice how much easier it was to compute $\Gamma_{11}{}^2$ extrinsically than intrinsically.

EXAMPLE 4.6. Consider the upper hemisphere $\mathbf{x}(r, s) = (r, s, \sqrt{1 - r^2 - s^2})$. Let $\boldsymbol{\gamma}(t) = (\sin t, 0, \cos t)$. This is a unit speed curve in $\mathbf{x}$ for $-\pi/2 < t < \pi/2$.

$$\mathbf{T} = (\cos t, 0, -\sin t), \mathbf{T}' = (-\sin t, 0, -\cos t),$$

and

$$\mathbf{n} = (r, s, \sqrt{1 - r^2 - s^2}).$$

At the point $(\sin t, 0, \cos t)$, $\mathbf{n}$ is $(\sin t, 0, \cos t)$. Then

$$\kappa_n = \langle \mathbf{T}', \mathbf{n} \rangle = -\sin^2 t - \cos^2 t = -1.$$

Since $\kappa \equiv 1$, and $\kappa^2 = \kappa_n{}^2 + \kappa_g{}^2$, we have $\kappa_g \equiv 0$ for this curve. Note that the image of this curve is part of a great circle.

PROBLEMS

***4.1.** Prove $\kappa^2 = \kappa_n{}^2 + \kappa_g{}^2$.

***4.2.** Show that the matrix (L_{ij}) for a surface of revolution (Problem 1.2) is

$$\frac{1}{\sqrt{\dot{r}^2 + \dot{z}^2}} \begin{pmatrix} \dot{r}\ddot{z} - \dot{z}\ddot{r} & 0 \\ 0 & r\dot{z} \end{pmatrix}.$$

4.3. Prove that for a surface of revolution $\det(L_{ij}) \equiv 0$ if and only if each meridian is a straight line.

4.4. Let $\bar{L}_{\alpha\beta}$ be the corresponding expression to L_{ij} in a coordinate system $\mathcal{V}$. Let $f\colon \mathcal{V} \longrightarrow \mathcal{U}$ be a coordinate transformation. Show that

$$L_{ij} = \pm \sum \bar{L}_{\alpha\beta} \frac{\partial v^\alpha}{\partial u^i} \frac{\partial v^\beta}{\partial u^j},$$

where the sign is that of $\det(\partial v^\alpha / \partial u^i)$. This shows that *up to sign*, the functions L_{ij} transform covariantly, (i.e., just like g_{ij}).

4.5. In the Monge patch $\mathbf{x}(u, v) = (u, v, u^2 + v^2)$ find the normal curvature of the curve $\boldsymbol{\gamma}(t) = \mathbf{x}(t^2, t)$ at $t = 1$.

***4.6.** The plane is a simple surface: $\mathbf{x}(r, s) = (r, s, 0)$. Show that the geodesic curvature of a curve in the plane is its plane curvature: $\kappa_g = k$.

4.7. The sphere is a surface of revolution. Find the geodesic curvature of a circle of latitude $w^1 =$ constant in Example 1.9.

***4.8.** Prove that if $\mathbf{x}$ is a coordinate patch of the sphere and $\mathbf{n}$ is the unit normal, then at any point $\mathbf{x}(r, s)$, $\mathbf{n}(r, s) = \pm\mathbf{x}(r, s)$ as vectors.

4.9. Let $\boldsymbol{\gamma}$ be a curve on the sphere. Prove κ_n is constant.

4.10. Let $\boldsymbol{\gamma}$ be a curve on the sphere with κ_g constant. Prove $\boldsymbol{\gamma}$ is a circle. (*Hint:* Problem 5.3 of Chapter 2 and Problem 4.9.)

4.11. Let $\bar{\Gamma}_{\alpha\beta}{}^\gamma$ denote the Christoffel symbols for a coordinate system $\mathcal{V}$, and $f\colon \mathcal{V} \longrightarrow \mathcal{U}$ a coordinate transformation. Use Gauss's formulas to prove

$$\bar{\Gamma}_{\alpha\beta}{}^\gamma = \sum \left(\sum \Gamma_{ij}{}^k \frac{\partial u^i}{\partial v^\alpha} \frac{\partial u^j}{\partial v^\beta} + \frac{\partial^2 u^k}{\partial v^\alpha \partial v^\beta} \right) \frac{\partial v^\gamma}{\partial u^k}.$$

This formula is the first example of a transformation law that is not well behaved (because of the term $(\partial^2 u^k / \partial v^\alpha\, \partial v^\beta)$).

4.12. Prove that

$$\frac{1}{2}\frac{\partial(\ln g)}{\partial u^1} = \Gamma_{11}{}^1 + \Gamma_{12}{}^2 \quad \text{and} \quad \frac{1}{2}\frac{\partial(\ln g)}{\partial u^2} = \Gamma_{21}{}^1 + \Gamma_{22}{}^2.$$

4.13. Does the sign of the geodesic curvature have any meaning in a coordinate patch? In a surface?

4–5. GEODESICS

In plane geometry, the straight lines play a very important role as the basis for most constructions and for the formation of most figures studied. We would like to find curves on an arbitrary surface that play an analogous role. However, there are several properties of straight lines and it is not clear which is the most important. That is, it is not clear which of these properties should be taken as the definition of a "straight line" on an arbitrary surface. For example:

SL1 Straight lines have (plane) curvature zero.
SL2 Straight lines give the shortest path between two points.
SL3 Given two points there is a unique straight line joining them.
SL4 The tangent vectors to a straight line are all parallel.

We shall generalize each of these properties to curves on surfaces and explore their interrelationships.

DEFINITION. A *geodesic* on a surface M is a unit speed curve on M with geodesic curvature equal to zero everywhere.

This is an immediate translation of property SL1 above. Since the geodesic curvature of a plane curve is its plane curvature (Problem 4.6), this means that straight lines in the plane are geodesics.

PROPOSITION 5.1. A unit speed curve $\boldsymbol{\gamma}(s)$ in M is a geodesic if and only if $[\mathbf{n}, \mathbf{T}, \mathbf{T}'] \equiv 0$.

Proof: By Equation (4-17), $\kappa_g = [\mathbf{n}, \mathbf{T}, \mathbf{T}']$ is the geodesic curvature. ∎

PROPOSITION 5.2. Let $\boldsymbol{\gamma}(s)$ be a unit speed curve, $\mathbf{x}$ a coordinate patch, and write $\boldsymbol{\gamma}(s) = \mathbf{x}(\gamma^1(s), \gamma^2(s))$. $\boldsymbol{\gamma}$ is a geodesic if only if

$$\gamma^{k\prime\prime} + \sum \Gamma_{ij}{}^k \gamma^{i\prime}\gamma^{j\prime} \equiv 0 \quad \text{for } k = 1, 2. \tag{5-1}$$

Proof: By Equation (4-10) $\kappa_g \mathbf{S} = \sum (\gamma^{k\prime\prime} + \sum \Gamma_{ij}{}^k \gamma^{i\prime}\gamma^{j\prime})\mathbf{x}_k$. Since $\mathbf{x}_1$ and $\mathbf{x}_2$ are independent, the result follows. ∎

PROPOSITION 5.3. A unit speed curve $\gamma(s)$ on a surface M is a geodesic if and only if γ'' is everywhere normal to the surface (i.e., is a multiple of the normal to M).

Proof: $\gamma'' = \kappa\mathbf{N} = \kappa_g\mathbf{S} + \kappa_n\mathbf{n}$. γ'' is normal to the surface if and only if $\kappa_g = 0$. ▮

EXAMPLE 5.4. Consider the great circle $\gamma(s) = (\sin s, 0, \cos s)$ on S^2. $\gamma' = (\cos s, 0, -\sin s)$, $\gamma'' = (-\sin s, 0, -\cos s)$. Then $\gamma''(s)$ is the inward pointing normal to S^2 at $\gamma(s)$. By Proposition 5.3, γ is a geodesic since a normal to S^2 at $\gamma(s)$ is $\pm\gamma(s)$. Since there is nothing geometrically special about this particular great circle, *every great circle of S^2 is a geodesic.* The converse is true: if γ is a geodesic of S^2 then γ is an arc of a great circle (see Problem 5.3).

PROPOSITION 5.5. Let M be a surface of revolution generated by the unit speed curve $(r(t), z(t))$. Then
(a) every meridian is a geodesic; and
(b) a circle of latitude is a geodesic if and only if the tangent $\mathbf{x}_1$ to the meridians is parallel to the axis of revolution at all points on the circle of latitude.

Proof: M may be parametrized by $\mathbf{x}(t, \theta) = (r(t)\cos\theta, r(t)\sin\theta, z(t))$ as in Problem 1.2. Since t is arc length along the generating curve, the metric matrix is, according to Problem 3.1,

$$(g_{ij}) = \begin{pmatrix} \dot{r}^2 + \dot{z}^2 & 0 \\ 0 & r^2 \end{pmatrix} = \begin{pmatrix} 1 & 0 \\ 0 & r^2 \end{pmatrix}.$$

Let $u^1 = t$, $u^2 = \theta$ so that the Christoffel symbols can be computed by Proposition 4.3: $\Gamma_{12}{}^2 = \Gamma_{21}{}^2 = \dot{r}/r$, $\Gamma_{22}{}^1 = -r\dot{r}$, and all other $\Gamma_{ij}{}^k$ are zero. By Proposition 5.2, a unit speed curve $\gamma(s) = \mathbf{x}(t(s), \theta(s))$ is a geodesic if and only if it is a solution of the differential equations

$$t'' - r\dot{r}\theta'\theta' = 0 \tag{5-2}$$

$$\theta'' + 2\left(\frac{\dot{r}}{r}\right)t'\theta' = 0. \tag{5-3}$$

(a) A meridian is given by $\theta =$ constant. Then θ' and θ'' are zero and (5-3) is satisfied. Along a meridian $t = s$, so that $t' = 1$ and $t'' \equiv 0$. Thus (5-2) is satisfied. Hence each meridian is a geodesic.

(b) A circle of latitude is given by $t =$ constant. Then $t'' = t' \equiv 0$. Since $\gamma(s) = \mathbf{x}(t(s), \theta(s))$ has unit speed,

$$1 = |\gamma'(s)|^2 = \left|\frac{\partial \mathbf{x}}{\partial t}t' + \frac{\partial \mathbf{x}}{\partial \theta}\theta'\right|^2 = g_{22}(\theta')^2.$$

We therefore have

$$1 = r^2(\theta')^2.$$

This last equation implies that $0 \neq \theta' = \pm 1/r$. (Note r is constant if t is.) Thus $\theta'' \equiv 0$ and a circle of latitude satisfies (5-3). Since $\theta' \neq 0$ and $r > 0$, a circle of latitude satisfies (5-2) if and only if $\dot{r} = 0$. This occurs if and only if $\mathbf{x}_1 = (\dot{r}\cos\theta, \dot{r}\sin\theta, \dot{z})$ is parallel to the axis of rotation (0, 0, 1). See Figure 4.14. ∎

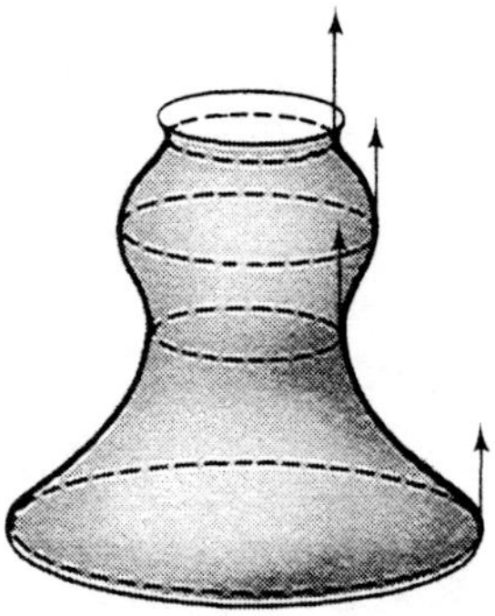

FIGURE 4.14

EXAMPLE 5.6. Let M be the Monge patch $\mathbf{x}(u, v) = (u, v, u^2 - v^2)$. Let $\boldsymbol{\gamma}(t)$ be the (non-unit speed) curve $\boldsymbol{\gamma}(t) = (t, 0, t^2)$ in M. Now

$$\boldsymbol{\gamma}' = \dot{\boldsymbol{\gamma}}\frac{dt}{ds} \quad\text{and}\quad \boldsymbol{\gamma}'' = \ddot{\boldsymbol{\gamma}}\left(\frac{dt}{ds}\right)^2 + \dot{\boldsymbol{\gamma}}\frac{d^2t}{ds^2}$$

so that $[\mathbf{n}, \boldsymbol{\gamma}', \boldsymbol{\gamma}''] = [\mathbf{n}, \dot{\boldsymbol{\gamma}}, \ddot{\boldsymbol{\gamma}}]\,(dt/ds)^3$.

$$\mathbf{n} = \frac{(-2u, 2v, 1)}{\sqrt{1 + 4u^2 + 4v^2}}$$

$$\mathbf{n}(u(t), v(t)) = \frac{(-2t, 0, 1)}{\sqrt{1 + 4t^2}}$$

$$\dot{\boldsymbol{\gamma}} = (1, 0, 2t) \quad\text{and}\quad \ddot{\boldsymbol{\gamma}} = (0, 0, 2)$$

so that $[\mathbf{n}, \dot{\boldsymbol{\gamma}}, \ddot{\boldsymbol{\gamma}}] = 0$. Hence $[\mathbf{n}, \boldsymbol{\gamma}', \boldsymbol{\gamma}''] = 0$ and $\boldsymbol{\gamma}$ is a geodesic by Proposition 5.1.

In the plane, a straight line is determined once a point on the line and the direction of that line (at that point) are given. The next theorem says that this is true for geodesics in general.

THEOREM 5.7. Let P be a point on a surface M and let $\mathbf{X}$ be a unit tangent vector at P. Then, if $s_0 \in R$ is given, there exists a unique geodesic $\boldsymbol{\gamma}$ with $\boldsymbol{\gamma}(s_0) = P$, $\boldsymbol{\gamma}'(s_0) = \mathbf{X}$.

Proof: Let $\mathbf{x}$ be a patch about P with $P = \mathbf{x}(0, 0)$, and $\mathbf{X} = \sum X^i\mathbf{x}_i$. If there is a geodesic $\boldsymbol{\gamma}(s) = \mathbf{x}(\gamma^1(s), \gamma^2(s))$, then we must have $\gamma^{k''} = -\sum \Gamma_{ij}{}^k \gamma^{i'}\gamma^{j'}$ with initial conditions $\gamma^i(s_0) = 0$, $\gamma^{i'}(s_0) = X^i$. Conversely, any solution to this initial value problem is a geodesic with the required properties, *if the solution is a unit speed curve.* Picard's theorem (Theorem 5.1 of Chapter 2) implies there is a unique solution to this initial value problem, for values of s near s_0. We must show that the solution $\boldsymbol{\gamma}(s) = \mathbf{x}(\gamma^1(s), \gamma^2(s))$ to the differential equation is unit speed.

Let $f(s) = |\boldsymbol{\gamma}'|^2 = \sum g_{ij}\gamma^{i\prime}\gamma^{j\prime}$. Then

$$\begin{aligned} f'(s) &= \sum \frac{\partial g_{ij}}{\partial u^k}\gamma^{k\prime}\gamma^{i\prime}\gamma^{j\prime} + \sum g_{ij}\gamma^{i\prime\prime}\gamma^{j\prime} + \sum g_{ij}\gamma^{i\prime}\gamma^{j\prime\prime} \\ &= \sum g_{il}\Gamma_{jk}{}^{l}\gamma^{k\prime}\gamma^{i\prime}\gamma^{j\prime} + \sum g_{jl}\Gamma_{ik}{}^{l}\gamma^{k\prime}\gamma^{j\prime}\gamma^{i\prime} \\ &\quad + \sum g_{il}\gamma^{l\prime\prime}\gamma^{i\prime} + \sum g_{jl}\gamma^{l\prime\prime}\gamma^{j\prime} \\ &= \sum g_{il}(\gamma^{l\prime\prime} + \Gamma_{jk}{}^{l}\gamma^{j\prime}\gamma^{k\prime})\gamma^{i\prime} + \sum g_{jl}(\gamma^{l\prime\prime} + \Gamma_{ik}{}^{l}\gamma^{i\prime}\gamma^{k\prime})\gamma^{j\prime}, \end{aligned}$$

which is zero since (γ^1, γ^2) solves the differential equation. Thus $f(s)$ is constant. Since $f(s_0) = |\mathbf{X}|^2 = 1$, f must be identically 1. Thus $\boldsymbol{\gamma}'$ is a unit vector and $\boldsymbol{\gamma}$ is a unit speed curve, hence in this case a geodesic.

Repeated application of this gives $\boldsymbol{\gamma}(s)$ defined for s in some open interval (a, b), even though the image of $\boldsymbol{\gamma}$ may not be all in one coordinate patch. The proof shows that $\boldsymbol{\gamma}$ is unique at all points where it is defined. ▮

EXAMPLE 5.8. Let M be the surface consisting of all the points in $\mathbf{R}^2$ except $(2, 0)$. Let $P = (0, 0)$ and let $\mathbf{X} = (1, 0)$ be the unit vector pointing in the direction of the positive x-axis. The associated geodesic is $\boldsymbol{\gamma}(s) = (s, 0)$ and is defined for all $s < 2$. Since $\boldsymbol{\gamma}$ is continuous, it is impossible to define $\boldsymbol{\gamma}(2)$ as anything but $(2, 0)$, which is not in M. It is therefore possible for the geodesic, which is the solution of the differential equation, to be defined only for $|s| < \epsilon < \infty$.

(The question as to whether a geodesic extends indefinitely, i.e., is defined for all $s \in (-\infty, \infty)$, is metrical. One calls a surface *complete* if every geodesic extends indefinitely. A famous theorem due to Hopf and Rinow says that M is complete if and only if it is complete as a metric space. See Hicks [1965].)

The next theorem tells us that geodesics also possess property SL2 of straight lines. The proof we give is a variant of Gauss's proof of Theorem 5.3. The technique of the proof is to start with a length-minimizing curve $\boldsymbol{\gamma}$ and assume its geodesic curvature is not zero. We "wiggle" the curve to form a family $\boldsymbol{\alpha}_t$ of curves with the same end points as $\boldsymbol{\gamma}$ and with $\boldsymbol{\alpha}_0 = \boldsymbol{\gamma}$. The function $L(t)$ which gives the length of $\boldsymbol{\alpha}_t$ must have a minimum at $t = 0$ so that $L'(0) = 0$. This fact together with integration by parts yields a contradiction. The idea of differentiating with respect to the "wiggle" parameter t is due to Gauss. A proof like this is called "variational" and is a cornerstone in the calculus of variations.

THEOREM 5.9. Let $\boldsymbol{\gamma}$ be a unit speed curve in a surface M between points $P = \boldsymbol{\gamma}(a)$ and $Q = \boldsymbol{\gamma}(b)$. If $\boldsymbol{\gamma}$ is the shortest curve between P and Q, then $\boldsymbol{\gamma}$ is a geodesic.

Proof: Let $a < s_0 < b$ and κ_g be the geodesic curvature of $\boldsymbol{\gamma}$. We shall show that $\kappa_g(s_0) = 0$. Suppose $\kappa_g(s_0) \neq 0$. Then there exist numbers c and d with $a < c < s_0 < d < b$, $\kappa_g \neq 0$ on $[c, d]$, and the image of $[c, d]$ under $\boldsymbol{\gamma}$ contained in a coordinate patch $\mathbf{x}$. Note that the segment of $\boldsymbol{\gamma}$ from $\boldsymbol{\gamma}(c)$ to $\boldsymbol{\gamma}(d)$ must be the shortest curve joining $\boldsymbol{\gamma}(c)$ and $\boldsymbol{\gamma}(d)$ or else there is a piecewise regular curve from $\boldsymbol{\gamma}(a)$ to $\boldsymbol{\gamma}(c)$ to $\boldsymbol{\gamma}(d)$ to $\boldsymbol{\gamma}(b)$ that is shorter than $\boldsymbol{\gamma}$. But $\boldsymbol{\gamma}$ is assumed to give the shortest curve from $\boldsymbol{\gamma}(a)$ to $\boldsymbol{\gamma}(b)$.

Let $\lambda(s)$ be a C^2 function defined for $c \leq s \leq d$ such that

$$\lambda(c) = \lambda(d) = 0,\ \lambda(s_0) \neq 0, \quad \text{and} \quad \lambda(s)\kappa_g(s) \geq 0 \qquad \text{for } c \leq s \leq d.$$

(If $\boldsymbol{\gamma}$ is C^4, then $\lambda(s) = (s - c)(d - s)\kappa_g(s)$ will work.) If $\mathbf{S} = \mathbf{n} \times \boldsymbol{\gamma}'$, then in the patch $\mathbf{x}$ we have $\lambda(s)\mathbf{S} = \sum v^i(s)\mathbf{x}_i$ for some $v^i\colon [c, d] \longrightarrow \mathbf{R}$. Let $\boldsymbol{\gamma}(s)$ be given by $\boldsymbol{\gamma}(s) = \mathbf{x}(\gamma^1(s), \gamma^2(s))$ and define a family of curves by

$$\boldsymbol{\alpha}_t(s) = \mathbf{x}(\gamma^1(s) + tv^1(s), \gamma^2(s) + tv^2(s)),$$

where $|t|$ is small enough. $\boldsymbol{\alpha}_t$ is a curve from $\boldsymbol{\gamma}(c)$ to $\boldsymbol{\gamma}(d)$ for each choice of t with $\boldsymbol{\alpha}_0 = \boldsymbol{\gamma}$. We may also write $\boldsymbol{\alpha}_t(s) = \boldsymbol{\alpha}(s; t)$. (See Figure 4.15.)

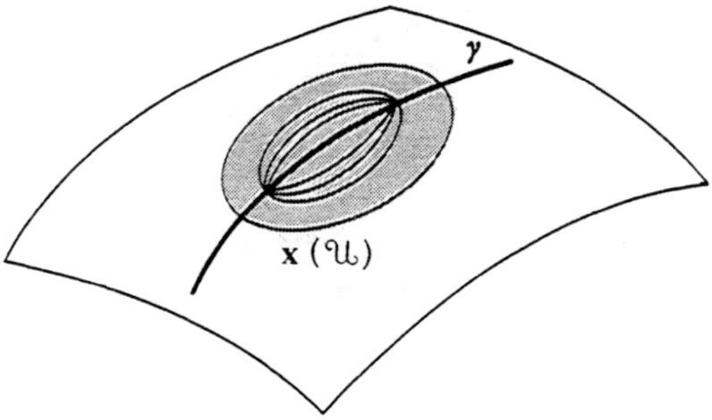

FIGURE 4.15

The length of $\boldsymbol{\alpha}(s; t)$ is $L(t) = \int_c^d \langle \partial\boldsymbol{\alpha}/\partial s, \partial\boldsymbol{\alpha}/\partial s\rangle^{1/2}\, ds$. $L(t)$ has a minimum for $t = 0$ since $\boldsymbol{\alpha}(s; 0) = \boldsymbol{\gamma}(s)$ gives the shortest path.

$$L'(t) = \frac{d}{dt}\int_c^d \left\langle \frac{\partial\boldsymbol{\alpha}}{\partial s}, \frac{\partial\boldsymbol{\alpha}}{\partial s}\right\rangle^{1/2} ds = \int_c^d \frac{\partial}{\partial t}\left\langle \frac{\partial\boldsymbol{\alpha}}{\partial s}, \frac{\partial\boldsymbol{\alpha}}{\partial s}\right\rangle^{1/2} ds$$

$$= \int_c^d \frac{1}{2}\,\frac{2\left\langle \dfrac{\partial^2\boldsymbol{\alpha}}{\partial t\,\partial s}, \dfrac{\partial\boldsymbol{\alpha}}{\partial s}\right\rangle}{\left\langle \dfrac{\partial\boldsymbol{\alpha}}{\partial s}, \dfrac{\partial\boldsymbol{\alpha}}{\partial s}\right\rangle^{1/2}}\, ds = \int_c^d \frac{\left\langle \dfrac{\partial^2\boldsymbol{\alpha}}{\partial s\,\partial t}, \dfrac{\partial\boldsymbol{\alpha}}{\partial s}\right\rangle}{\left\langle \dfrac{\partial\boldsymbol{\alpha}}{\partial s}, \dfrac{\partial\boldsymbol{\alpha}}{\partial s}\right\rangle^{1/2}}\, ds.$$

At $t = 0$,

$$\left\langle \frac{\partial\boldsymbol{\alpha}}{\partial s}, \frac{\partial\boldsymbol{\alpha}}{\partial s}\right\rangle = \left\langle \frac{d\boldsymbol{\gamma}}{ds}, \frac{d\boldsymbol{\gamma}}{ds}\right\rangle = 1.$$

Thus

$$L'(0) = \int_c^d \left\langle \frac{\partial^2 \boldsymbol{\alpha}}{\partial s\, \partial t}, \frac{\partial \boldsymbol{\alpha}}{\partial s} \right\rangle \bigg|_{t=0} ds$$

$$= \int_c^d \left[\frac{d}{ds} \left\langle \frac{\partial \boldsymbol{\alpha}}{\partial t}, \frac{\partial \boldsymbol{\alpha}}{\partial s} \right\rangle \bigg|_{t=0} - \left\langle \frac{\partial \boldsymbol{\alpha}}{\partial t}, \frac{\partial^2 \boldsymbol{\alpha}}{\partial s^2} \right\rangle \bigg|_{t=0} \right] ds$$

$$= \left\langle \frac{\partial \boldsymbol{\alpha}}{\partial t}, \frac{\partial \boldsymbol{\alpha}}{\partial s} \right\rangle \bigg|_{t=0} \bigg|_c^d - \int_c^d \left\langle \frac{\partial \boldsymbol{\alpha}}{\partial t}, \frac{\partial^2 \boldsymbol{\alpha}}{\partial s^2} \right\rangle \bigg|_{t=0} ds.$$

$$\frac{\partial \boldsymbol{\alpha}}{\partial t} \bigg|_{t=0} = \sum v^i(s) \mathbf{x}_i = \lambda(s) \mathbf{S}.$$

But λ was constructed so that $\lambda(c) = \lambda(d) = 0$. Thus

$$0 = L'(0) = 0 - \int_c^d \langle \lambda(s)\mathbf{S}, \kappa_g(s)\mathbf{S} + \kappa_n(s)\mathbf{n} \rangle\, ds$$

$$= -\int_c^d \lambda(s) \kappa_g(s)\, ds < 0.$$

This contradiction implies $\kappa_g(s_0) = 0$. ∎

The above theorem is true if $\boldsymbol{\gamma}$ is only assumed to be piecewise regular unit speed (see Hicks [1965]). The converse is false. A geodesic need not minimize distances. Let P and Q be two points on S^2 with $P \neq \pm Q$. There are two geodesics of different lengths joining P to Q, corresponding to the two arcs of the great circle through P and Q. The longer geodesic does not minimize length.

The next two examples show that, in general, property SL3 of straight lines is false for geodesics.

Example 5.10. Let M be the surface of Example 5.8. Then there is no geodesic joining the points $(0, 0)$ and $(4, 0)$.

Example 5.11. Let M be the unit sphere with P and Q the two poles. There are an infinite number of great circles through the poles. Hence there is not a unique geodesic from P to Q.

The closest we can come to having property SL3 is the following theorem which is local in nature (i.e., it says nothing about geodesics between "distant" points). To get stronger, global theorems requires further topological assumptions (e.g., simple connectivity) and further geometric assumptions (e.g., on curvature). The reader is referred to M. Berger [1965]. The proof of Whitehead's theorem below is beyond the scope of this book. It may be found in Hicks [1965].

THEOREM 5.12 (J. H. C. Whitehead, 1932). Let P be a point on a surface M. Then there is an ϵ-neighborhood $\mathcal{U}$ of P such that any two points of $\mathcal{U}$ can be joined by a unique geodesic of shortest length, and this geodesic is contained in $\mathcal{U}$.

PROBLEMS

5.1. Show that a meridian of a surface of revolution is a geodesic *without* solving the differential equations as was done in Proposition 5.5. Also, determine which circles of latitude are geodesics. (*Hint:* Proposition 5.3.)

5.2. Let M be a surface and Π a plane that intersects M in a curve $\boldsymbol{\gamma}$. Show that $\boldsymbol{\gamma}$ is a geodesic if Π is a plane of symmetry of M, i.e., the two sides are mirror images.

***5.3.** Combine Problems 4.9 and 4.10 to show that any geodesic on S^2 is a great circle.

5.4. Let $\boldsymbol{\gamma}$ be a straight line in a surface M. Prove $\boldsymbol{\gamma}$ is a geodesic.

***5.5.** Suppose $\mathbf{x}$ is a coordinate patch such that $g_{11} \equiv 1$ and $g_{12} \equiv 0$. Prove that the u^1-curves are geodesics. (Such a patch is called a *geodesic coordinate patch.*)

†5.6. Prove that if M is a surface of revolution and $\boldsymbol{\gamma}$ is a geodesic, then $r \cos \beta(s) =$ constant, where $\beta(s)$ is the angle between $\boldsymbol{\gamma}'(s)$ and the circle of latitude (of radius r) through $\boldsymbol{\gamma}(s)$.

5.7. Let $\boldsymbol{\gamma}(t)$ be a geodesic not parametrized by arc length. Prove

$$\frac{d^2\gamma^i}{dt^2} + \sum \Gamma_{jk}{}^i \frac{d\gamma^j}{dt}\frac{d\gamma^k}{dt} = -\frac{d\gamma^i}{dt}\frac{d^2t}{ds^2}\left(\frac{ds}{dt}\right)^2 \qquad \text{for } i = 1 \text{ and } 2.$$

†5.8. Prove that a regular curve $\boldsymbol{\gamma}(t)$ is a geodesic if and only if

$$\frac{d^2\gamma^1}{dt^2}\frac{d\gamma^2}{dt} - \frac{d^2\gamma^2}{dt^2}\frac{d\gamma^1}{dt} + \sum \left(\frac{d\gamma^2}{dt}\Gamma_{ij}{}^1 - \frac{d\gamma^1}{dt}\Gamma_{ij}{}^2\right)\frac{d\gamma^i}{dt}\frac{d\gamma^j}{dt} \equiv 0.$$

5.9. Let M be the Monge patch $\mathbf{x}(u, v) = (u, v, uv)$. Show that the non-unit speed curve $\boldsymbol{\gamma}(t) = (t, -t, -t^2)$ is a geodesic when parametrized by arc length. Can you find other geodesics in M?

5.10. Let $\boldsymbol{\alpha}(s) = (f(s), g(s))$ be a simple unit speed plane curve. Let $\mathbf{x}(s, t)$ be the surface $\mathbf{x}(s, t) = (f(s), g(s), t)$. (This is called a *cylinder* over $\boldsymbol{\alpha}(s)$.) Let β be a fixed constant and let $\boldsymbol{\gamma}(\theta) = (f(\theta), g(\theta), \theta \tan \beta)$. Prove $\boldsymbol{\gamma}$ is a geodesic (θ is not arc length). Prove $\boldsymbol{\gamma}$ is a helix.

5.11. Let M be the surface given by $x^2 + y^2 - z^2 = 1$ (Example 2.8). Find as many geodesics as you can.

5.12. Find as many geodesics as you can on the Monge patch $(u, v, u^2 - v^2)$. (*Hint:* Problems 5.2, 5.4)

5.13. Find as many geodesics as you can on the surface

$$x^2 + \frac{y^2}{4} - z^2 = 1.$$

†5.14. Let **x** and **y** be two different coordinate patches for part of a surface M. Let $\mathbf{X} = \sum X^j \mathbf{x}_j = \sum \bar{X}^\beta \mathbf{y}_\beta$ and $\mathbf{Y} = \sum Y^i \mathbf{x}_i = \sum \bar{Y}^\alpha \mathbf{y}_\alpha$ be two vector fields. Define symbols Z^k and $\bar{Z}^\gamma$ by

$$Z^k = \sum \frac{\partial Y^k}{\partial u^j} X^j + \sum \Gamma_{ij}{}^k Y^i X^j$$

and

$$\bar{Z}^\gamma = \sum \frac{\partial \bar{Y}^\gamma}{\partial v^\beta} \bar{X}^\beta + \sum \bar{\Gamma}_{\alpha\beta}{}^\gamma \bar{Y}^\alpha \bar{X}^\beta.$$

Prove that $\bar{Z}^\gamma = \sum Z^k(\partial v^\gamma/\partial u^k)$. (*Hint:* Problem 4.11.) This proves that $\sum Z^k \mathbf{x}_k = \sum \bar{Z}^\gamma \mathbf{y}_\gamma$ defines a vector field $\mathbf{Z} = \nabla_{\mathbf{X}}\mathbf{Y}$, called the *covariant derivative* of **Y** with respect to **X**. This is one of the most fundamental concepts of modern differential geometry. It is due to Levi-Civita (1917) and will be used in Chapter 7.

†5.15. Let **x** be a patch where $g_{11} = g_{22} = a(u^1) + b(u^2)$ and $g_{12} \equiv 0$. Let $\boldsymbol{\gamma}(s)$ be a geodesic and let θ be the angle between $\boldsymbol{\gamma}$ and the u^1-curves (i.e., between $\boldsymbol{\gamma}'$ and $\mathbf{x}_1$). Prove $a \sin^2 \theta - b \cos^2 \theta =$ constant. (*Hint:* Equation (4-16).)

4–6. PARALLEL VECTOR FIELDS ALONG A CURVE AND PARALLELISM

Finally we would like to generalize property SL4 which involves the concept of "parallel." In plane geometry there are concepts of both parallel lines and parallel vectors. We shall treat the concept of parallel vectors as the more primitive of the two. Indeed, there are geometries (e.g., spherical geometry) in which parallel lines do not exist.

DEFINITION. A *vector field along a curve* $\boldsymbol{\gamma}: [a, b] \longrightarrow M$ is a function **X** which assigns to each $t \in [a, b]$ a tangent vector $\mathbf{X}(t)$ to M at $\boldsymbol{\gamma}(t)$.

EXAMPLE 6.1. Let $\boldsymbol{\gamma}(s)$ be a unit speed curve in M. Then $\mathbf{T}(s)$ is a vector field along $\boldsymbol{\gamma}$.

EXAMPLE 6.2. Let $\boldsymbol{\gamma}(s)$ be a unit speed curve in a coordinate patch **x**. Then $\mathbf{S}(s) = \mathbf{n} \times \mathbf{T}$ is a vector field along $\boldsymbol{\gamma}$. Since **T** and **S** are linearly independent, any vector field **X** along $\boldsymbol{\gamma}$ has the form

$$\mathbf{X}(s) = a(s)\mathbf{T}(s) + b(s)\mathbf{S}(s).$$

DEFINITION. A vector field $\mathbf{X}(t)$ along $\boldsymbol{\gamma}(t)$ is *differentiable* if as a function $\mathbf{X}: [a, b] \longrightarrow \mathbf{R}^3$ it is differentiable.

That $d\mathbf{X}/dt$ is normal to the surface means there is no tangential component in the change of $\mathbf{X}$ as t varies. Thus an "intrinsic" being living on the surface could detect no change in $\mathbf{X}$ along $\boldsymbol{\gamma}$. The being would view the vectors $\mathbf{X}(t_1)$ and $\mathbf{X}(t_2)$ (for t_1 near t_2) as parallel if $d\mathbf{X}/dt$ is normal. Hence the following:

DEFINITION. A differentiable vector field $\mathbf{X}(t)$ along $\boldsymbol{\gamma}(t)$ is *parallel along* $\boldsymbol{\gamma}(t)$ if $d\mathbf{X}/dt$ is perpendicular to M.

EXAMPLE 6.3. Let $\boldsymbol{\gamma}(t) = (a(t), b(t), 0)$ be a plane curve. Let

$$\mathbf{X}(t) = (A(t), B(t), 0)$$

be a vector field along $\boldsymbol{\gamma}$. $d\mathbf{X}/dt = (dA/dt, dB/dt, 0)$. The normal to the surface is $(0, 0, 1)$. Hence $d\mathbf{X}/dt$ is perpendicular to the surface (plane) if and only if $dA/dt = 0 = dB/dt$. Therefore $\mathbf{X}$ is parallel along $\boldsymbol{\gamma}$ if and only if A and B are constants. This is the usual notion of parallel vectors in the plane. Note that in this case, the notion of parallel along $\boldsymbol{\gamma}$ is independent of $\boldsymbol{\gamma}$. This is atypical and in fact characterizes the plane locally.

EXAMPLE 6.4. Let M be the unit sphere, $\boldsymbol{\gamma}(t)$ the equator, and $\mathbf{X}(t)$ the unit vector pointing north at each point on $\boldsymbol{\gamma}$. Then $\mathbf{X}$ is parallel since $\mathbf{X}(t) = (0, 0, 1)$ and $d\mathbf{X}/dt = (0, 0, 0)$ is certainly perpendicular to the surface.

EXAMPLE 6.5. Let M be the unit sphere, and $\boldsymbol{\gamma}(t)$ the circle of latitude

$$\boldsymbol{\gamma}(t) = \left(\frac{\sqrt{2}}{2}\cos t, \frac{\sqrt{2}}{2}\sin t, \frac{\sqrt{2}}{2}\right).$$

Let $\mathbf{X}(t)$ be the unit vector field that points toward the north pole at each point of $\boldsymbol{\gamma}$. Analytically

$$\mathbf{X}(t) = \left(\frac{-\sqrt{2}}{2}\cos t, \frac{-\sqrt{2}}{2}\sin t, \frac{\sqrt{2}}{2}\right)$$

so that

$$\frac{d\mathbf{X}}{dt} = \left(\frac{\sqrt{2}}{2}\sin t, \frac{-\sqrt{2}}{2}\cos t, 0\right),$$

which is not normal to the surface. $\mathbf{X}$ is not parallel along $\boldsymbol{\gamma}$.

PROPOSITION 6.6. If $\boldsymbol{\gamma}(t) = \mathbf{x}(\gamma^1(t), \gamma^2(t))$ is a regular curve in a coordinate patch $\mathbf{x}$ and $\mathbf{X}(t)$ is a differentiable vector field along $\boldsymbol{\gamma}$ with $\mathbf{X} = \sum X^i \mathbf{x}_i$, then $\mathbf{X}(t)$ is parallel along $\boldsymbol{\gamma}$ if and only if

$$0 = \frac{dX^k}{dt} + \sum \Gamma_{ij}{}^k X^i \frac{d\gamma^j}{dt}, \qquad k = 1 \text{ and } 2.$$

Proof: $\mathbf{X}$ is parallel if and only if for all l,

$$0 \equiv \left\langle \frac{d\mathbf{X}}{dt}, \mathbf{x}_l \right\rangle = \left\langle \sum \frac{dX^i}{dt} \mathbf{x}_i, \mathbf{x}_l \right\rangle + \left\langle \sum X^i \mathbf{x}_{ij} \frac{d\gamma^j}{dt}, \mathbf{x}_l \right\rangle;$$

that is, if and only if

$$0 = \sum \frac{dX^i}{dt} g_{il} + \sum \langle \mathbf{x}_{ij}, \mathbf{x}_l \rangle X^i \frac{d\gamma^j}{dt}, \qquad l = 1, 2. \tag{6-1}$$

If these equations are multiplied by g^{lk} and summed over l, the result is $0 = \sum (dX^i/dt)\, \delta_i^k + \sum \Gamma_{ij}{}^k X^i\, (d\gamma^j/dt)$ or

$$0 = \frac{dX^k}{dt} + \sum \Gamma_{ij}{}^k X^i \frac{d\gamma^j}{dt}, \qquad k = 1, 2. \tag{6-2}$$

Conversely, if Equations (6-2) are multiplied by g_{kl} and summed over k, the result is (6-1). Thus $\mathbf{X}(t)$ is parallel if and only if (6-2) holds. ∎

Notice that we did not assume that t is arc length along $\boldsymbol{\gamma}$. Further note that the differential equation that $\mathbf{X}$ must satisfy depends only on the given curve $\boldsymbol{\gamma}$ and the intrinsic quantities $\Gamma_{ij}{}^k$. Hence the concept of parallel along a curve is intrinsic.

THEOREM 6.7. Let $\boldsymbol{\gamma}(t)$ be a regular curve on a C^2 surface M. Let $\tilde{\mathbf{X}}$ be a vector tangent to M at $\boldsymbol{\gamma}(t_0)$. Then there exists a unique vector field $\mathbf{X}(t)$ that is parallel along $\boldsymbol{\gamma}(t)$ with $\mathbf{X}(t_0) = \tilde{\mathbf{X}}$.

Proof: Let $\mathbf{x}$ be a patch about $\boldsymbol{\gamma}(t_0)$. $\boldsymbol{\gamma}(t) = \mathbf{x}(\gamma^1(t), \gamma^2(t))$. Consider the initial value problem

$$\frac{dX^k}{dt} = -\sum \Gamma_{ij}{}^k(\gamma^1(t), \gamma^2(t)) X^i(t) \frac{d\gamma^j}{dt}, \qquad k = 1, 2,$$

$$X^k(t_0) = \tilde{X}^k.$$

By Picard's Theorem, this has a unique solution for values of t near t_0. By Proposition 6.6 this solution is parallel along $\boldsymbol{\gamma}$ (where defined) and any parallel field along $\boldsymbol{\gamma}$ whose value at t_0 is $\tilde{\mathbf{X}}$ must solve this problem.

Repeated application of this gives a unique $\mathbf{X}$ defined along all of $\boldsymbol{\gamma}$. ∎

DEFINITION. The unique vector field $\mathbf{X}(t)$ parallel along $\boldsymbol{\gamma}(t)$ such that $\mathbf{X}(t_0) = \tilde{\mathbf{X}}$ is called the *parallel translate* of $\tilde{\mathbf{X}}$ along $\boldsymbol{\gamma}$.

EXAMPLE 6.8. Let $M = S^2$ and $\boldsymbol{\gamma}$ be a circle of latitude $\phi = \phi_0$ so that $\boldsymbol{\gamma}(t) = (\sin \phi_0 \cos t, \sin \phi_0 \sin t, \cos \phi_0)$. We shall work in the coordinate patch $\mathbf{x}(\phi, \theta) = (\sin \phi \cos \theta, \sin \phi \sin \theta, \cos \phi)$. Let $\tilde{\mathbf{X}} = \mathbf{x}_1(\phi_0, 0)$ so that $\tilde{\mathbf{X}}$ is a unit vector pointing south. We now compute (using Theorem 6.7) the parallel translate of $\tilde{\mathbf{X}}$.

Recall that by Problem 3.1

$$(g_{ij}) = \begin{pmatrix} 1 & 0 \\ 0 & \sin^2 \phi \end{pmatrix}$$

so that $\Gamma_{12}{}^2 = \Gamma_{21}{}^2 = \cot \phi$, $\Gamma_{22}{}^1 = -\sin \phi \cos \phi$, and all other $\Gamma_{ij}{}^k$ are zero. Now $\boldsymbol{\gamma}(t) = \mathbf{x}(\gamma^1(t), \gamma^2(t))$ where $\gamma^1(t) = \phi_0$ and $\gamma^2(t) = t$ so that the equations for parallel translation (Equation (6-2)) become

$$\frac{dX^1}{dt} - \sin \phi_0 \cos \phi_0 \, X^2 = 0 \quad \text{and} \quad \frac{dX^2}{dt} + \cot \phi_0 X^1 = 0^{\cdot}$$

with $X^1(0) = 1$ and $X^2(0) = 0$ (i.e., $\mathbf{X}(0) = \tilde{\mathbf{X}}$). It is easy to check that $X^1(t) = \cos((\cos \phi_0)t)$ and $X^2(t) = (-\sin((\cos \phi_0)t))/\sin \phi_0$ are solutions to the system of differential equations which satisfy the initial conditions. Thus the parallel translate of $\tilde{\mathbf{X}}$ is

$$\mathbf{X}(t) = \cos((\cos \phi_0)t)\mathbf{x}_1 - \frac{\sin((\cos \phi_0)t)}{\sin \phi_0}\mathbf{x}_2.$$

Note that $X(0) \neq X(2\pi)$. See Figure 4.16.

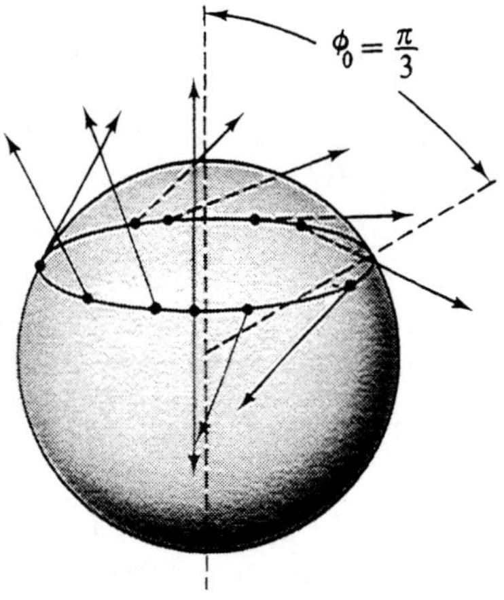

FIGURE 4.16

PROPOSITION 6.9. Let $\mathbf{X}(t)$ and $\mathbf{Y}(t)$ both be parallel along a regular curve $\boldsymbol{\gamma}$ in M. Then $|\mathbf{X}(t)|$ is constant and so is the angle between $\mathbf{X}(t)$ and $\mathbf{Y}(t)$.

Proof: Let $f(t) = \langle \mathbf{X}(t), \mathbf{Y}(t) \rangle$. $df/dt = \langle d\mathbf{X}/dt, \mathbf{Y} \rangle + \langle \mathbf{X}, d\mathbf{Y}/dt \rangle = 0 + 0$, so f is constant. If $\mathbf{X} = \mathbf{Y}$, this implies $|\mathbf{X}|$ is constant. The cosine of the angle between $\mathbf{X}$ and $\mathbf{Y}$ is $f(t)/|\mathbf{X}||\mathbf{Y}|$, which is then constant and so is the angle. ▮

Comment: This theorem says that when we parallel translate vectors, angles and lengths are preserved, just as in plane geometry. However, we shall see below in Example 6.11 that if two different curves join P to Q and $\tilde{\mathbf{X}}$ is a tangent vector at P, the value of the vector field at Q depends on which curve is used. This is one of the critical differences between geometry in the plane and the geometry on an arbitrary surface. It was T. Levi-Civita (1917) who first realized the importance of parallel translation.

DEFINITION. A regular curve $\boldsymbol{\gamma}(t)$ on a surface M is *maximally straight* if $d\boldsymbol{\gamma}/dt$ is parallel along $\boldsymbol{\gamma}$.

This is the generalization of property SL4 of straight lines in terms of the given notion of parallel.

PROPOSITION 6.10. A regular curve $\boldsymbol{\gamma}(t)$ on a surface M is maximally straight if and only if dt/ds is constant and $\boldsymbol{\gamma}(t(s))$ is a geodesic, where s is the arc length.

Proof: Assume $\boldsymbol{\gamma}(t(s))$ is a geodesic and $dt/ds \equiv c$. Then

$$d\boldsymbol{\gamma}/dt = \mathbf{T}\, ds/dt = \mathbf{T}/c.$$

$\boldsymbol{\gamma}(t)$ is maximally straight if $d\boldsymbol{\gamma}/dt$ is parallel along $\boldsymbol{\gamma}$. $d^2\boldsymbol{\gamma}/dt^2 = \mathbf{T}'/c^2$. Since $\boldsymbol{\gamma}(t(s))$ is a geodesic $\mathbf{T}' = \boldsymbol{\gamma}''$ is normal to the surface. Thus $d\boldsymbol{\gamma}/dt$ is parallel, and $\boldsymbol{\gamma}(t)$ is maximally straight.

Now assume that $\boldsymbol{\gamma}(t)$ is maximally straight. $d^2\boldsymbol{\gamma}/dt^2$ is normal to M. We need to show that $dt/ds = \text{constant}$ and $d\mathbf{T}/ds$ is normal to M. $d\boldsymbol{\gamma}/dt = (d\boldsymbol{\gamma}/ds)(ds/dt) = \mathbf{T}(ds/dt)$. Since $d\boldsymbol{\gamma}/dt$ is parallel, its length $|ds/dt|$ is constant by Proposition 6.9. Since ds/dt is continuous, ds/dt is a constant $\neq 0$. Thus dt/ds, which is the reciprocal of ds/dt, is a constant.

$$\frac{d\mathbf{T}}{ds} = \frac{d^2\boldsymbol{\gamma}}{ds^2} = \left(\frac{d^2\boldsymbol{\gamma}}{dt^2}\right)\left(\frac{dt}{ds}\right)^2 + \left(\frac{d\boldsymbol{\gamma}}{dt}\right)\left(\frac{d^2t}{ds^2}\right).$$

Since dt/ds is constant, $d^2t/ds^2 = 0$ and $d\mathbf{T}/ds$ is normal because $d^2\boldsymbol{\gamma}/dt^2$ is. ∎

We might note that if $\boldsymbol{\gamma}(t)$ is maximally straight, then to an intrinsic being $\boldsymbol{\gamma}(t)$ has no acceleration (the acceleration vector $d^2\boldsymbol{\gamma}/dt^2$ is normal to the surface) and hence is traveling in uniform or "linear" motion. When differential geometry is applied to physics, and particularly to general relativity, the concept of uniform motion mentioned in Newton's first law of motion (the law of inertia) is translated to motion along a geodesic because the Euclidean notion of straight line has been lost. (See Misner, Thorne, and Wheeler [1973].)

We shall now give two examples to show that the notion of parallel translation depends upon the path. In the first example we take two curves with the same end points and parallel translate a given vector along each curve and get different answers. In the second example we translate a vector around a "geodesic triangle" but do not return to the original vector.

EXAMPLE 6.11. On S^2 consider two different meridians $\boldsymbol{\gamma}, \boldsymbol{\alpha}$ from the north to the south pole. Let $\tilde{\mathbf{X}}$ be the unit tangent to $\boldsymbol{\gamma}$ at the north pole. If $\tilde{\mathbf{X}}$ is parallel translated along $\boldsymbol{\gamma}$ to the south pole, we get the unit tangent to $\boldsymbol{\gamma}$ at the south pole since $\boldsymbol{\gamma}$ is a geodesic. If $\tilde{\mathbf{X}}$ is parallel translated along $\boldsymbol{\alpha}$, it must keep constant angle with $\boldsymbol{\alpha}'$, which is also parallel. Let θ be the angle between $\boldsymbol{\gamma}'$ and $\boldsymbol{\alpha}'$ at the north pole. The angle between the two translates of $\tilde{\mathbf{X}}$ at the south pole is 2θ. Note that 2θ is the area of the region bounded by $\boldsymbol{\gamma}$ and $\boldsymbol{\alpha}$. See Figure 4.17.

EXAMPLE 6.12. On S^2 consider a great circle segment $\boldsymbol{\gamma}_1$ from the north pole N to point A on the equator, followed by part of the equator $\boldsymbol{\gamma}_2$ to B, then a great circle segment $\boldsymbol{\gamma}_3$ back to N. Let $\tilde{\mathbf{X}}$ be the unit vector pointing south along $\boldsymbol{\gamma}_1$ at N. This may be parallel translated along $\boldsymbol{\gamma}_1$ to A, where it points south, then along $\boldsymbol{\gamma}_2$ to B, where it points south, and then along $\boldsymbol{\gamma}_3$ to N, where it points south along $\boldsymbol{\gamma}_3$. The final result is a vector different from $\tilde{\mathbf{X}}$. In fact, the angle between the two vectors at N is $\theta =$ angle between $\boldsymbol{\gamma}_1$ and $\boldsymbol{\gamma}_3$. θ also equals the area of the region surrounded by $\boldsymbol{\gamma}_1, \boldsymbol{\gamma}_2$ and $\boldsymbol{\gamma}_3$. See Figure 4.18.

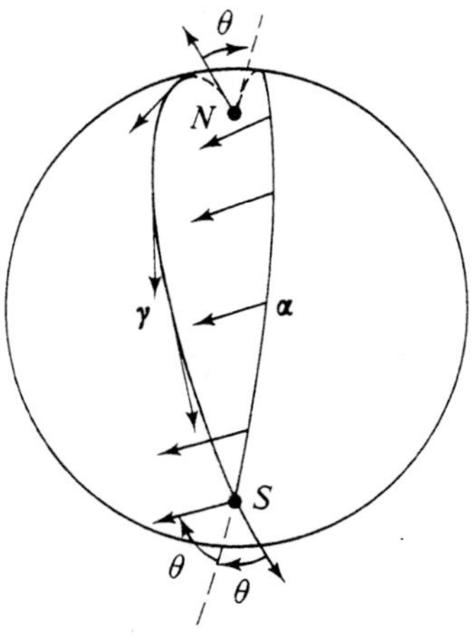

FIGURE 4.17

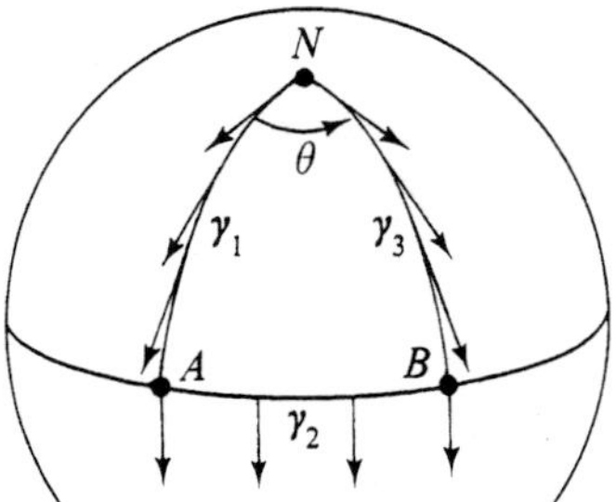

FIGURE 4.18

PROBLEMS

Equation (4-6) will be useful in many of these problems.

6.1. Let $\boldsymbol{\gamma}(t)$ be a maximally straight curve. Prove there are constants $a, b \in \mathbf{R}$ such that $s = at + b$, where s is arc length.

6.2. Let $\boldsymbol{\gamma}(s)$ be a unit speed curve on a surface M and let $\mathbf{S}$ be its intrinsic normal. Prove that $\mathbf{S}$ is parallel along $\boldsymbol{\gamma}$ if and only if $\boldsymbol{\gamma}$ is a geodesic.

6.3. Let $\mathbf{X}_N$ be the tangential component of the normal vector $\mathbf{N}$ of a unit speed curve $\boldsymbol{\gamma}$ on a surface M.
(a) Prove that $\mathbf{X}_N = \mathbf{N} - \langle \mathbf{N}, \mathbf{n} \rangle \mathbf{n}$ and that $\mathbf{X}_N$ is a vector field along $\boldsymbol{\gamma}$.
(b) Prove that the following are equivalent:
  (i) $\mathbf{X}_N = \mathbf{0}$;
  (ii) $\boldsymbol{\gamma}$ is a geodesic;
  (iii) $\mathbf{X}_N$ is parallel along $\boldsymbol{\gamma}$.

6.4. Let $\mathbf{X}_B$ be the tangential component of the binormal vector $\mathbf{B}$ of a unit speed curve $\boldsymbol{\gamma}$ on a surface M.
(a) Prove $\mathbf{X}_B = -(\kappa_n/\kappa)\mathbf{S}$.
(b) Prove that the following are equivalent:
  (i) $\mathbf{X}_B = \mathbf{B}$;
  (ii) $\boldsymbol{\gamma}$ is a geodesic;
  (iii) $\mathbf{X}_B$ is not zero and is parallel along $\boldsymbol{\gamma}$.

6.5. Let $\boldsymbol{\gamma}$ be a curve with $\boldsymbol{\gamma}(0) = P$, $\boldsymbol{\gamma}(1) = Q$. If $\tilde{\mathbf{X}} \in T_P M$, let $\mathbf{X}$ be the parallel translate of $\tilde{\mathbf{X}}$ along $\boldsymbol{\gamma}$. Define $\boldsymbol{\gamma}^{\#}: T_P M \longrightarrow T_Q M$ by $\boldsymbol{\gamma}^{\#}(\tilde{\mathbf{X}}) = \mathbf{X}(1)$.
(a) Prove that $\boldsymbol{\gamma}^{\#}$ is a linear transformation.
(b) Prove that $\boldsymbol{\gamma}^{\#}$ is an isometry, that is, that

$$\langle \boldsymbol{\gamma}^{\#}(\tilde{\mathbf{X}}), \boldsymbol{\gamma}^{\#}(\tilde{\mathbf{Y}}) \rangle = \langle \tilde{\mathbf{X}}, \tilde{\mathbf{Y}} \rangle.$$

(c) Prove that $\boldsymbol{\gamma}^{\#}$ is an isomorphism. $\boldsymbol{\gamma}^{\#}$ is called the *parallelism* defined by $\boldsymbol{\gamma}$.
(This problem requires an understanding of the proof of Theorem 6.7.)

4–7. THE SECOND FUNDAMENTAL FORM AND THE WEINGARTEN MAP

In the previous sections we considered concepts related to the geodesic curvature of a curve $\boldsymbol{\gamma}(s)$ on a surface M. We now turn our attention to the normal curvature of the curve. It will tell us how M is curving in the direction of $\mathbf{T}$ (see Proposition 7.2). In order to study how M is curving at a point, without reference to a direction, we define the Weingarten map L. This map will be essentially the directional derivative of the normal vector. (If $\nu: M \longrightarrow S^2 \subset \mathbf{R}^3$ is the normal spherical image given locally by $\nu(P) = \mathbf{n}(P)$, then L can be viewed as the derivative of ν.) It will be the eigenvalues of L at a point which will tell us how M curves at that point, as we shall see in Section 4-8.

As before, let $\boldsymbol{\gamma}$ be a unit speed curve on M whose image is contained in a coordinate patch $\mathbf{x}(\mathcal{U})$. Let $\boldsymbol{\gamma}(s) = \mathbf{x}(\gamma^1(s), \gamma^2(s))$. We have (from (4-6) and (4-8))

$$\boldsymbol{\gamma}'' = \kappa_n \mathbf{n} + \kappa_g \mathbf{S} \tag{7-1}$$

(7-2) $$\mathbf{x}_{ij} = L_{ij}\mathbf{n} + \sum \Gamma_{ij}{}^{k}\mathbf{x}_k.$$

Recall that the L_{ij} were called the coefficients of the second fundamental form.

DEFINITION. The *second fundamental form* II on M is the bilinear form on T_PM (for each $P \in M$) given by $\text{II}(\mathbf{X}, \mathbf{Y}) = \sum L_{ij}X^iY^j$, where

$$\mathbf{X} = \sum X^i\mathbf{x}_i, \qquad \mathbf{Y} = \sum Y^j\mathbf{x}_j \in T_PM.$$

PROPOSITION 7.1. Let M be a surface. Then

(a) II is a symmetric bilinear form on T_PM for each $P \in M$;

(b) if $\boldsymbol{\gamma}$ is a unit speed curve with tangent $\mathbf{T}$, then $\kappa_n = \text{II}(\mathbf{T}, \mathbf{T})$;

(c) if $\boldsymbol{\alpha}$ and $\boldsymbol{\beta}$ are regular curves with $\boldsymbol{\alpha}(0) = \boldsymbol{\beta}(0)$ and whose velocity vectors are dependent at $t = 0$, then $\boldsymbol{\alpha}$ and $\boldsymbol{\beta}$ have the same normal curvature at $t = 0$.

Proof:

(a) It is easy to show that

$$\text{II}(\mathbf{X}_1 + \mathbf{X}_2, \mathbf{Y}) = \text{II}(\mathbf{X}_1, \mathbf{Y}) + \text{II}(\mathbf{X}_2, \mathbf{Y}),$$

$$\text{II}(\mathbf{X}, \mathbf{Y}_1 + \mathbf{Y}_2) = \text{II}(\mathbf{X}, \mathbf{Y}_1) + \text{II}(\mathbf{X}, \mathbf{Y}_2)$$

and

$$\text{II}(\mathbf{X}, r\mathbf{Y}) = \text{II}(r\mathbf{X}, \mathbf{Y}) = r\text{II}(\mathbf{X}, \mathbf{Y})$$

(for all $r \in \mathbf{R}$ and $\mathbf{X}, \mathbf{X}_1, \mathbf{X}_2, \mathbf{Y}, \mathbf{Y}_1, \mathbf{Y}_2 \in T_PM$), which is what is meant by bilinearity. Since $L_{ij} = L_{ji}$, $\text{II}(\mathbf{X}, \mathbf{Y}) = \text{II}(\mathbf{Y}, \mathbf{X})$ and II is symmetric.

(b) This is precisely Equation (4-9).

(c) Let $\mathbf{T}$ and $\bar{\mathbf{T}}$ denote the (unit) tangent vectors to $\boldsymbol{\alpha}$ and $\boldsymbol{\beta}$ at $\boldsymbol{\alpha}(0) = \boldsymbol{\beta}(0)$. Since $\dot{\boldsymbol{\alpha}}(0)$ and $\dot{\boldsymbol{\beta}}(0)$ are dependent, $\mathbf{T} = \pm\bar{\mathbf{T}}$. Hence, the normal curvature of $\boldsymbol{\alpha}$ at $t = 0$ is $\text{II}(\mathbf{T}, \mathbf{T}) = \text{II}(\pm\bar{\mathbf{T}}, \pm\bar{\mathbf{T}}) = \text{II}(\bar{\mathbf{T}}, \bar{\mathbf{T}})$, which is the normal curvature of $\boldsymbol{\beta}$ at $t = 0$. ∎

PROPOSITION 7.2. Let $\boldsymbol{\gamma}(s)$ be a unit speed curve in a surface M with normal curvature κ_n at P. Let $\bar{\boldsymbol{\gamma}}$ be the curve formed by the intersection of M with the plane Π through P spanned by $\mathbf{n}$ and $\boldsymbol{\gamma}'$. (See Figure 4.19.) Then $|\kappa_n|$ is the curvature $\bar{\kappa}$ of the plane curve $\bar{\boldsymbol{\gamma}}$.

Proof: Part (c) of Proposition 7.1 shows that $\boldsymbol{\gamma}$ and $\bar{\boldsymbol{\gamma}}$ have the same normal curvatures. However, $\bar{\boldsymbol{\gamma}}$ is a plane curve with normal $\pm\mathbf{n}$ at P. Its plane curvature $\bar{k}$ satisfies $|\bar{k}| = \bar{\kappa}$. On the other hand, $\bar{k}$ is $\pm$(normal curvature of $\bar{\boldsymbol{\gamma}}$) $= \pm\kappa_n$. Thus $|\kappa_n| = |\bar{k}| = \bar{\kappa}$. ∎

We now discuss directional derivatives in preparation for the Weingarten map.

A subset $\mathcal{R}$ of a surface M is *open* if for each $P \in \mathcal{R}$ there is an ϵ-neighborhood of P in M contained in $\mathcal{R}$. (The image of a proper coordinate patch

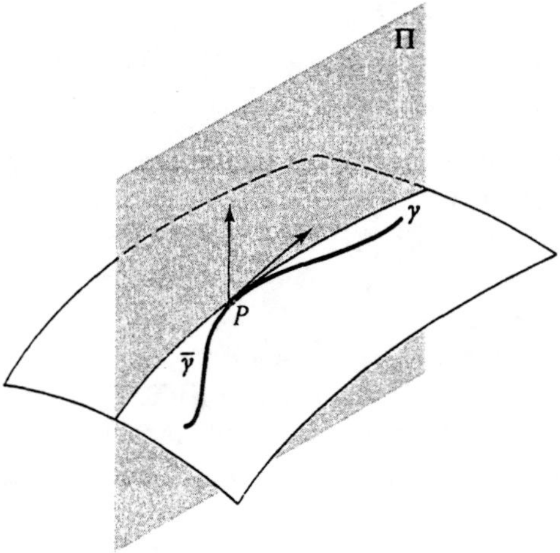

FIGURE 4.19

is open.) A function $f: \mathcal{R} \to \mathbf{R}$ is *differentiable* if for every C^1 curve $\boldsymbol{\alpha}(t)$ with $\boldsymbol{\alpha}(0) \in \mathcal{R}$ the derivative $(d(f \circ \boldsymbol{\alpha})/dt)(0)$ exists.

DEFINITION. Let f be a differentiable function defined on an ϵ-neighborhood of $P \in M$. Let $\mathbf{X} \in T_PM$ and let $\boldsymbol{\alpha}(t)$ be a curve on M such that $\boldsymbol{\alpha}(0) = P$ and $\mathbf{X} = (d\boldsymbol{\alpha}/dt)(0)$. The *directional derivative of f in the direction* $\mathbf{X}$ is $\mathbf{X}f = (d(f \circ \boldsymbol{\alpha})/dt)(0)$.

Note that $\mathbf{X}f$ is a number for each $\mathbf{X} \in T_PM$ and each differentiable function f. We will show that $\mathbf{X}f$ is well defined (i.e., independent of choice of $\boldsymbol{\alpha}$) and give an easier way to compute $\mathbf{X}f$ after an example.

EXAMPLE 7.3. Let

$$M = S^2, P = (1, 0, 0), \mathbf{X} = (0, 1, 0), \quad \text{and} \quad f(x, y, z) = x^2 + y.$$

If we choose $\boldsymbol{\alpha}(s) = (\cos s, \sin s, 0)$, then $f \circ \boldsymbol{\alpha} = \cos^2 s + \sin s$ so that $(f \circ \boldsymbol{\alpha})'(s) = -2 \cos s \sin s + \cos s$. Hence $\mathbf{X}f = (f \circ \boldsymbol{\alpha})'(0) = 1$.

PROPOSITION 7.4. Let $\mathbf{x}: \mathcal{U} \to M$ be a coordinate patch for M about $P = \mathbf{x}(0, 0)$. If $\mathbf{X} = \sum X^i\mathbf{x}_i$, then $\mathbf{X}f = \sum_{i=1}^{2} X^i(\partial(f \circ \mathbf{x})/\partial u^i)(0, 0)$. In particular, $\mathbf{X}f$ does not depend on the choice of $\boldsymbol{\alpha}$ such that $\mathbf{X} = \dot{\boldsymbol{\alpha}}(0)$ and $\boldsymbol{\alpha}(0) = P$.

Proof: We may write $\boldsymbol{\alpha}(t) = \mathbf{x}(\alpha^1(t), \alpha^2(t))$. Therefore

$$(f \circ \boldsymbol{\alpha})(t) = (f \circ \mathbf{x})(\alpha^1(t), \alpha^2(t)).$$

The chain rule implies

$$\frac{d}{dt}(f \circ \boldsymbol{\alpha})(t) = \sum_i \frac{\partial(f \circ \mathbf{x})}{\partial u^i}\frac{d\alpha^i}{dt}.$$

Since $X^i = (d\alpha^i/dt)(0)$ (Equation (1-8)), the above expression when evaluated at $t = 0$ completes the proof. ▮

Using Proposition 7.4 it is easy to show:

LEMMA 7.5. Let f be a differentiable function in an ϵ-neighborhood of $P \in M$. If $\mathbf{X}, \mathbf{Y} \in T_PM$ and $r \in \mathbf{R}$, then $(\mathbf{X} + \mathbf{Y})f = \mathbf{X}f + \mathbf{Y}f$ and $(r\mathbf{X})f = r(\mathbf{X}f)$.

If $\mathbf{f} = (f^1, f^2, f^3)$ is a vector-valued function defined on an open set in M, we can let $\mathbf{X}\mathbf{f} = (\mathbf{X}f^1, \mathbf{X}f^2, \mathbf{X}f^3)$ be the definition of the directional derivative of a vector-valued function. In particular, this may be done for the function $\mathbf{f} = \mathbf{n}$. (Remember that $\mathbf{n}$ is really only defined up to sign: in a coordinate patch $\mathbf{n}$ is determined and if a change of coordinates is made, the worst that can happen is that the sign of $\mathbf{n}$ is changed.)

DEFINITION. The *Weingarten map* L is, for each $P \in M$, the function $\mathsf{L}: T_PM \to \mathbf{R}^3$ given by $\mathsf{L}(\mathbf{X}) = -\mathbf{X}\mathbf{n}$. (The minus sign is for future convenience.)

Because $\mathbf{n}$ is only determined up to sign, the same is true of L.

PROPOSITION 7.6. Let M be a surface. Then

(a) L is a linear transformation from T_PM to T_PM;

(b) If $\mathsf{L}(\mathbf{x}_k) = \sum L^l{}_k \mathbf{x}_l$, then $L^l{}_k = \sum L_{ik} g^{il}$.

Proof:

(a) It is immediate from Lemma 7.5 that $\mathsf{L}(\mathbf{X} + \mathbf{Y}) = \mathsf{L}(\mathbf{X}) + \mathsf{L}(\mathbf{Y})$ and $\mathsf{L}(r\mathbf{X}) = r\mathsf{L}(\mathbf{X})$. Thus L is a linear transformation. We must show that $\mathsf{L}(\mathbf{X}) \in T_PM$ for each $\mathbf{X} \in T_PM$. Since L is linear, we need only check this on the basis $\{\mathbf{x}_1, \mathbf{x}_2\}$. Let $\boldsymbol{\alpha}_i$ be the u^i-curve through P. Then

$$\mathsf{L}(\mathbf{x}_i) = -\mathbf{x}_i(\mathbf{n}) = -\frac{d}{du^i}(\mathbf{n} \circ \boldsymbol{\alpha}_i) = -\frac{\partial \mathbf{n}}{\partial u^i}.$$

Since $1 = \langle \mathbf{n}, \mathbf{n} \rangle$, $0 = \langle \partial\mathbf{n}/\partial u^i, \mathbf{n} \rangle$ and $\partial\mathbf{n}/\partial u^i$ is tangent to M at P. Hence

$$\mathsf{L}(\mathbf{x}_i) = -\frac{\partial \mathbf{n}}{\partial u^i} = -\mathbf{n}_i \tag{7-3}$$

and $\mathsf{L}(\mathbf{x}_i)$ is tangent to M. Note that (7-3) defines $\mathbf{n}_i$.

(b) Since $\mathbf{x}_i$ is tangent to M, $\langle \mathbf{n}, \mathbf{x}_i \rangle = 0$. Hence

$$\begin{aligned} 0 = \frac{\partial \langle \mathbf{n}, \mathbf{x}_i \rangle}{\partial u^k} &= \left\langle \frac{\partial \mathbf{n}}{\partial u^k}, \mathbf{x}_i \right\rangle + \langle \mathbf{n}, \mathbf{x}_{ik} \rangle = -\langle \mathsf{L}(\mathbf{x}_k), \mathbf{x}_i \rangle + L_{ik} \\ &= L_{ik} - \langle \textstyle\sum L^j{}_k \mathbf{x}_j, \mathbf{x}_i \rangle = L_{ik} - \sum L^j{}_k \langle \mathbf{x}_j, \mathbf{x}_i \rangle \\ &= L_{ik} - \textstyle\sum L^j{}_k g_{ji}. \end{aligned}$$

Thus

(7-4) $$L_{ik} = \sum L^j{}_k g_{ji}.$$

Therefore $\sum L_{ik}g^{il} = \sum L^j{}_k g_{ji} g^{il} = \sum L^j{}_k \delta_j{}^l = L^l{}_k$. That is,

(7-5) $$L^l{}_k = \sum L_{ik} g^{il}. \quad ▮$$

A classical geometer would say that $L^l{}_k$ is obtained from L_{ik} by raising an index. L (or $L^l{}_k$) is a "tensor of type (1, 1)" because it has one upper and one lower index and behaves in a certain fashion under change of coordinates (see Problem 7.6). Note that $(L^l{}_k)$ is the matrix representing L with respect to the basis $\{\mathbf{x}_1, \mathbf{x}_2\}$.

PROPOSITION 7.7. On a surface M we have Weingarten's equations

(7-6) $$\mathbf{n}_j = -\sum L^k{}_j \mathbf{x}_k.$$

Proof: By Equation (7-3) $L(\mathbf{x}_j) = -\partial\mathbf{n}/\partial u^j$ so that

$$\mathbf{n}_j = \frac{\partial \mathbf{n}}{\partial u^j} = -\mathsf{L}(\mathbf{x}_j) = -\sum L^k{}_j \mathbf{x}_k. \quad ▮$$

For a surface M, the basis $\{\mathbf{x}_1, \mathbf{x}_2, \mathbf{n}\}$ of $\mathbf{R}^3$ plays a role analogous to the Frenet-Serret frame $\{\mathbf{T}, \mathbf{N}, \mathbf{B}\}$ of a curve. Gauss's formulas (7-2) and the Weingarten equations (7-6) give the analogues of the Frenet-Serret equations:

(7-2) $$\mathbf{x}_{ij} = L_{ij}\mathbf{n} + \sum \Gamma_{ij}{}^k \mathbf{x}_k$$

(7-6) $$\mathbf{n}_j = -\sum L^k{}_j \mathbf{x}_k$$

EXAMPLE 7.8. Let $M = \mathbf{R}^2$. We show that L is the zero transformation (at each point) in two ways. First by the definition of L: $\mathbf{n} \equiv (0, 0, 1)$ is independent of P so that $\mathsf{L}(\mathbf{X}) = \mathbf{0}$ for all $\mathbf{X} \in T_PM$. The second way is to compute the L_{ik} for M: $\mathbf{x}(u^1, u^2) = (u^1, u^2, 0)$ and

$$L_{ik} = \langle \mathbf{x}_{ik}, \mathbf{n} \rangle = \langle \mathbf{0}, \mathbf{n} \rangle = 0$$

for all i and k. Hence raising an index by (7-5) gives $L^l{}_k = 0$ so that $\mathsf{L} = 0$.

EXAMPLE 7.9. Let $M = S^2$ and choose $\mathbf{n}$ to be the outward pointing normal. Let $\mathbf{X} \in T_PS^2$ and choose $\boldsymbol{\alpha}$ to be a curve on M such that $\dot{\boldsymbol{\alpha}}(0) = \mathbf{X}$, $\boldsymbol{\alpha}(0) = P$, and $\boldsymbol{\alpha}(t) = (x(t), y(t), z(t))$. At $\boldsymbol{\alpha}(t)$, $\mathbf{n} = \boldsymbol{\alpha}(t)$ so that

$$\begin{aligned} \mathsf{L}(\mathbf{X}) &= -\mathbf{X}\mathbf{n} = -(\mathbf{X}x, \mathbf{X}y, \mathbf{X}z) \\ &= -(\dot{x}(0), \dot{y}(0), \dot{z}(0)) = -\dot{\boldsymbol{\alpha}}(0) = -\mathbf{X}. \end{aligned}$$

That is, L is the negative of the identity. If we had chosen the inward pointing normal, we would have had L equals the identity.

The following result says that L is a self-adjoint (or symmetric) linear transformation.

LEMMA 7.10. If $P \in M$ and $\mathbf{X}, \mathbf{Y} \in T_P M$, then

$$\mathrm{II}(\mathbf{X}, \mathbf{Y}) = \langle \mathsf{L}(\mathbf{X}), \mathbf{Y} \rangle = \langle \mathbf{X}, \mathsf{L}(\mathbf{Y}) \rangle.$$

Proof: Problem 7.4. ▮

PROBLEMS

7.1. Show that for $M = S^2$, L is a plus or minus the identity by computing the L_{ik} in a coordinate patch and raising an index.

7.2. Show that for $M = S^1 \times (0, 1)$ (Example 2.3) L can be represented by the matrix $\begin{pmatrix} 1 & 0 \\ 0 & 0 \end{pmatrix}$.

7.3. Find L for the torus in Problem 1.1.

***7.4.** Prove Lemma 7.10.

7.5. Find the matrix $(L^i{}_j)$ for a surface of revolution. (Compare your answer with 7.3)

7.6. Let $f: \mathcal{V} \longrightarrow \mathcal{U}$ be a coordinate transformation. How are the $\bar{L}^\alpha{}_\beta$ related to the $L^i{}_j$? Compare your result with Equation (2-1) of Chapter 1.

7.7. Let $\boldsymbol{\gamma}(s) = \mathbf{x}(\gamma^1(s), \gamma^2(s))$ be a unit speed curve in a surface M. Note that $\{\mathbf{T}, \mathbf{S}, \mathbf{n}\}$ give a right-handed orthonormal basis of $\mathbf{R}^3$ at each point of $\boldsymbol{\gamma}$. View $\mathbf{n}$ as $\mathbf{n}(s) = \mathbf{n}(\gamma^1(s), \gamma^2(s))$ and prove the following analogues of the Frenet-Serret equations:
(a) $\mathbf{T}' = \mathrm{II}(\mathbf{T}, \mathbf{T})\mathbf{n} + \kappa_g \mathbf{S}$;
(b) $\mathbf{S}' = -\kappa_g \mathbf{T} + \mathrm{II}(\mathbf{T}, \mathbf{S})\mathbf{n}$;
(c) $\mathbf{n}' = -\mathrm{II}(\mathbf{T}, \mathbf{T})\mathbf{T} - \mathrm{II}(\mathbf{T}, \mathbf{S})\mathbf{S}$.

4–8. PRINCIPAL, GAUSSIAN, MEAN, AND NORMAL CURVATURES

We now have two ways to measure how a surface is curving. The first is through the normal curvature of curves. This is aesthetically unsatisfactory because it forces us to break up a surface into infinitely many curves. The second measurement involves the change of the normal, which is given by L. Since L is a linear transformation, there are two associated numerical invariants: the determinant of $(L^i{}_j)$ (the *Gaussian curvature*) and one-half the trace of $(L^i{}_j)$ (the *mean curvature*). This section is devoted to understanding the relationship between these different kinds of curvatures.

We know that the normal curvature of γ at P depends only on the unit tangent of γ at P. If we knew all the possible values that κ_n takes on at P we would know how M curves. One step in this direction would be to find the maximum and minimum values that κ_n takes on. That is, determine the maximum and minimum of $\mathrm{II}(\mathbf{X}, \mathbf{X})$ as $\mathbf{X}$ runs over all unit vectors in T_PM. This means we are maximizing (and minimizing) $\mathrm{II}(\mathbf{X}, \mathbf{X})$ subject to the constraint $\langle \mathbf{X}, \mathbf{X} \rangle = 1$. The method of Lagrange multipliers tells us to find the critical values of

$$\begin{aligned} f(\mathbf{X}, \lambda) &= \mathrm{II}(\mathbf{X}, \mathbf{X}) - \lambda(\langle \mathbf{X}, \mathbf{X}\rangle - 1) = \langle \mathsf{L}(\mathbf{X}), \mathbf{X}\rangle - \lambda\langle \mathbf{X}, \mathbf{X}\rangle + \lambda \\ &= \langle \mathsf{L}(\mathbf{X}) - \lambda \mathbf{X}, \mathbf{X}\rangle + \lambda \end{aligned}$$

at P.

In terms of a coordinate patch $\mathbf{x}$ we consider the function

$$f(X^1, X^2, \lambda) = \lambda + \sum (L^i{}_j - \lambda\delta^i{}_j)X^jX^kg_{ik}.$$

Following the method of Lagrange multipliers we must have $\partial f/\partial X^1 = 0$, $\partial f/\partial X^2 = 0$, and $\partial f/\partial \lambda = 0$. Now $\partial f/\partial \lambda = \langle -\mathbf{X}, \mathbf{X}\rangle + 1$, so $\partial f/\partial \lambda = 0$ is the equation $\langle \mathbf{X}, \mathbf{X}\rangle = 1$.

$$\frac{\partial f}{\partial X^l} = 2\sum (L^i{}_j - \lambda\delta^i{}_j)g_{il}X^j.$$

Hence $\partial f/\partial X^l = 0$ implies $\sum (L^i{}_j - \lambda\delta^i{}_j)g_{il}X^jY^l = 0$ for all choices of Y^l. That is, $\langle \mathsf{L}(\mathbf{X}) - \lambda\mathbf{X}, \mathbf{Y}\rangle = 0$ for all $\mathbf{Y}$, or $\mathsf{L}(\mathbf{X}) = \lambda\mathbf{X}$. Thus λ is an eigenvalue of L, and $\mathbf{X}$ is a corresponding unit eigenvector. L has real eigenvalues since L is self-adjoint (Lemma 7.10) and so the Lagrange problem has a solution. (Alternatively, the problem has a solution since $\mathrm{II}(\mathbf{X}, \mathbf{X})$ does have a maximum and minimum: the set of unit vectors in T_PM is closed and bounded, i.e., compact.) The eigenvalues are the roots of

$$0 = \det(\mathsf{L} - \lambda\mathsf{I}) = \lambda^2 - (\operatorname{trace}\mathsf{L})\lambda + \det\mathsf{L}.$$

Denote these roots by κ_1 and κ_2 with $\kappa_1 \geq \kappa_2$.

LEMMA 8.1. Let $P \in M$ and λ, $\mathbf{X}$ be an eigenvalue-eigenvector pair for L at P. Let $\mathbf{Y}$ be a unit vector in T_PM such that $\langle \mathbf{X}, \mathbf{Y}\rangle = 0$. Then $\mathbf{Y}$ is also an eigenvector.

Proof: Note that $\{\mathbf{X}, \mathbf{Y}\}$ is a basis for T_PM.

$$0 = \langle \lambda\mathbf{X}, \mathbf{Y}\rangle = \langle \mathsf{L}(\mathbf{X}), \mathbf{Y}\rangle = \langle \mathbf{X}, \mathsf{L}(\mathbf{Y})\rangle.$$

Hence $\mathsf{L}(\mathbf{Y})$ is orthogonal to $\mathbf{X}$ and $\mathsf{L}(\mathbf{Y}) = \mu\mathbf{Y}$ for some μ since T_PM has dimension 2. Thus $\mathbf{Y}$ is also an eigenvector. ∎

PROPOSITION 8.2. At each point of a surface M there are two orthogonal directions such that the normal curvature takes its maximum value in one direction and its minimum along the other.

Proof: By the preceding discussion we know that the maximum and minimum values of κ_n are taken in the direction of eigenvectors of L. Note that if $\mathbf{X}_{(i)}$ is a unit eigenvector of L with eigenvalue κ_i, then the value of κ_n in the direction of $\mathbf{X}_{(i)}$ is

$$\text{II}(\mathbf{X}_{(i)}, \mathbf{X}_{(i)}) = \langle \mathsf{L}(\mathbf{X}_{(i)}), \mathbf{X}_{(i)} \rangle = \kappa_i \langle \mathbf{X}_{(i)}, \mathbf{X}_{(i)} \rangle = \kappa_i.$$

Thus the maximum value of κ_n is κ_1 and the minimum is κ_2. If $\kappa_1 \neq \kappa_2$, then $\kappa_1 \langle \mathbf{X}_{(1)}, \mathbf{X}_{(2)} \rangle = \langle \mathsf{L}(\mathbf{X}_{(1)}), \mathbf{X}_{(2)} \rangle = \langle \mathbf{X}_{(1)}, \mathsf{L}(\mathbf{X}_{(2)}) \rangle = \kappa_2 \langle \mathbf{X}_{(1)}, \mathbf{X}_{(2)} \rangle$ implies that $\langle \mathbf{X}_{(1)}, \mathbf{X}_{(2)} \rangle = 0$. If $\kappa_1 = \kappa_2$, Lemma 8.1 implies that a unit vector orthogonal to $\mathbf{X}_{(1)}$ is also an eigenvector. Hence it may be chosen as $\mathbf{X}_{(2)}$. In either case $\langle \mathbf{X}_{(1)}, \mathbf{X}_{(2)} \rangle = 0$. ∎

DEFINITION. The *principal curvatures* of a surface M at a point P are the eigenvalues of L there (κ_1 and κ_2). Corresponding unit eigenvectors are called *principal directions* at P.

DEFINITION. An *umbilic* is a point where $\kappa_1 = \kappa_2$.

EXAMPLE 8.3. Every point of either S^2 or $\mathbf{R}^2$ is an umbilic. We shall see that the converse is true in Section 6-1.

DEFINITION. A *line of curvature* on a surface M is a curve whose tangent vector at each point is a principal direction at that point.

In Problem 8.5 you will show that the meridians and circles of latitude of a surface of revolution are lines of curvature.

THEOREM 8.4 (Euler). Let $\mathbf{Y}$ be a unit vector tangent to M at P. Then

$$\text{II}(\mathbf{Y}, \mathbf{Y}) = \kappa_1 \cos^2 \theta + \kappa_2 \sin^2 \theta,$$

where θ is the angle between $\mathbf{Y}$ and the principal direction $\mathbf{X}_{(1)}$ corresponding to κ_1.

Proof: Let $\mathbf{X}_{(2)}$ be the principal direction corresponding to κ_2.

$$\mathsf{L}(\mathbf{X}_{(i)}) = \kappa_i \mathbf{X}_{(i)}.$$

Thus in terms of the orthonormal basis $\mathbf{X}_{(1)}$, $\mathbf{X}_{(2)}$, L is represented by $\begin{pmatrix} \kappa_1 & 0 \\ 0 & \kappa_2 \end{pmatrix}$.

$$\mathbf{Y} = \cos \theta \mathbf{X}_{(1)} + \sin \theta \mathbf{X}_{(2)}.$$

$$\begin{aligned}\text{II}(\mathbf{Y}, \mathbf{Y}) &= \langle \mathsf{L}(\mathbf{Y}), \mathbf{Y} \rangle \\ &= \langle \kappa_1 \cos \theta \mathbf{X}_{(1)} + \kappa_2 \sin \theta \mathbf{X}_{(2)}, \cos \theta \mathbf{X}_{(1)} + \sin \theta \mathbf{X}_{(2)} \rangle \\ &= \kappa_1 \cos^2 \theta + \kappa_2 \sin^2 \theta. \quad ∎\end{aligned}$$

Note that this theorem can be used to compute normal curvatures.

EXAMPLE 8.5. Consider the Monge patch $(u, v, v^2 - u^2)$ defined for $u^2 + v^2 < 1$. The curves $u = 0$ and $v = 0$ are lines of curvature. See Figure 4.20.

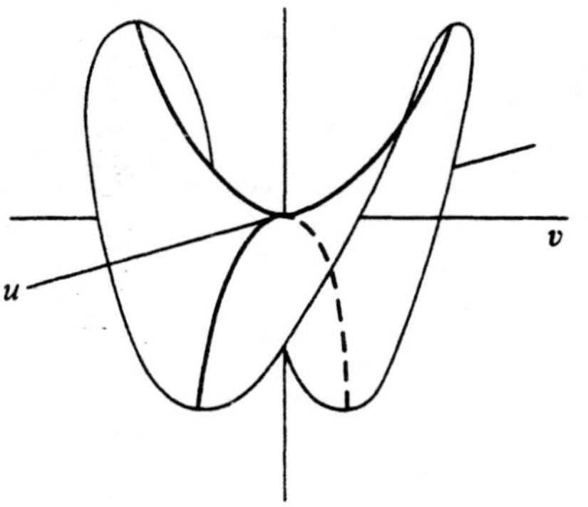

FIGURE 4.20

DEFINITION. The *Gaussian curvature* of M at P is $K = \kappa_1\kappa_2 = \det \mathsf{L}$. The *mean curvature* of M at P is $H = \frac{1}{2}(\kappa_1 + \kappa_2) = \frac{1}{2}\operatorname{trace}(\mathsf{L})$.

The terminology "mean curvature" reflects the fact that H is the average normal curvature (see Problems 8.3 and 8.11). Surfaces for which $H \equiv 0$ are called *minimal surfaces*. A minimal surface has the local property that if a "small" region of the surface is deformed slightly without changing the boundary curve, the area of the region is increased. That is, the given region has *minimal* area among all nearby regions. (The concept of area is developed below.) Such a surface may be made experimentally by dipping a simple closed curve made out of wire into some liquid soap. The resulting soap film bounded by the wire frame is a minimal surface. See do Carmo [1976] for a more extensive study of the differential geometry of minimal surfaces. See Almgren [1966] for a detailed study of soap films and their geometric properties.

We shall now give a geometric interpretation of the Gaussian curvature. This will require the concept of integration on a surface.

DEFINITION. If $x : \mathcal{U} \longrightarrow \mathbf{R}^3$ is a parametrized surface, the *area of a subset* $\mathcal{R} \subset \mathbf{x}(\mathcal{U})$ is $\iint_{x^{-1}(\mathcal{R})} [\mathbf{x}_1, \mathbf{x}_2, \mathbf{n}]\, du^1\, du^2$ or

(8-1) $$A(\mathcal{R}) = \iint_{x^{-1}(\mathcal{R})} \sqrt{g}\, du^1\, du^2.$$

$\sqrt{g}\, du^1\, du^2$ is called the *area element* and is denoted dA.

If it is necessary to integrate a function, such as K, over a surface, it is done with respect to dA: $\iint_{\mathcal{R}} K\, dA = \iint_{x^{-1}(\mathcal{R})} K(u^1, u^2)\sqrt{g}\, du^1\, du^2$. See the digression below for a justification of this definition. In this chapter, the concept of area is used only in Proposition 8.6, which can be omitted.

DEFINITION. The *normal spherical image* (or *Gauss map*) of a surface is the function $\nu : M \longrightarrow S^2$ which sends each point of M to the normal at M. (Actually ν is only defined on a coordinate patch, where there is a well defined normal.)

PROPOSITION 8.6. The Gaussian curvature K at a point P is the limit of the ratio $A(\nu(\mathcal{R}))/A(\mathcal{R})$ as the region $\mathcal{R}$ shrinks to the point P and A denotes (signed) area.

Proof: $A(\mathcal{R}) = \iint_{x^{-1}(\mathcal{R})} [\mathbf{x}_1, \mathbf{x}_2, \mathbf{n}]\, du^1\, du^2 = \iint \sqrt{g}\, du^1\, du^2$. The function $\mathbf{n}: x^{-1}(\mathcal{R}) \longrightarrow S^2$ has image $\nu(\mathcal{R})$ and is a parametrized surface that might not be regular ($\mathbf{n}_1 \times \mathbf{n}_2 = \mathbf{0}$?). This lack of regularity does not affect the formula for the absolute area of $\nu(\mathcal{R})$: $|A(\nu(\mathcal{R}))| = \iint_{x^{-1}(\mathcal{R})} [\mathbf{n}_1, \mathbf{n}_2, \mathbf{m}]\, du^1\, du^2$, where $\mathbf{m} = \mathbf{n}_1 \times \mathbf{n}_2/|\mathbf{n}_1 \times \mathbf{n}_2| = \pm\mathbf{n}$. The plus sign holds if ν preserves orientation, the minus if it reverses it. Thus signed area is given by

$$A(\nu(\mathcal{R})) = \iint [\mathbf{n}_1, \mathbf{n}_2, \mathbf{n}]\, du^1\, du^2.$$

Then

$$C(P) = \lim_{\mathcal{R} \to P} \frac{A(\nu(\mathcal{R}))}{A(\mathcal{R})} = \frac{[\mathbf{n}_1, \mathbf{n}_2, \mathbf{n}]}{[\mathbf{x}_1, \mathbf{x}_2, \mathbf{n}]}.$$

By Problem 8.9 $\mathbf{n}_1 \times \mathbf{n}_2 = K\sqrt{g}\,\mathbf{n}$. Thus $[\mathbf{n}_1, \mathbf{n}_2, \mathbf{n}] = K\sqrt{g}$ and $C(P) = K$. ∎

This theorem gives geometric significance to the sign of K: K negative means ν reverses orientation at P.

Note that neither the matrix L_{ij} nor the mean curvature is intrinsic. It is a remarkable fact that the Gaussian curvature is intrinsic, as we shall see in Theorem 9.2.

We have given some geometric meaning to the Weingarten map L in terms of its trace $2H$ and determinant K. We shall now give a further geometric interpretation to the second fundamental form. We know that $\mathrm{II}(\mathbf{X}, \mathbf{Y})$ is a symmetric bilinear form on T_PM for each $P \in M$. If we think of II as a generalization of the length of a vector, then we can ask which vectors have "generalized length" $+1$ and -1 with respect to II. This information (which is classically called the Dupin indicatrix) provides information about the geometry of M near P.

DEFINITION. The *Dupin indicatrix* D of M at $P \in M$ is the subset of T_PM given by

$$D = \{\mathbf{X} \in T_PM \mid \mathrm{II}(\mathbf{X}, \mathbf{X}) = 1\} \cup \{\mathbf{X} \in T_PM \mid \mathrm{II}(\mathbf{X}, \mathbf{X}) = -1\}.$$

We shall write $D = D^+ \cup D^-$. In terms of local coordinates, $D^+ = \{\sum X^i \mathbf{x}_i \mid \sum L_{ij}X^iX^j = 1\}$ and $D^- = \{\sum X^i \mathbf{x}_i \mid \sum L_{ij}X^iX^j = -1\}$. (Since the sign of $\mathbf{n}$ is not well defined, neither is that of II. A change in the sign of $\mathbf{n}$ will interchange D^+ and D^- but does not, of course, change their union which is D.) The local coordinate description shows that D^+ and D^- are (possibly degenerate) conic sections when viewed as graphs in the tangent plane. Note that the axes $(\mathbf{x}_1, \mathbf{x}_2)$ that we use for T_PM may not be orthogonal since g_{12} may not be zero.

EXAMPLE 8.7. Let $M = S^2$. By Problem 7.1 L is the identity transformation. Hence $\text{II}(\mathbf{X}, \mathbf{X}) = \langle \mathsf{L}(\mathbf{X}), \mathbf{X}\rangle = \langle \mathbf{X}, \mathbf{X}\rangle$. Then $D = D^+ =$ the unit circle in T_PM. See also Problem 8.20.

Depending on the signs of the principal curvatures and whether they vanish or not, the Dupin indicatrix may be an ellipse, two conjugate pairs of hyperbolas (i.e., having the same asymptotes and semi-axes), two parallel lines, or empty. We leave the proof of the precise statement (below) as Problem 8.21.

PROPOSITION 8.3. Let P be a point on M with Gaussian curvature K at P. Then
(a) if $K > 0$, the Dupin indicatrix is an ellipse;
(b) if $K < 0$, the Dupin indicatrix is two conjugate pairs of hyperbolas; and
(c) if $K = 0$, the Dupin indicatrix is two parallel lines if one principal curvature is not zero and is empty if both principal curvatures are zero.

Because of this proposition, the point P is called *elliptic* if $K > 0$, *hyperbolic* if $K < 0$, *parabolic* if only one principal curvature vanishes, and *flat* (or *planar*) if both principal curvatures are zero.

The geometric significance of the Dupin indicatrix comes from the classical interpretation we now describe. The basic idea is to shift the tangent plane parallel to itself in the direction of the normal to the surface. The curve of intersection of this new plane and the surface is "approximately the Dupin indicatrix." See Figures 4.21, 4.22, 4.23 for the case of elliptic, hyperbolic, and parabolic points, respectively.

We now make the classical notion that the intersection is "approximately the Dupin indicatrix" more precise. Let $\mathbf{x}\colon \mathcal{U} \longrightarrow \mathbf{R}^3$ be a coordinate patch about P with $\mathbf{x}(0, 0) = \mathbf{x}_0 = P$. The tangent plane at P is given by

$$T_PM = \{\mathbf{y} \in \mathbf{R}^3 \mid \langle \mathbf{y} - \mathbf{x}_0, \mathbf{n}\rangle = 0\},$$

where $\mathbf{n}$ is the normal to M at P. Let ϵ be any (possibly negative) number and $\Pi_\epsilon = \{\mathbf{y} \in \mathbf{R}^3 \mid \langle \mathbf{y} - \mathbf{x}_0, \mathbf{n}\rangle = \epsilon\}$. Π_ϵ is the plane parallel to T_PM at a

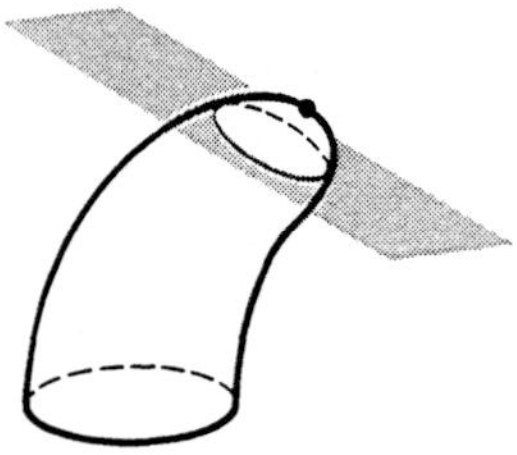

FIGURE 4.21

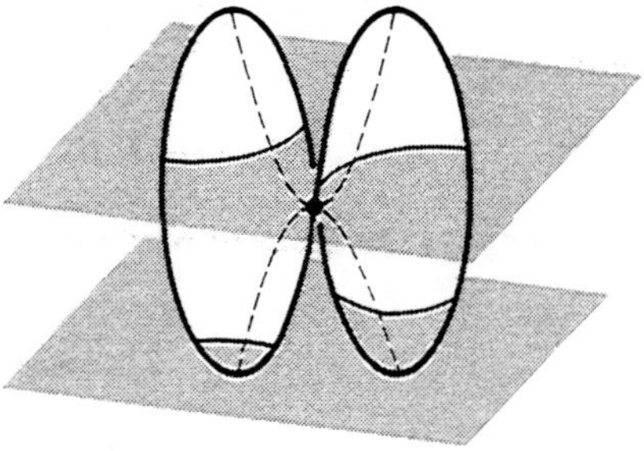

FIGURE 4.22

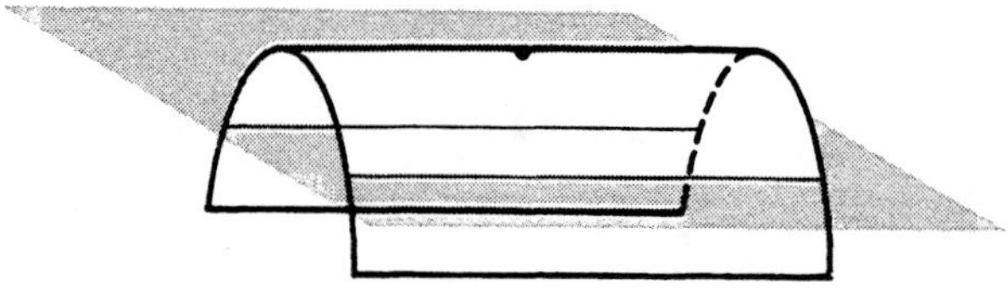

FIGURE 4.23

signed distance ϵ from it. Since we wish to know what $M \cap \Pi_\epsilon$ is for small ϵ, we shall examine

$$\mathbf{x}(\mathcal{U}) \cap \Pi_\epsilon = \{\mathbf{x}(u^1, u^2) \mid (u^1, u^2) \in \mathcal{U} \quad \text{and} \quad \langle \mathbf{x}(u^1, u^2) - \mathbf{x}_0, \mathbf{n} \rangle = \epsilon\}.$$

To this end we expand $\mathbf{x}$ in a Taylor series about $(0, 0)$:

$$\mathbf{x}(u^1, u^2) = \mathbf{x}_0 + \sum \mathbf{x}_i(0, 0)u^i + \tfrac{1}{2} \sum \mathbf{x}_{ij}(0, 0)u^i u^j + \mathbf{r}(u^1, u^2).$$

Then

$$\begin{aligned} \epsilon &= \langle \mathbf{x}(u^1, u^2) - \mathbf{x}_0, \mathbf{n} \rangle \\ &= \sum \langle \mathbf{x}_i(0, 0), \mathbf{n} \rangle u^i + \tfrac{1}{2} \sum \langle \mathbf{x}_{ij}(0, 0), \mathbf{n} \rangle u^i u^j + \langle \mathbf{r}, \mathbf{n} \rangle \end{aligned}$$

or

$$\epsilon = \tfrac{1}{2} \sum L_{ij} u^i u^j + \langle \mathbf{r}, \mathbf{n} \rangle \tag{8-2}$$

Providing $\mathbf{x}$ is sufficiently differentiable we may assume that the remainder $\mathbf{r}(u^1, u^2)$ satisfies

$$\lim \frac{\mathbf{r}(u^1, u^2)}{(u^1)^2 + (u^2)^2} = \mathbf{0} \quad \text{as} \quad (u^1, u^2) \longrightarrow (0, 0).$$

Thus for u^1 and u^2 small, $\mathbf{r}$ is quite small and the intersection is approximated by the conic section $\epsilon = \frac{1}{2} \sum L_{ij} u^i u^j$. We shall make a change of coordinates that will have the effect of normalizing this conic section, and will stretch it as $\epsilon \longrightarrow 0$. Currently we are using u^1 and u^2 as coordinates in the plane, with axes given by $\mathbf{x}_1(0, 0)$ and $\mathbf{x}_2(0, 0)$, which may not be perpendicular. Define new coordinates here by $\lambda^i = u^i/\sqrt{2|\epsilon|}$. Notice that as $\epsilon \longrightarrow 0$ this has the effect of stretching the graph of $M \cap \Pi_\epsilon$ relative to a fixed (λ^1, λ^2) coordinate system.

If we substitute $u^i = \lambda^i\sqrt{2|\epsilon|}$ into Equation (8-2) and divide by $|\epsilon|$ we obtain

$$\pm 1 = \sum L_{ij}\lambda^i\lambda^j + \frac{\langle \mathbf{r}(\sqrt{2|\epsilon|}\lambda^1, \sqrt{2|\epsilon|}\lambda^2), \mathbf{n}\rangle}{|\epsilon|}.$$

We compute the limiting position of this locus as $\epsilon \longrightarrow 0$. By assumption,

$$\lim_{\epsilon\to 0} \frac{\mathbf{r}(\sqrt{2|\epsilon|}\lambda^1, \sqrt{2|\epsilon|}\lambda^2)}{2|\epsilon|((\lambda^1)^2 + (\lambda^2)^2)} = \mathbf{0}.$$

Hence

$$\lim_{\epsilon\to 0} \frac{\mathbf{r}(\sqrt{2|\epsilon|}\lambda^1, \sqrt{2|\epsilon|}\lambda^2)}{|\epsilon|} = \mathbf{0}.$$

Thus the limiting position as $\epsilon \longrightarrow 0$ of these "stretched conic sections" is given by

$$\{\sum \lambda^i\mathbf{x}_i \mid \pm 1 = \sum L_{ij}\lambda^i\lambda^j\} = D.$$

This is the classical interpretation of the Dupin indicatrix.

DEFINITION. Two non-zero vectors $\mathbf{X}$ and $\mathbf{Y}$ are called *conjugate directions* (or *conjugates*) if $\mathrm{II}(\mathbf{X}, \mathbf{Y}) = 0$.

DEFINITION. A tangent vector $\mathbf{X}$ at P is an *asymptotic direction* (or is *self-conjugate*) if $\mathrm{II}(\mathbf{X}, \mathbf{X}) = 0$. An *asymptotic curve* is one whose tangent vector is an asymptotic direction at each point.

The terminology asymptotic direction is due to the fact that the asymptotic directions at a hyperbolic point are the asymptotes of the Dupin indicatrix, which is an hyperbola. See Example 8.9 below. Various properties of asymptotic and conjugate directions are developed in Problems 8.22 through 8.35.

EXAMPLE 8.9. Let $\mathbf{x}$ be the Monge patch $\mathbf{x}(u, v) = (u, v, u^2 - 2v^2)$. Then $\mathbf{x}_1 = (1, 0, 2u)$, $\mathbf{x}_2 = (0, 1, -4v)$, $\mathbf{n} = (-2u, 4v, 1)/\sqrt{1 + 4u^2 + 16v^2}$, $\mathbf{x}_{11} = (0, 0, 2)$, $\mathbf{x}_{12} = (0, 0, 0) = \mathbf{x}_{21}$, $\mathbf{x}_{22} = (0, 0, -4)$,

$$L_{11} = \frac{2}{\sqrt{1 + 4u^2 + 16v^2}},$$

$$L_{12} = L_{21} = 0,$$

and

$$L_{22} = \frac{-4}{\sqrt{1 + 4u^2 + 16v^2}}.$$

At $\mathbf{x}(0, 0)$ the coordinate axes $\mathbf{x}_1$ and $\mathbf{x}_2$ are orthogonal. The Dupin indicatrix there is given by

$$\pm 1 = 2(X^1)^2 - 4(X^2)^2,$$

and consists of two conjugate pairs of hyperbolas, as pictured in Figure 4.24. The asymptotic directions are given by

$$0 = \mathrm{II}(\mathbf{X}, \mathbf{X}) = 2(X^1)^2 - 4(X^2)^2 = 0$$

or $X^1 = \pm\sqrt{2}\,X^2$. Note that these give the asymptotes of the hyperbolas in Figure 4.24. Note that $\mathbf{x}_1$ and $\mathbf{x}_2$ are conjugates. If $\mathbf{X} = \mathbf{x}_1 + \mathbf{x}_2$ then $\mathbf{Y} = Y^1\mathbf{x}_1 + Y^2\mathbf{x}_2$ is conjugate to $\mathbf{X}$ if and only if

$$0 = \langle \mathrm{L}(\mathbf{X}), \mathbf{Y} \rangle = 2X^1Y^1 - 4X^2Y^2 = 2Y^1 - 4Y^2$$

or $Y^1 = 2Y^2$. Hence $2\mathbf{x}_1 + \mathbf{x}_2$ is conjugate to $\mathbf{x}_1 + \mathbf{x}_2$.

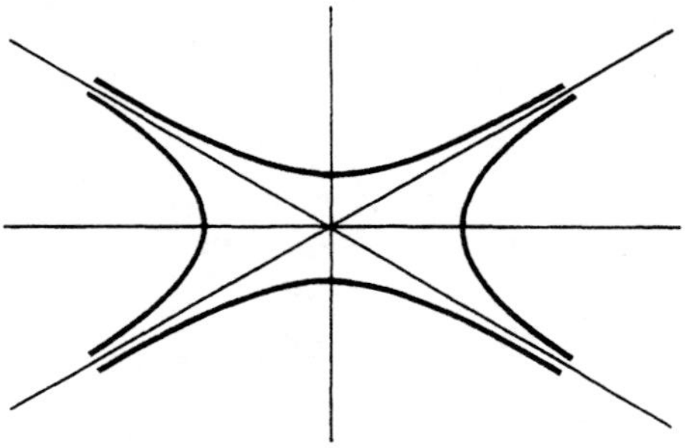

FIGURE 4.24

DIGRESSION

We indicate here in a very brief fashion, the origin of the area formula in a coordinate patch. For a more rigorous foundation see Fulks [1969].

Consider a region $\mathcal{A}$ contained in the image of a coordinate patch $\mathbf{x}: \mathcal{U} \longrightarrow \mathbf{R}^3$ bounded by four parametric curves: $u^1 = a$, $u^1 = a + \Delta u^1$, $u^2 = b$, $u^2 = b + \Delta u^2$. In the tangent plane at $\mathbf{x}(a, b)$ there is a parallelogram $\mathcal{B}$ spanned by $\Delta u^1\, \mathbf{x}_1$ and $\Delta u^2\, \mathbf{x}_2$. If Δu^1 and Δu^2 are quite small, the area of $\mathcal{B}$ and $\mathcal{A}$ should be almost the same. See Figure 4.25.

The area of $\mathcal{B}$ is $|\Delta u^1\, \mathbf{x}_1 \times \Delta u^2 \mathbf{x}_2| = \Delta u^1\, \Delta u^2\, |\mathbf{x}_1 \times \mathbf{x}_2| = \Delta u^1\, \Delta u^2 \sqrt{g}$. The area of a large region $\mathcal{R}$ in $\mathbf{x}(\mathcal{U})$ should be the sum of lots of small areas of regions like $\mathcal{A}$ and thus should almost equal the sum of terms $\Delta u^1\, \Delta u^2 \sqrt{g}$. In the limit this says

$$\text{area of } \mathcal{R} = \iint_{x^{-1}(\mathcal{R})} \sqrt{g}\; du^1\, du^2.$$

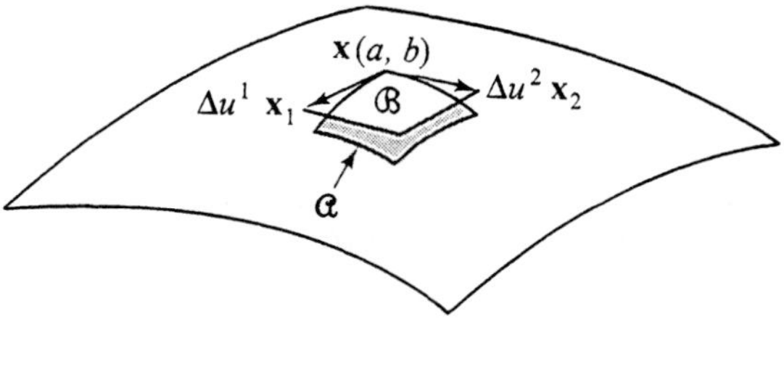

FIGURE 4.25

If $\mathcal{R}$ is also contained in the $\mathcal{V}$ coordinate patch we must have area of $\mathcal{R} = \iint_{y^{-1}(\mathcal{R})} \sqrt{\bar{g}}\, dv^1\, dv^2$, where $\bar{g} = \det(\bar{g}_{\alpha\beta})$. Since $(g_{ij}) = (J^{-1})^t(\bar{g}_{\alpha\beta})J^{-1}$,

$$g = \det(g_{ij}) = \det(J^{-1})\det(\bar{g}_{\alpha\beta})\det(J^{-1}) = \bar{g}(\det(J^{-1}))^2.$$

Thus $\sqrt{g} = \sqrt{\bar{g}}\,|\det(J^{-1})|$. Thus

$$\iint_{x^{-1}(\mathcal{R})} \sqrt{g}\, du^1\, du^2 = \iint_{x^{-1}(\mathcal{R})} (\sqrt{\bar{g}} \circ f^{-1})|\det(J^{-1})|\, du^1\, du^2,$$

which, by Jacobi's rule for transformation of integrals (the substitution or change of variables formula for multiple integrals), equals $\iint_{y^{-1}(\mathcal{R})} \sqrt{\bar{g}}\, dv^1\, dv^2$. ($f\colon \mathcal{V} \longrightarrow \mathcal{U}$ is the transformation of coordinates.)

EXAMPLE 8.10. For a Monge patch $\mathbf{x}(r, s) = (r, s, f(r, s))$ we have

$$g = (1 + f_r^2)(1 + f_s^2) - (f_r f_s)^2 = 1 + f_r^2 + f_s^2.$$

The area is given by $\iint \sqrt{1 + f_r^2 + f_s^2}\, dr\, ds$. This should be a familiar formula from Calculus III.

EXAMPLE 8.11. For the upper hemisphere (Example 1.7 of Chapter 4) we have $f(r, s) = \sqrt{1 - r^2 - s^2}$. $f_r = -r/\sqrt{1 - r^2 - s^2}$, $f_s = -s/\sqrt{1 - r^2 - s^2}$, and $\sqrt{1 + f_r^2 + f_s^2} = \sqrt{1/(1 - r^2 - s^2)}$. Hence the area of a hemisphere must be

$$\begin{aligned}\iint_{\mathcal{U}} \frac{dr\, ds}{\sqrt{1 - r^2 - s^2}} &= \int_{-1}^{1}\int_{-\sqrt{1-s^2}}^{\sqrt{1-s^2}} \frac{1}{\sqrt{1 - r^2 - s^2}}\, dr\, ds \\ &= \int_{-1}^{1} \left(\arcsin \frac{r}{\sqrt{1 - s^2}}\right)\Bigg|_{-\sqrt{1-s^2}}^{\sqrt{1-s^2}} ds \\ &= \int_{-1}^{1} \left(\frac{\pi}{2} - \left(-\frac{\pi}{2}\right)\right) ds = \int_{-1}^{1} \pi\, ds = 2\pi.\end{aligned}$$

Recall from Calculus II that if a function f defined in $\mathbf{R}^2$ is given in terms of polar coordinates, then the integral of f over a region is given by $\iint f(r, \theta)\, r\, dr\, d\theta$, not $\iint f(r, \theta)\, dr\, d\theta$. Furthermore, the expression $r\, dr\, d\theta$ is precisely what is integrated to find area: area of $\mathcal{R} = \iint_{\mathcal{R}} r\, dr\, d\theta$. Similarly, if we have a function f defined on a surface, its integral over a region $\mathcal{R}$ contained in a coordinate patch will be defined as $\iint f(u^1, u^2)\sqrt{g}\, du^1\, du^2$.

PROBLEMS

General

8.1. Find the Gaussian and mean curvatures of $\mathbf{R}^2$, S^2, T^2 (Problems 1.1 and 7.3), and $S^1 \times (0, 1)$ (Problem 7.2).

8.2. Prove $H^2 \geq K$. When does equality hold?

8.3. Let $\mathbf{X}$ and $\mathbf{Y}$ be orthonormal vectors at P. Prove

$$H = \tfrac{1}{2}(\mathrm{II}(\mathbf{X}, \mathbf{X}) + \mathrm{II}(\mathbf{Y}, \mathbf{Y})).$$

8.4. Prove that the u^i-curves are lines of curvature if and only if

$$g_{12} \equiv L_{12} \equiv 0.$$

8.5. Prove that the circles of latitude and the meridians of a surface of revolution are lines of curvature.

†8.6. Suppose $g_{12} \equiv 0$. Prove

$$K = -\frac{1}{2\sqrt{g}}\left(\frac{\partial}{\partial u^2}\frac{\frac{\partial g_{11}}{\partial u^2}}{\sqrt{g}} + \frac{\partial}{\partial u^1}\frac{\frac{\partial g_{22}}{\partial u^1}}{\sqrt{g}}\right).$$

***8.7.** Suppose $g_{11} \equiv 1$, $g_{12} \equiv 0$. Prove

$$\frac{\partial^2\sqrt{g_{22}}}{(\partial u^1)^2} + K\sqrt{g_{22}} = 0.$$

8.8. Determine which points of the Monge patch $\mathbf{x}(u, v) = (u, v, u^3 + v^3)$ have $K > 0$ and which have $K < 0$.

***8.9.** Prove $\mathbf{n}_1 \times \mathbf{n}_2 = K\sqrt{g}\,\mathbf{n}$.

8.10. What are the principal curvatures for a surface of revolution?

8.11. Prove $H = (1/2\pi)\int_0^{2\pi} \kappa_n \, d\theta$, where θ is as in Euler's Theorem (8.4). This shows that the mean curvature is the average of the normal curvature over all directions.

†8.12. Suppose two surfaces M_1 and M_2 intersect in a curve $\mathcal{C}$. Let κ be the curvature of $\mathcal{C}$, λ_i the normal curvature of $\mathcal{C}$ in M_i, and θ the angle between the normals of M_1 and M_2. Prove

$$\kappa^2 \sin^2\theta = \lambda_1{}^2 + \lambda_2{}^2 - 2\lambda_1\lambda_2 \cos\theta.$$

8.13. Let $\gamma(s)$ be a unit speed curve on a surface M. The *geodetic torsion* of γ is $\tau_g = -\langle \mathbf{S}, d\mathbf{n}/ds \rangle$ (where $\mathbf{S} = \mathbf{n} \times \mathbf{T}$).
(a) Prove $\tau_g \equiv 0$ if and only if γ is a line of curvature.
(b) Prove if γ is a geodesic then $\tau = \tau_g$.
(c) Does $\tau = \tau_g$ imply γ is a geodesic? (Give proof or counterexample.)

†8.14. Suppose that in a coordinate patch $\mathbf{x}: \mathcal{U} \longrightarrow M$ the Weingarten map satisfies $\mathsf{L} = f\,\mathsf{I}$ where $f: \mathcal{U} \longrightarrow \mathbf{R}$ and I is the identity linear transformation. Prove f is a constant in a neighborhood of each point.

Set A—Minimal Surfaces

8.15. Let M be the surface of revolution generated by the non-unit speed curve $\boldsymbol{\alpha}(t) = ((1/a)\cosh(at + b), t)$. Show that M is minimal ($H \equiv 0$). M is called a *catenoid*.

8.16. Let M be the surface $\mathbf{x}(u^1, u^2) = (u^2 \cos u^1, u^2 \sin u^1, pu^1)$. Show that M is minimal. M is called a *helicoid.*

8.17. Prove that a surface is minimal if and only if

$$g_{11}L_{22} - 2g_{12}L_{12} + g_{22}L_{11} \equiv 0.$$

†8.18. Prove that Problem 8.15, together with the plane, gives all surfaces of revolution that are minimal.

Set B—Dupin Indicatrix, Asymptotic and Conjugate Directions

8.19. Classify all the points of the Monge patch $\mathbf{x}(u, v) = (u, v, u^3 + v^2)$ as to being elliptic, hyperbolic, parabolic, or flat.

8.20. Let $P = (1/2, 1/2, 1/\sqrt{2}) \in S^2$. Graph the Dupin indicatrix of S^2 at P, using axes parallel to $\mathbf{x}_1$ and $\mathbf{x}_2$, where the coordinate patch is the upper hemisphere (Example 1.7). Be careful, $\mathbf{x}_1$ and $\mathbf{x}_2$ are not orthogonal. Your result should still be a circle as in Example 8.7.

8.21. Prove Proposition 8.3. (*Hint:* Write elements of T_PM in terms of a basis of principal (orthogonal) directions.)

8.22. Show that there are 0, 1, or 2 linearly independent asymptotic directions at P, depending on whether P is elliptic, parabolic, or hyperbolic.

8.23. Prove that the principal directions bisect the asymptotic directions at each hyperbolic point.

8.24. Prove that the parametric curves are asymptotic curves if and only if $L_{11} = L_{22} = 0$.

8.25. Let $\boldsymbol{\gamma}$ be a curve on M. Suppose that the curvature κ of $\boldsymbol{\gamma}$ is nonzero. Prove that $\boldsymbol{\gamma}$ is an asymptotic curve if and only if $\mathbf{B}$ is normal to the surface, i.e., the osculating plane of $\boldsymbol{\gamma}$ coincides with the tangent plane of M at each point of $\boldsymbol{\gamma}$.

8.26. Let $\boldsymbol{\gamma}$ be a straight line lying on a surface M. Prove that $\boldsymbol{\gamma}$ is asymptotic.

†8.27. Let $\mathbf{x}$ be the Monge patch $\mathbf{x}(u, v) = (u, v, (u^2/4) - v^2)$. Prove that all asymptotic curves are straight. (*Hint:* Problem 8.26 may be useful.)

8.28. Prove that a geodesic is an asymptotic curve if and only if it is a straight line.

8.29. Two surfaces $\mathbf{x}(u^1, u^2)$ and $\mathbf{y}(u^1, u^2)$ are said to have *contact of order r* at (a, b) if $\mathbf{x} - \mathbf{y}$ and all its derivatives up through order r are zero at (a, b). If $\mathbf{x}\colon \mathfrak{U} \longrightarrow \mathbf{R}^3$ is a surface with $P = \mathbf{x}(0, 0)$, determine under what conditions $\mathbf{y} = \mathbf{a} + \sum \mathbf{b}_i u^i + \sum \mathbf{c}_{ij} u^i u^j$ has second order contact with $\mathbf{x}$ at $(0, 0)$. The resulting surface $\mathbf{y}$ is called the *osculating paraboloid.* Show that $\mathbf{x}$ and $\mathbf{y}$ have the same normal vector at P.

Consider a plane parallel to the tangent plane at P. How is its intersection with the osculating paraboloid related to the Dupin indicatrix? (This problem is an analog for surfaces of Problem Set D of Section 2-4 which deals with spherical contact for curves.)

8.30. Prove that if $K \neq 0$ and $\mathbf{X} \neq 0$ then there are exactly two unit vectors $\mathbf{Y}_1$ and $\mathbf{Y}_2$ such that $\mathbf{X}$ and $\mathbf{Y}_i$ are conjugates for $i = 1$ and 2.

8.31. Prove that the tangents to the parametric curves are conjugates at each point if and only if $L_{12} \equiv 0$. In this case we say that the parametric curves form a *conjugate family of curves.*

8.32. Let M be a surface with no umbilics. Prove that the parametric curves form an orthogonal conjugate family of curves if and only if they are lines of curvature.

8.33. Can a line of curvature ever be an asymptotic curve?

8.34. Prove that $\langle \mathsf{L}(\mathbf{X}), \mathsf{L}(\mathbf{Y})\rangle - 2H\langle \mathsf{L}(\mathbf{X}), \mathbf{Y}\rangle + K\langle \mathbf{X}, \mathbf{Y}\rangle \equiv 0$. (*Hint:* Write $\mathbf{X}$ and $\mathbf{Y}$ as linear combinations of the principal directions or use the Cayley-Hamilton Theorem from linear algebra.)

†8.35. Prove the Beltrami-Enneper Theorem: If $\boldsymbol{\gamma}$ has non-zero curvature, torsion τ, and is an asymptotic curve, then $\tau^2 = -K$, where K is the Gaussian curvature. (The proof will make use of Problem 8.34.)

Set C—Ruled Surfaces

A ruled surface is a surface swept out by a moving straight line, much as a curve is swept out by a moving point. Such a surface can be parametrized in the form $\mathbf{x}(s, t) = \boldsymbol{\alpha}(s) + t\boldsymbol{\beta}(s)$, where $\boldsymbol{\alpha}(s)$ is a unit speed curve and $|\boldsymbol{\beta}(s)| = 1$. However, not every function of this form gives a surface. There are minor problems with self intersections if $\mathbf{x}$ is not one-to-one. These will be ignored. However, there may be *singular points*, that is, points where $\mathbf{x}_1 \times \mathbf{x}_2 = \mathbf{0}$. We allow this below but will keep track of any singular points.

8.36. Prove that the surfaces in Problems 1.4, 1.9, and 1.14 of Section 4-1 are ruled.

†8.37. Prove that the surface $z = x^2 - y^2$ is doubly ruled; that is, through each point of the surface there are two straight lines that lie on the surface.

†8.38. Prove that the surface

$$\frac{x^2}{a^2} + \frac{y^2}{b^2} - \frac{z^2}{c^2} = 1$$

is doubly ruled.

8.39. Prove that the curvature of a ruled surface is never positive.

A ruled surface for which $\boldsymbol{\beta}' \equiv \mathbf{0}$ is called a *cylinder*. (See Problem 5.10 for an example.) A ruled surface for which $\boldsymbol{\beta}'(s) \neq \mathbf{0}$ for all s is called *noncylindrical.*

8.40. Let $\mathbf{x}(s, t) = \boldsymbol{\alpha}(s) + t\boldsymbol{\beta}(s)$ be a noncylindrical ruled surface. Prove there exists a unique curve $\boldsymbol{\gamma}(s) = \boldsymbol{\alpha}(s) + r(s)\boldsymbol{\beta}(s)$, which in general is not unit speed, such that $\langle \boldsymbol{\gamma}', \boldsymbol{\beta}' \rangle \equiv 0$. This is called the *line of striction*.

8.41. Find the line of striction for the surfaces in Problem 8.36.

8.42. If $\boldsymbol{\gamma}(s)$ is the line of striction of the ruled surface $\mathbf{x}(s, t) = \boldsymbol{\alpha}(s) + t\boldsymbol{\beta}(s)$, prove that $\mathbf{y}(s, u) = \boldsymbol{\gamma}(s) + u\boldsymbol{\beta}(s)$ is a reparametrization, where $u = t - r(s)$.

8.43. Show that any point on the line of striction where $\boldsymbol{\gamma}' = 0$ is a singular point.

8.44. Show that every singular point of a ruled surface lies on the line of striction. (*Hint:* Parametrize as $\mathbf{y}(s, u) = \boldsymbol{\gamma}(s) + u\boldsymbol{\beta}(s)$.)

Set D—Developable Surfaces

A ruled surface is called *developable* if along each line of the ruling the tangent planes are all parallel.

8.45. Let $\mathbf{x}(s, t) = \boldsymbol{\alpha}(s) + t\boldsymbol{\beta}(s)$ be a developable surface. Show that the normal vector $\mathbf{n}$ does not depend upon the parameter t.

8.46. Let $\mathbf{x}(s, t) = \boldsymbol{\alpha}(s) + t\boldsymbol{\alpha}'(s)$ be the *tangent developable surface* of a unit speed curve $\boldsymbol{\alpha}(s)$ (Problem 1.14). Show that $\mathbf{x}$ is developable.

8.47. Compute the metric matrix for the surface in Problem 8.46.

8.48. Show that a surface of revolution generated by a straight line is developable.

8.49. Prove that $\mathbf{x}(s, t) = \boldsymbol{\alpha}(s) + t\boldsymbol{\beta}(s)$ is developable if and only if $[\boldsymbol{\alpha}', \boldsymbol{\beta}, \boldsymbol{\beta}'] \equiv 0$.

8.50. If $\boldsymbol{\alpha}(s)$ is a unit speed curve, under what conditions is

$$\mathbf{x}(s, t) = \boldsymbol{\alpha}(s) + t\mathbf{N}(s)$$

developable?

8.51. When is $\mathbf{y}(s, t) = \boldsymbol{\alpha}(s) + t\mathbf{B}(s)$ developable?

8.52. Show that a developable surface has Gaussian curvature equal to zero.

8.53. Show that each line in the ruling of a developable surface is both an asymptotic curve and a line of curvature.

8.54. Let M be a developable surface without umbilics. Let $\boldsymbol{\alpha}(s)$ be a (unit speed) line of curvature corresponding to the nonzero principal curvature. Show that $\boldsymbol{\alpha}(s)$ is orthogonal to each line in the ruling.

8.55. Let $\boldsymbol{\alpha}(s)$ be chosen as in Problem 8.54 and parametrize the developable surface M by $\mathbf{x}(s, t) = \boldsymbol{\alpha}(s) + t\boldsymbol{\beta}(s)$ with $|\boldsymbol{\beta}| = 1$. Show that $\boldsymbol{\beta}'(s) = \lambda(s)\boldsymbol{\alpha}'(s)$ for some function $\lambda(s)$.

8.56. If $\lambda(s)$ is as in Problem 8.55, show that
(a) $\lambda \equiv 0$ implies M is a cylinder;
(b) $\lambda \equiv$ constant $\neq 0$ implies all lines of the ruling have a point in common and in this case that point is the line of striction (M is a cone, but not necessarily circular);
(c) λ and λ' both nonzero implies M is the tangent developable surface of its line of striction.

4–9. RIEMANNIAN CURVATURE AND GAUSS'S THEOREMA EGREGIUM

In this section we introduce the Riemannian curvature tensor $(R_i{}^j{}_{kl})$, which is an intrinsic invariant. We then obtain Gauss's equations and the Codazzi-Mainardi equations and use them to prove the surprising fact that the Gaussian curvature is $\sum R_1{}^l{}_{21}g_{l2}/g$ and hence intrinsic.

Let $\mathbf{x}: \mathfrak{U} \longrightarrow \mathbf{R}^3$ be a coordinate patch on M with Christoffel symbols $\Gamma_{ij}{}^k$ and second fundamental form coefficients L_{ij}.

DEFINITION. The *Riemannian curvature tensor* with index (i, l, j, k) is given by

$$R_i{}^l{}_{jk} = \frac{\partial \Gamma_{ik}{}^l}{\partial u^j} - \frac{\partial \Gamma_{ij}{}^l}{\partial u^k} + \sum (\Gamma_{ik}{}^p \Gamma_{pj}{}^l - \Gamma_{ij}{}^p \Gamma_{pk}{}^l) \tag{9-1}$$

for all $1 \leq i, l, j, k \leq 2$.

Most readers complain about the definition of $R_i{}^l{}_{jk}$ because it is so replete with symbols that they feel (correctly) that the geometric meaning of the Riemannian curvature tensor is lost. In order to define $R_i{}^l{}_{jk}$ more geometrically we would have to take a much more abstract approach using the formalism of covariant derivatives. We will do this in Chapter 7.

PROPOSITION 9.1. For all $1 \leq i, l, j, k \leq 2$
(a) (Gauss's equations)

$$R_i{}^l{}_{jk} = L_{ik}L^l{}_j - L_{ij}L^l{}_k \tag{9-2}$$

(b) (Codazzi-Mainardi equations)

$$\frac{\partial L_{ij}}{\partial u^k} - \frac{\partial L_{ik}}{\partial u^j} = \sum (\Gamma_{ik}{}^l L_{lj} - \Gamma_{ij}{}^l L_{lk}) \tag{9-3}$$

Proof: We prove both equations simultaneously by computing $\partial \mathbf{x}_{ij}/\partial u^k = \mathbf{x}_{ijk}$ two ways and comparing the tangential and normal components of the result. Using Gauss's formulas (4-8), we have $\mathbf{x}_{ijk} = \partial(L_{ij}\mathbf{n} + \sum \Gamma_{ij}{}^l \mathbf{x}_l)/\partial u^k$. Therefore

$$\begin{aligned}\mathbf{x}_{ijk} &= \frac{\partial L_{ij}}{\partial u^k}\mathbf{n} + L_{ij}\mathbf{n}_k + \sum \frac{\partial \Gamma_{ij}{}^l}{\partial u^k}\mathbf{x}_l + \sum \Gamma_{ij}{}^l \mathbf{x}_{lk} \\ &= \frac{\partial L_{ij}}{\partial u^k}\mathbf{n} - \sum L_{ij}L^l{}_k \mathbf{x}_l + \sum \frac{\partial \Gamma_{ij}{}^l}{\partial u^k}\mathbf{x}_l + \sum \Gamma_{ij}{}^l \Gamma_{lk}{}^m \mathbf{x}_m + \sum \Gamma_{ij}{}^l L_{lk}\mathbf{n} \\ &= \left(\frac{\partial L_{ij}}{\partial u^k} + \sum \Gamma_{ij}{}^l L_{lk}\right)\mathbf{n} + \sum \left(\frac{\partial \Gamma_{ij}{}^l}{\partial u^k} - L_{ij}L^l{}_k + \sum \Gamma_{ij}{}^p \Gamma_{pk}{}^l\right)\mathbf{x}_l.\end{aligned}$$

Similarly

$$\mathbf{x}_{ikj} = \left(\frac{\partial L_{ik}}{\partial u^j} + \sum \Gamma_{ik}{}^l L_{lj}\right)\mathbf{n} + \sum \left(\frac{\partial \Gamma_{ik}{}^l}{\partial u^j} - L_{ik}L^l{}_j + \sum \Gamma_{ik}{}^p \Gamma_{pj}{}^l\right)\mathbf{x}_l.$$

Since we assume that $\mathbf{x}$ is at least C^3,

$$\mathbf{x}_{ijk} = \frac{\partial^3 \mathbf{x}}{\partial u^k\, \partial u^j\, \partial u^i} = \frac{\partial^3 \mathbf{x}}{\partial u^j\, \partial u^k\, \partial u^i} = \mathbf{x}_{ikj}.$$

$\{\mathbf{x}_1, \mathbf{x}_2, \mathbf{n}\}$ is a basis of $\mathbf{R}^3$ at each point so that the various components of $\mathbf{x}_{ijk}$ with respect to this basis must equal those of $\mathbf{x}_{ikj}$. Hence

$$\left(\frac{\partial L_{ij}}{\partial u^k} + \sum \Gamma_{ij}{}^l L_{lk}\right) = \left(\frac{\partial L_{ik}}{\partial u^j} + \sum \Gamma_{ik}{}^l L_{lj}\right) \tag{9-4}$$

$$\left(\frac{\partial \Gamma_{ij}{}^l}{\partial u^k} - L_{ij}L^l{}_k + \sum \Gamma_{ij}{}^p \Gamma_{pk}{}^l\right) = \left(\frac{\partial \Gamma_{ik}{}^l}{\partial u^j} - L_{ik}L^l{}_j + \sum \Gamma_{ik}{}^p \Gamma_{pj}{}^l\right) \tag{9-5}$$

Equation (9-4) can be rewritten as

$$\frac{\partial L_{ij}}{\partial u^k} - \frac{\partial L_{ik}}{\partial u^j} = \sum (\Gamma_{ik}{}^l L_{lj} - \Gamma_{ij}{}^l L_{lk})$$

which is the Codazzi-Mainardi equation (9-3). Equation (9-5) may be rewritten as

$$\frac{\partial \Gamma_{ik}{}^l}{\partial u^j} - \frac{\partial \Gamma_{ij}{}^l}{\partial u^k} + \sum (\Gamma_{ik}{}^p \Gamma_{pj}{}^l - \Gamma_{ij}{}^p \Gamma_{pk}{}^l) = L_{ik}L^l{}_j - L_{ij}L^l{}_k.$$

Since the left-hand side of this equation is $R_i{}^l{}_{jk}$ (by the definition), we have Gauss's equation (9-2). ∎

The important point here is that the Riemann curvature tensor is defined intrinsically by Equation (9-1). Gauss's equation (9-2) gives an extrinsic description in terms of the second fundamental form and the Weingarten map.

We use the above to prove:

THEOREM 9.2 (Gauss's *Theorema Egregium.*) The Gaussian curvature K of a surface is intrinsic.

Proof: By Gauss's equation (9-2) $R_i{}^l{}_{jk} = L_{ik}L^l{}_j - L_{ij}L^l{}_k$. Hence

$$\sum R_i{}^l{}_{jk}\, g_{lm} = L_{ik}L_{mj} - L_{ij}L_{mk}.$$

Let $i = k = 1$ and $j = m = 2$.

$$\begin{aligned}\sum R_1{}^l{}_{21} g_{l2} &= L_{11}L_{22} - L_{12}L_{21} = \det(L_{ij}) = \det((L^k{}_j)(g_{ik})) \\ &= \det(L^k{}_j)\det(g_{ik}) = Kg.\end{aligned}$$

Hence $K = \sum R_1{}^l{}_{21} g_{l2}/g$. ∎

Gauss called this theorem *egregium* because it is truly remarkable. (Look up the meaning of egregious if it is unfamiliar.) K is defined very extrinsically—in terms of $\mathbf{n}$, or in terms of L, or in terms of ν, none of which are intrinsic. Yet K is intrinsic.

It can be shown that if a different coordinate patch is used, the coefficients $\bar{R}_\alpha{}^\delta{}_{\beta\gamma}$ defined there satisfy

$$R_i{}^l{}_{jk} = \sum \bar{R}_\alpha{}^\delta{}_{\beta\gamma} \frac{\partial v^\alpha}{\partial u^i}\frac{\partial u^l}{\partial v^\delta}\frac{\partial v^\beta}{\partial u^j}\frac{\partial v^\gamma}{\partial u^k}.$$

From this it follows that there is a trilinear function mapping

$$T_pM \times T_pM \times T_pM \longrightarrow T_pM,$$

called the *Riemann curvature tensor*, by $\mathsf{R}(\mathbf{X},\mathbf{Y})\mathbf{Z} = \sum R_i{}^l{}_{jk} X^j Y^k Z^i \mathbf{x}_l$. We then note that $\langle \mathsf{R}(\mathbf{x}_2, \mathbf{x}_1)\mathbf{x}_1, \mathbf{x}_2\rangle / |\mathbf{x}_2 \times \mathbf{x}_1|^2 = K$. In modern differential geometry (of higher dimensions) the expression $\langle \mathsf{R}(\mathbf{X},\mathbf{Y})\mathbf{Y},\mathbf{X}\rangle$, where $\mathbf{X}$ and $\mathbf{Y}$ are orthonormal, is called a sectional curvature. It plays a significant role in describing the geometry of M. This will be covered in more depth in Chapter 7.

One important aspect of the Codazzi-Mainardi equations (9-3) is that they give integrability conditions. The Fundamental Theorem of Surfaces states that if g_{ij} and L_{ij} are symmetric functions such that $g_{11} > 0$, $\det(g_{ij}) > 0$, and both Equations (9-2) and (9-3) hold, then there is a surface in $\mathbf{R}^3$, unique up to position in space, having the g_{ij} and L_{ij} as coefficients of the first and second fundamental form. See Section 4-10. Thus the situation for surfaces is a bit different than that for curves, where only the (natural) restriction $\kappa > 0$ is placed upon κ and τ in order to determine a curve.

PROBLEMS

9.1. Show that if $M = \mathbf{R}^2$, then $R_i{}^l{}_{jk} = 0$ for all $1 \leq i, l, j, k \leq 2$ both intrinsically and extrinsically.

9.2. Prove that the Riemann curvature tensor has the following symmetry properties:
(a) $R_i{}^l{}_{jk} = -R_i{}^l{}_{kj}$, hence $R_i{}^l{}_{jj} = 0$;
(b) $R_i{}^l{}_{jk} + R_j{}^l{}_{ki} + R_k{}^l{}_{ij} \equiv 0$;
(c) if $R_{imjk} = \sum R_i{}^l{}_{jk} g_{lm}$ then $R_{imjk} = -R_{mijk}$; and
(d) $R_{imjk} = R_{jkim}$.

9.3. Compute the Riemann curvature tensor for $M = S^2$
(a) extrinsically;
(b) intrinsically. (*Hint:* Use Problem 9.2 to cut down on the calculations.)

9.4. Compute the $R_i{}^l{}_{jk}$ for
(a) T^2 (Problem 1.1);
(b) $S^1 \times (0, 1)$ (Example 2.3).

The following problems are on the whole more difficult than most of our exercises. They are intended to introduce some of the modern terminology to help bridge the gap between this text and more advanced texts. They may be omitted if Chapter 7 is not covered.

9.5. Let $\mathbf{X}$ and $\mathbf{Y}$ be vector fields defined on a surface M. In a coordinate patch $\mathbf{x}: \mathcal{U} \longrightarrow M$ define

$$Z^i = \sum_j \left(X^j \frac{\partial Y^i}{\partial u^j} - Y^j \frac{\partial X^i}{\partial u^j}\right),$$

where $\mathbf{X} = \sum X^i \mathbf{x}_i$ and $\mathbf{Y} = \sum Y^j \mathbf{x}_j$. Similarly, in the patch $\mathbf{y}: \mathcal{V} \longrightarrow M$ define

$$\bar{Z}^\alpha = \sum_\beta \left(\bar{X}^\beta \frac{\partial \bar{Y}^\alpha}{\partial v^\beta} - \bar{Y}^\beta \frac{\partial \bar{X}^\alpha}{\partial v^\beta}\right).$$

Prove that $\bar{Z}^i = \sum \bar{Z}^\alpha \partial u^i / \partial v^\alpha$. This means that $\sum Z^i \mathbf{x}_i = \sum \bar{Z}^\alpha \mathbf{y}_\alpha = \mathbf{Z}$ is a vector field. $\mathbf{Z}$ is denoted $[\mathbf{X}, \mathbf{Y}]$ and is called the *Lie bracket* of $\mathbf{X}$ and $\mathbf{Y}$. Prove that $[\mathbf{x}_i, \mathbf{x}_j] = \mathbf{0}$. Prove that $[\mathbf{X}, \mathbf{Y}] = -[\mathbf{Y}, \mathbf{X}]$. Show that $[u^1\mathbf{x}_2, u^1u^2\mathbf{x}_1] = (u^1)^2\mathbf{x}_1 - u^1u^2\mathbf{x}_2$.

9.6. Recall that in Problem 5.14 we defined the vector field

$$\nabla_{\mathbf{X}}\mathbf{Y} = \sum X^i\left(\frac{\partial Y^k}{\partial u^i} + \sum \Gamma_{ij}{}^k Y^j\right)\mathbf{x}_k$$

in a coordinate patch $\mathbf{x}: \mathcal{U} \longrightarrow M$. Prove that

$$\nabla_{\mathbf{X}}\mathbf{Y} - \nabla_{\mathbf{Y}}\mathbf{X} - [\mathbf{X}, \mathbf{Y}] \equiv \mathbf{0}.$$

(*Hint:* $\Gamma_{ij}{}^{k} = \Gamma_{ji}{}^{k}$ (Why?).)

9.7. In this problem we interpret a vector field as a "directional derivative." Let $f: M \longrightarrow \mathbf{R}$ be a function defined on all of M. We say that f is *differentiable* if for every coordinate patch $\mathbf{x}: \mathcal{U} \longrightarrow M$, $f \circ \mathbf{x}: \mathcal{U} \longrightarrow \mathbf{R}$ is differentiable. If $\mathbf{X}$ is a vector field on M, then in the patch $\mathbf{x}$ set $\mathbf{X}(f) = \sum X^i\, \partial f/\partial u^i$. Prove that in terms of another patch $\mathbf{y}: \mathcal{V} \longrightarrow M$ we have $\sum \bar{X}^\alpha\, \partial f/\partial v^\alpha = \sum X^i\, \partial f/\partial u^i$. Prove that if $a \in \mathbf{R}$, and f and g are differentiable functions on M then: (1) $\mathbf{X}(f+g) = \mathbf{X}(f) + \mathbf{X}(g)$; (2) $\mathbf{X}(f \cdot g) = g \cdot (\mathbf{X}(f)) + f \cdot (\mathbf{X}(g))$, where $\cdot$ denotes multiplication; and (3) $\mathbf{X}(a) \equiv 0$, where a denotes the constant function $h(P) \equiv a$. It can be shown that any operator $\mathbf{X}$ that satisfies (1), (2), and (3) actually is a vector field. This will be the definition in Chapter 7.

9.8. Prove $[\mathbf{X}, \mathbf{Y}](f) = \mathbf{X}(\mathbf{Y}(f)) - \mathbf{Y}(\mathbf{X}(f))$. Let $f \cdot \mathbf{X}$ denote the vector field which at P is $f(P)$ times $\mathbf{X}$ at P. Prove $[f \cdot \mathbf{X}, \mathbf{Y}] = f \cdot [\mathbf{X}, \mathbf{Y}] - (\mathbf{Y}(f)) \cdot \mathbf{X}$. Prove $[\mathbf{X} + \mathbf{Y}, \mathbf{Z}] = [\mathbf{X}, \mathbf{Z}] + [\mathbf{Y}, \mathbf{Z}]$.

9.9. Prove $\nabla_{\mathbf{X}}(\mathbf{Y} + \mathbf{Z}) = \nabla_{\mathbf{X}}\mathbf{Y} + \nabla_{\mathbf{X}}\mathbf{Z}$. Prove $\nabla_{\mathbf{X}+\mathbf{Y}}(\mathbf{Z}) = \nabla_{\mathbf{X}}\mathbf{Z} + \nabla_{\mathbf{Y}}\mathbf{Z}$. Prove $\nabla_{f \cdot \mathbf{X}}(\mathbf{Y}) = f \cdot \nabla_{\mathbf{X}}\mathbf{Y}$. Prove $\nabla_{\mathbf{X}}(f \cdot \mathbf{Y}) = \mathbf{X}(f) \cdot \mathbf{Y} + f \cdot \nabla_{\mathbf{X}}\mathbf{Y}$.

9.10. If $\mathbf{X}, \mathbf{Y}, \mathbf{Z}$ are vector fields on M, set

$$\mathsf{R}(\mathbf{X}, \mathbf{Y})\mathbf{Z} = \nabla_{\mathbf{X}}(\nabla_{\mathbf{Y}}\mathbf{Z}) - \nabla_{\mathbf{Y}}(\nabla_{\mathbf{X}}\mathbf{Z}) - \nabla_{[\mathbf{X}, \mathbf{Y}]}\mathbf{Z}.$$

Prove that in a patch $\mathbf{x}: \mathcal{U} \longrightarrow M$ we have $\mathsf{R}(\mathbf{X}, \mathbf{Y})\mathbf{Z} = \sum W^l \mathbf{x}_l$ with $W^l = \sum R_i{}^l{}_{jk} Z^i X^j Y^k$ and $R_i{}^l{}_{jk}$ is defined by (9-1). This is the same concept as that at the end of this section.

9.11. Prove $\mathsf{R}(\mathbf{X}, \mathbf{Y})\mathbf{Z} = -\mathsf{R}(\mathbf{Y}, \mathbf{X})\mathbf{Z}$.
Prove $\mathbf{X}(\langle \mathbf{Y}, \mathbf{Z} \rangle) = \langle \nabla_{\mathbf{X}}\mathbf{Y}, \mathbf{Z} \rangle + \langle \mathbf{Y}, \nabla_{\mathbf{X}}\mathbf{Z} \rangle$.
Prove $\langle \mathsf{R}(\mathbf{X}, \mathbf{Y})\mathbf{Z}, \mathbf{Z} \rangle \equiv 0$.
Prove $\langle \mathsf{R}(\mathbf{X}, \mathbf{Y})\mathbf{Z}, \mathbf{W} \rangle = -\langle \mathsf{R}(\mathbf{X}, \mathbf{Y})\mathbf{W}, \mathbf{Z} \rangle$.
Prove $\mathsf{R}(\mathbf{X}, \mathbf{Y})\mathbf{Z} + \mathsf{R}(\mathbf{Y}, \mathbf{Z})\mathbf{X} + \mathsf{R}(\mathbf{Z}, \mathbf{X})\mathbf{Y} \equiv \mathbf{0}$.
Prove $\langle \mathsf{R}(\mathbf{X}, \mathbf{Y})\mathbf{Z}, \mathbf{W} \rangle = \langle \mathsf{R}(\mathbf{Z}, \mathbf{W})\mathbf{X}, \mathbf{Y} \rangle$.

9.12. Prove $\langle \mathsf{R}(\mathbf{X}, \mathbf{Y})\mathbf{Y}, \mathbf{X} \rangle = |\mathbf{X} \times \mathbf{Y}|^2 K$, where K is the Gaussian curvature.

9.13. Let $\mathbf{X}$ and $\mathbf{Y}$ be two fixed vector fields. If $\mathbf{Z}$ is a vector field define $\mathsf{S}(\mathbf{Z}) = \mathsf{R}(\mathbf{Z}, \mathbf{X})\mathbf{Y}$. Prove $\mathsf{S}(f \cdot \mathbf{Z}) = f \cdot \mathsf{S}(\mathbf{Z})$ and

$$\mathsf{S}(\mathbf{Z} + \mathbf{W}) = \mathsf{S}(\mathbf{Z}) + \mathsf{S}(\mathbf{W}).$$

In particular, this means that S is a linear transformation from T_PM to T_PM. Prove that the trace of S is $|\mathbf{X}, \mathbf{Y}|^2 K$.

4–10. ISOMETRIES AND THE FUNDAMENTAL THEOREM OF SURFACES

In this section we shall discuss the question as to when two surfaces are geometrically the same. There will be two concepts: isometry, which refers to intrinsic geometry, and rigidity, which refers to both the intrinsic and extrinsic geometry. In the latter case, two surfaces will be considered the same if they differ only in their position in space, as for example the two spheres $\{\mathbf{x} \in \mathbf{R}^3 \mid |\mathbf{x}| = 1\}$ and $\{\mathbf{x} \in \mathbf{R}^3 \mid |\mathbf{x} - (1, 0, 0)| = 1\}$. It is the former concept that shall be our primary interest.

DEFINITION. Let $f\colon M \longrightarrow N$ be a function between surfaces. f is *differentiable* if for each $P \in M$ there are coordinate patches $\mathbf{x}\colon \mathcal{U} \longrightarrow M$ and $\mathbf{y}\colon \mathcal{V} \longrightarrow N$ about P and $f(P)$ respectively such that $y^{-1} \circ f \circ \mathbf{x}\colon \mathcal{U} \longrightarrow \mathcal{V}$ is differentiable as a function of two variables.

In practice, f is often given in local coordinates by giving $y^{-1} \circ f \circ \mathbf{x}$, as in the following example.

EXAMPLE 10.1. Let M be the surface of revolution

$$\mathbf{x}(u, \theta) = (\cosh(u)\cos\theta, \cosh(u)\sin\theta, u)$$

for $-\sinh^{-1}(1) < u < \sinh^{-1}(1)$ and $0 < \theta < 2\pi$. Let N be the helicoid $\mathbf{y}(v, \phi) = (v\cos\phi, v\sin\phi, \phi)$ for $-1 < v < 1$ and $0 < \phi < 2\pi$. The function $y^{-1} \circ f \circ \mathbf{x}(u, \theta) = (\sinh(u), \theta)$ is clearly differentiable and gives a differentiable function $f\colon M \longrightarrow N$ by $f(\mathbf{x}(u, \theta)) = \mathbf{y}(\sinh(u), \theta)$. Note that this particular function is also one-to-one and onto. See Figure 4.26. The meridians of M are sent to straight lines on N and the circles of latitude are sent to circular helices.

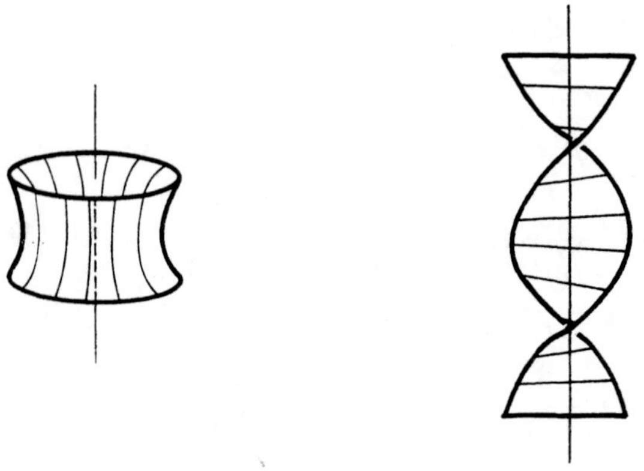

FIGURE 4.26

DEFINITION. An *isometry* from M to N is a one-to-one, onto, differentiable function $f: M \longrightarrow N$ such that for any curve $\boldsymbol{\gamma}: [c, d] \longrightarrow M$, the length of $\boldsymbol{\gamma}$ equals the length of $f \circ \boldsymbol{\gamma}$. M and N are *isometric* if such an isometry exists.

In Problem 10.1 you will verify that a rotation of S^2 is an isometry. A less obvious example is the following.

EXAMPLE 10.2. If f, M, N are as in Example 10.1, then f is an isometry. This is seen as follows. With respect to the coordinates (u, θ), the metric matrix of M is

$$(g_{ij}) = \begin{pmatrix} \cosh^2(u) & 0 \\ 0 & \cosh^2(u) \end{pmatrix}.$$

With respect to the cordinates (v, ϕ), the metric matrix of N is

$$(h_{ij}) = \begin{pmatrix} 1 & 0 \\ 0 & 1 + v^2 \end{pmatrix}.$$

If $\boldsymbol{\gamma}: [c, d] \longrightarrow M$ is given by $\boldsymbol{\gamma}(t) = \mathbf{x}(u(t), \theta(t))$, then $d\boldsymbol{\gamma}/dt$ has length $\cosh(u(t))\sqrt{\dot{u}^2 + \dot{\theta}^2}$. Since $(f \circ \boldsymbol{\gamma})(t) = \mathbf{y}(\sinh(u(t)), \theta(t))$, $d(f \circ \boldsymbol{\gamma})/dt$ has length $\sqrt{\cosh^2(u(t))\dot{u}^2 + (1 + \sinh^2(u(t)))\dot{\theta}^2} = \cosh(u(t))\sqrt{\dot{u}^2 + \dot{\theta}^2}$ also. Hence $\boldsymbol{\gamma}$ and $f \circ \boldsymbol{\gamma}$ must have the same length and f is an isometry.

In the above example we saw that the vectors $d\boldsymbol{\gamma}/dt$ and $d(f \circ \boldsymbol{\gamma})/dt$ had the same length for a particular isometry. This is true in general.

PROPOSITION 10.3. Let $f: M \longrightarrow N$ be an isometry and let $\boldsymbol{\gamma}: [c, d] \longrightarrow M$ be a regular curve. Then $d\boldsymbol{\gamma}/dt$ and $d(f \circ \boldsymbol{\gamma})/dt$ are vectors of the same length for each value of $t \in (c, d)$.

Proof: For each $t^* \in (c, d)$, the curve $\boldsymbol{\gamma}: [c, t^*] \longrightarrow M$ has the same length as $f \circ \boldsymbol{\gamma}: [c, t^*] \longrightarrow N$. Hence $\int_c^{t^*} |d\boldsymbol{\gamma}/dt|\, dt = \int_c^{t^*} |d(f \circ \boldsymbol{\gamma})/dt|\, dt$. If we differentiate with respect to t^* and evaluate at $t^* = t$ we obtain

$$\left|\frac{d\boldsymbol{\gamma}}{dt}\right| = \left|\frac{d(f \circ \boldsymbol{\gamma})}{dt}\right|. \quad \blacksquare$$

DEFINITION. Two surfaces M and N are *locally isometric* if for each $P \in M$ there are open sets $\mathcal{U}' \subset M$, $\mathcal{V}' \subset N$ with $P \in \mathcal{U}'$, and an isometry $f: \mathcal{U}' \longrightarrow \mathcal{V}'$ (and vice versa for $Q \in N$). Such an f is called a *local isometry*.

EXAMPLE 10.4. In Example 10.1 we now allow $0 \leq \theta \leq 2\pi$, so that M is the entire surface of revolution, and $-\infty < \phi < \infty$ so that N is an infinite staircase. f gives one local isometry. Choosing a second coordinate patch to cover M we can easily define another local isometry from M to N so that M and N are locally isometric. However, they are not isometric because M has finite area while N has infinite area and we shall see in Proposition 10.5 that isometries preserve the intrinsic geometry, in particular, the area. This proposition says that isometries preserve the first fundamental form.

PROPOSITION 10.5. Two surfaces M and N are locally isometric if and only if for each $P \in M$ there is an open set $\mathcal{U} \subset \mathbf{R}^2$ and coordinate patches $\mathbf{x}: \mathcal{U} \longrightarrow M$, $\mathbf{y}: \mathcal{U} \longrightarrow N$ with $P \in \mathbf{x}(\mathcal{U})$ such that the metric coefficients of $\mathbf{x}$ and $\mathbf{y}$ are the same. Hence locally isometric surfaces have the same local intrinsic geometry at corresponding points.

Proof: First assume that $f: \mathcal{U}' \longrightarrow \mathcal{V}'$ is a local isometry with $P \in \mathcal{U}'$. Let $\mathbf{x}: \mathcal{U} \longrightarrow M$ be a coordinate patch about P with $\mathbf{x}(\mathcal{U}) \subset \mathcal{U}'$. Let $\mathbf{y} = f \circ \mathbf{x}: \mathcal{U} \longrightarrow N$, which is clearly a coordinate patch about $f(P)$. (g_{ij}) will denote the metric tensor of $\mathbf{x}$ and (h_{ij}) will denote the metric tensor of $\mathbf{y}$. We shall first show that $g_{ij}(a, b) = h_{ij}(a, b)$ for all $(a, b) \in \mathcal{U}$ when $i = j$. Let $\boldsymbol{\gamma}(t) = \mathbf{x}(a + t, b)$ (i.e., a u^1-curve). By Proposition 10.3, $|\dot{\boldsymbol{\gamma}}(t)| = |(f \circ \boldsymbol{\gamma})^{\boldsymbol{\cdot}}(t)|$ for all t. Since $\dot{\boldsymbol{\gamma}}(0) = \mathbf{x}_1(a, b)$ and $(f \circ \boldsymbol{\gamma})^{\boldsymbol{\cdot}}(0) = \mathbf{y}_1(a, b)$, $|\mathbf{x}_1(a, b)| = |\mathbf{y}_1(a, b)|$ and $g_{11}(a, b) = h_{11}(a, b)$. Similarly $g_{22}(a, b) = h_{22}(a, b)$ by using a u^2-curve.

Now we show that $g_{12}(a, b) = h_{12}(a, b)$. Let $\boldsymbol{\alpha}(t) = \mathbf{x}(a + t, b + t)$. $\dot{\boldsymbol{\alpha}}(0) = \mathbf{x}_1(a, b) + \mathbf{x}_2(a, b)$ and $(f \circ \boldsymbol{\alpha})^{\boldsymbol{\cdot}}(0) = \mathbf{y}_1(a, b) + \mathbf{y}_2(a, b)$. Proposition 10.3 implies $|\dot{\boldsymbol{\alpha}}(0)| = |(f \circ \boldsymbol{\alpha})^{\boldsymbol{\cdot}}(0)|$ so that

$$\langle \mathbf{x}_1 + \mathbf{x}_2, \mathbf{x}_1 + \mathbf{x}_2 \rangle = \langle \mathbf{y}_1 + \mathbf{y}_2, \mathbf{y}_1 + \mathbf{y}_2 \rangle$$

or

$$\langle \mathbf{x}_1, \mathbf{x}_1 \rangle + 2\langle \mathbf{x}_1, \mathbf{x}_2 \rangle + \langle \mathbf{x}_2, \mathbf{x}_2 \rangle = \langle \mathbf{y}_1, \mathbf{y}_1 \rangle + 2\langle \mathbf{y}_1, \mathbf{y}_2 \rangle + \langle \mathbf{y}_2, \mathbf{y}_2 \rangle$$

or

$$g_{11} + 2g_{12} + g_{22} = h_{11} + 2h_{12} + h_{22}.$$

By the previous paragraph $g_{11}(a, b) = h_{11}(a, b)$ and $g_{22}(a, b) = h_{22}(a, b)$ so that $g_{12}(a, b) = h_{12}(a, b)$ also. Hence $(g_{ij}) = (h_{ij})$.

On the other hand, if there exists an open set $\mathcal{U}$ and coordinate patches $\mathbf{x}: \mathcal{U} \longrightarrow M$ and $\mathbf{y}: \mathcal{U} \longrightarrow N$ with the same metric coefficients, let $\mathcal{U}' = \mathbf{x}(\mathcal{U})$, $\mathcal{V}' = \mathbf{y}(\mathcal{U})$ and define $f: \mathcal{U}' \longrightarrow \mathcal{V}'$ by $f(\mathbf{x}(u^1, u^2)) = \mathbf{y}(u^1, u^2)$. Then $y^{-1} \circ f \circ \mathbf{x}(u^1, u^2) = (u^1, u^2)$ is clearly one-to-one, onto, and differentiable. If $\boldsymbol{\gamma}: [c, d] \longrightarrow \mathcal{U}'$, $\boldsymbol{\gamma}(t) = \mathbf{x}(\gamma^1(t), \gamma^2(t))$ and $(f \circ \boldsymbol{\gamma})(t) = \mathbf{y}(\gamma^1(t), \gamma^2(t))$, then $\boldsymbol{\gamma}$ and $f \circ \boldsymbol{\gamma}$ have the same length $\int_c^d \sqrt{\dot{\gamma}^i \dot{\gamma}^j g_{ij}}\, dt = \int_c^d \sqrt{\dot{\gamma}^i \dot{\gamma}^j h_{ij}}\, dt$ since $g_{ij} = h_{ij}$. Hence f is a local isometry. ∎

EXAMPLE 10.6. Let f, M, N be as in Examples 10.1 and 10.2 so that $y^{-1} \circ f \circ \mathbf{x}(u, \theta) = (\sinh(u), \theta)$. Then the Jacobian matrix of $y^{-1} \circ f \circ \mathbf{x}$ is

$$\begin{pmatrix} \dfrac{\partial v}{\partial u} & \dfrac{\partial v}{\partial \theta} \\ \dfrac{\partial \phi}{\partial u} & \dfrac{\partial \phi}{\partial \theta} \end{pmatrix} = \begin{pmatrix} \cosh(u) & 0 \\ 0 & 1 \end{pmatrix}.$$

The metric matrix for N in the (u, θ) coordinates is then

$$\begin{pmatrix} \cosh u & 0 \\ 0 & 1 \end{pmatrix}\begin{pmatrix} 1 & 0 \\ 0 & 1 + v^2 \end{pmatrix}\begin{pmatrix} \cosh u & 0 \\ 0 & 1 \end{pmatrix}$$

$$= \begin{pmatrix} \cosh^2 u & 0 \\ 0 & 1 + v^2 \end{pmatrix}$$

$$= \begin{pmatrix} \cosh^2 u & 0 \\ 0 & 1 + \sinh^2 u \end{pmatrix}$$

$$= \begin{pmatrix} \cosh^2 u & 0 \\ 0 & \cosh^2 u \end{pmatrix},$$

which is the metric matrix for M in the (u, θ) coordinates.

By using Proposition 10.5 we shall show in the next section that any two surfaces of revolution having the same positive constant curvature are locally isometric. In fact, in Section 6-2, after we have introduced the concept of a geodesic coordinate patch, we shall prove the more powerful theorem stated below.

THEOREM 10.7. Let M and N be two surfaces with the same constant curvature $K \equiv c$. Then for each $P \in M$, $\mathbf{X} \in T_P M$, $Q \in N$, $\mathbf{Y} \in T_Q N$ with $\mathbf{X}$ and $\mathbf{Y}$ unit vectors, there is a local isometry f such that $f(P) = Q$ and $f \circ \boldsymbol{\alpha} = \boldsymbol{\gamma}$, where $\boldsymbol{\alpha}$ is the unique geodesic through P in the direction $\mathbf{X}$ and $\boldsymbol{\gamma}$ is the unique geodesic through Q in the direction $\mathbf{Y}$.

Note that the above theorem must hold in particular for $M = N$ and $P = Q$. Thus for each P on a surface of constant curvature there is a local isometry that acts like a rotation about P. Also, if $M = N$ there must be a local isometry that sends any given point P to any other point Q. This is a situation that is clear on S^2 and in the plane, where each local isometry is actually global, i.e., defined on all of S^2 or $\mathbf{R}^2$. (The isometries of S^2 consist of compositions of rotations about various axes and reflections. The isometries of $\mathbf{R}^2$ are compositions of translations, rotations, and reflections. See Problem 10.7.)

Gauss's *Theorema Egregium* may be stated in the language of isometries (and in fact originally was, using the word "developed" which means mapped by an isometry):

THEOREM 10.8. If two surfaces are locally isometric, their Gaussian curvatures at corresponding points are equal.

The analogue of Theorem 10.7 for mean curvature is false (see Problem 10.6). Further, the condition of constant curvature cannot be dropped from Theorem 10.7. Problem 10.10 gives an example of two surfaces and a function from one to the other which is not a local isometry yet corresponding points have the same *nonconstant* Gaussian curvature.

We now turn to the stronger notion of equivalence of surfaces called rigidity.

DEFINITION. An $n \times n$ matrix A is a *special orthogonal matrix* if $A^T = A^{-1}$ and $\det A = 1$. When thought of as a linear transformation from $\mathbf{R}^n$ to $\mathbf{R}^n$, A is called a *rotation*.

The terminology rotation comes from the situation when $n = 2$. If $A = \begin{pmatrix} a & b \\ c & d \end{pmatrix}$, then $A^{-1} = \begin{pmatrix} d & -b \\ -c & a \end{pmatrix} \Big/ \det(A)$. If $\det A = 1$ and $A^T = A^{-1}$, then $a = d$ and $b = -c$. $1 = \det A = ad - bc = a^2 + b^2$ implies we may write $a = \cos\phi$ and $b = -\sin\phi$ with $0 \leq \phi \leq 2\pi$. Hence

$$A = \begin{pmatrix} \cos\phi & -\sin\phi \\ \sin\phi & \cos\phi \end{pmatrix}.$$

If $\mathbf{v} = \begin{pmatrix} r\cos\theta \\ r\sin\theta \end{pmatrix} \in \mathbf{R}^2$, then

$$\begin{aligned} A\mathbf{v} &= \begin{pmatrix} \cos\phi & -\sin\phi \\ \sin\phi & \cos\phi \end{pmatrix} \begin{pmatrix} r\cos\theta \\ r\sin\theta \end{pmatrix} \\ &= \begin{pmatrix} r\cos\phi\cos\theta - r\sin\phi\sin\theta \\ r\sin\phi\cos\theta + r\cos\phi\sin\theta \end{pmatrix} \\ &= \begin{pmatrix} r\cos(\theta + \phi) \\ r\sin(\theta + \phi) \end{pmatrix}. \end{aligned}$$

Hence the special orthogonal matrix A rotates $\mathbf{R}^2$ through an angle of ϕ radians.

DEFINITION. $f\colon \mathbf{R}^n \longrightarrow \mathbf{R}^n$ is called a *rigid motion* of $\mathbf{R}^n$ if there is a rotation A of $\mathbf{R}^n$ and a vector $\mathbf{b} \in \mathbf{R}^n$ such that $f(\mathbf{v}) = (A\mathbf{v}) + \mathbf{b}$.

DEFINITION. Two surfaces M and N are *rigidly equivalent* if there is a rigid motion $f\colon \mathbf{R}^3 \longrightarrow \mathbf{R}^3$ such that $f(M) = N$.

Note that a rigid motion is the composition of a rotation (by A) and a translation (by $\mathbf{b}$). Since a rotation preserves distances, as does a translation, a rigid motion must preserve distances and the length of curves. Thus the restriction of a rigid motion f to a surface M is an isometry from M to $f(M)$. However, the converse is false and an example is given by Example 10.2.

The two surfaces are isometric but not rigidly equivalent. It can be shown that rigidly equivalent surfaces have the same first and second fundamental forms at corresponding points so that they must have the same mean curvatures as well as the same Gaussian curvature. In Problem 10.6 you will give another example of isometric surfaces which are not rigidly equivalent by exploiting this idea.

Recall that the Fundamental Theorem of Curves (Theorem 5.2 of Chapter 2) said that the curvature and torsion determine a space curve up to position, that is, up to a rigid motion. Further, any differentiable function $\bar{\kappa} > 0$ and continuous function $\bar{\tau}$ determine a curve with curvature $\bar{\kappa}$ and torsion $\bar{\tau}$. We now state the analogous theorem for surfaces and give a brief discussion of its proof.

THEOREM 10.9 (Fundamental Theorem of Surfaces). Let $\mathcal{U}$ be an open set in $\mathbf{R}^2$ such that any two points of $\mathcal{U}$ may be joined by a curve in $\mathcal{U}$ and let $L_{ij}: \mathcal{U} \longrightarrow \mathbf{R}$ and $g_{ij}: \mathcal{U} \longrightarrow \mathbf{R}$ be differentiable functions for $i = 1, 2$ and $j = 1, 2$ such that

(a) $L_{12} = L_{21}$, $g_{12} = g_{21}$, $g_{11} > 0$, $g_{22} > 0$, and $g_{11}g_{22} - (g_{12})^2 > 0$, and

(b) L_{ij} and g_{ij} satisfy Gauss's Equations (9-2) and the Codazzi-Mainardi Equations (9-3) where the $\Gamma_{ij}{}^k$ are determined by Equation (4-11).

Then, if $P \in \mathcal{U}$, there is an open set $\mathcal{V} \subset \mathcal{U}$ with $P \in \mathcal{V}$ and a simple surface $\mathbf{x}: \mathcal{V} \longrightarrow \mathbf{R}^3$ such that (g_{ij}) and (L_{ij}) are the matrices of the first and second fundamental forms. Further, if $\mathbf{y}: \mathcal{V} \longrightarrow \mathbf{R}^3$ is another simple surface with first and second fundamental forms (g_{ij}) and (L_{ij}), then $\mathbf{y}(\mathcal{V})$ is rigidly equivalent to $\mathbf{x}(\mathcal{V})$.

It is clear that conditions (a) and (b) are necessary (this is Proposition 9.1). What is surprising is that these conditions are sufficient! A detailed proof may be found in the Appendix to Chapter 4 of Do Carmo [1976].

We discussed the analogy between the Frenet-Serret frame $\{\mathbf{T}, \mathbf{N}, \mathbf{B}\}$ and the "frame" $\{\mathbf{x}_1, \mathbf{x}_2, \mathbf{n}\}$ just before Proposition 1.13. Because of this, the ideas of the proof of Theorem 10.9 are very similar to those of the Fundamental Theorem of Curves, but the technical details are more formidable. The idea is to set up a system of differential equations for the derivatives of $\mathbf{x}_1$, $\mathbf{x}_2$, and $\mathbf{n}$ in terms of g_{ij} and L_{ij} and then solve. To actually carry this out for surfaces requires an existence and uniqueness theorem from partial differential equations (instead of ordinary differential equations). In general, one cannot always solve a system of partial differential equations. In the case at hand, the equations can be solved if and only if condition (b) of the theorem holds! This once again points out how the theory of differential equations lies at the heart of geometry.

The first proof of Theorem 10.9 was given by O. Bonnet in 1867.

PROBLEMS

10.1. Let $f_\theta: S^2 \longrightarrow S^2$ be given by

$$f_\theta(x, y, z) = \begin{pmatrix} \cos\theta & -\sin\theta & 0 \\ \sin\theta & \cos\theta & 0 \\ 0 & 0 & 1 \end{pmatrix} \begin{pmatrix} x \\ y \\ z \end{pmatrix}.$$

Show that $f_\theta(0, 0, 1) = (0, 0, 1)$ and that f_θ is a rotation around the z-axis through an angle θ. Prove that f_θ is an isometry.

10.2. Let $S^2(r) = \{\mathbf{x} \in \mathbf{R}^3 \,|\, |\mathbf{x}| = r\}$ and let $f: S^2 \longrightarrow S^2(r)$ by $f(\mathbf{x}) = r\mathbf{x}$. Prove that f is one-to-one and onto but not an isometry.

10.3. Show that $\mathbf{R}^2$ and the cylinder $S^1 \times (-\infty, \infty)$ are locally isometric.

10.4. Prove that $\mathbf{R}^2$ and the torus T^2 are not locally isometric. (*Hint:* Problem 8.1.)

10.5. If $f: M \longrightarrow N$ is an isometry and $\boldsymbol{\gamma}$ is a geodesic in M, show that $f \circ \boldsymbol{\gamma}$ is a geodesic in N.

10.6. Find an example to show that the following statement is false: If M is locally isometric to N, then the mean curvature of M is equal to the mean curvature of N at corresponding points.

10.7. Prove that any rigid motion of $\mathbf{R}^2$ is an isometry of $\mathbf{R}^2$ onto itself (where $\mathbf{R}^2 = \{(x, y, z) \in \mathbf{R}^3 \,|\, z = 0\}$).

†10.8. An isometry f is called *orientation preserving* if the Jacobian of f has positive determinant. Prove that $f: \mathbf{R}^2 \longrightarrow \mathbf{R}^2$ is an orientation preserving isometry if and only if f is a rigid motion. Do this by setting $f(u^1, u^2) = (f^1(u^1, u^2), f^2(u^1, u^2))$ and proceeding as follows.

(a) Assume that $f(0, 0) = (0, 0)$, $(\partial f^1/\partial u^1)(0, 0) = (\partial f^2/\partial u^2)(0, 0) = 1$ and $(\partial f^1/\partial u^2)(0, 0) = (\partial f^2/\partial u^1)(0, 0) = 0$. Show that

$$\left(\frac{\partial f^1}{\partial u^i}\right)^2 + \left(\frac{\partial f^2}{\partial u^i}\right)^2 = 1$$

for $i = 1$ and 2 at all points of $\mathbf{R}^2$. Differentiate this last expression to show that $\partial^2 f^k/\partial u^i\,\partial u^j = 0$ at all points in an open set about $(0, 0)$ so that $\partial f^k/\partial u^j$ is constant there. Use this to show that $f(u^1, u^2) = (u^1, u^2)$.

(b) Assume that $g: \mathbf{R}^2 \longrightarrow \mathbf{R}^2$ is an isometry. Show that there is a rigid motion $T: \mathbf{R}^2 \longrightarrow \mathbf{R}^2$ such that $f = g \circ T^{-1}$ satisfies the assumptions of part (a). (*Hint:* $\mathbf{b} = g(0, 0)$.)

(c) Use parts (a), (b) and Problem 10.7 to finish the proof.

Note that if $f: \mathbf{R}^2 \longrightarrow \mathbf{R}^2$ is an isometry that does not preserve orientation and if $S = \begin{pmatrix} 1 & 0 \\ 0 & -1 \end{pmatrix}$ is the reflection in the u^1 axis, then $S \circ f$

is an orientation preserving isometry. Hence every isometry of $\mathbf{R}^2$ is either a rigid motion or a rigid motion followed by a reflection.

†10.9. If M is a surface show that the set of (global) isometries from M to M forms a group under the operation of composition.

10.10. Let M be the surface of revolution

$$\mathbf{x}(t, \theta) = (t \cos \theta, t \sin \theta, \log t) \quad \text{for} \quad 0 < t, 0 < \theta < 2\pi$$

and let N be the helicoid

$$\mathbf{y}(t, \theta) = (t \cos \theta, t \sin \theta, \theta) \quad \text{for} \quad 0 < t, 0 < \theta < 2\pi.$$

(a) Show that the metric tensors for M and N are respectively

$$\begin{pmatrix} \dfrac{t^2+1}{t^2} & 0 \\ 0 & t^2 \end{pmatrix} \quad \text{and} \quad \begin{pmatrix} 1 & 0 \\ 0 & 1+t^2 \end{pmatrix}$$

so that the obvious function from M to N is not a local isometry.

(b) Show that the Gaussian curvature of both surfaces at (t, θ) is $K = -1/[(1 + t^2)^2]$. (Hint: Problem 8.6.)

4–11. SURFACES OF CONSTANT CURVATURE

In this last section we shall discuss certain surfaces of constant Gaussian curvature. Our goal is to determine and describe all such surfaces that are actually surfaces of revolution. This topic is a good one for the final section of this chapter because this description will require most of the ideas of this chapter. Unfortunately, it will also entail the use of integrals that are nonelementary, and hence not computable by elementary functions. We shall find that such surfaces of revolution fall into three general classes depending upon the sign of their curvature. Within each class all surfaces with the same curvature will have the same local intrinsic geometry although their global and extrinsic properties will differ; that is, each surface is locally isometric to every other surface in the class with the same curvature, but is not globally isometric to any of the others.

We recall briefly some facts about surfaces of revolution. Let

$$\boldsymbol{\alpha}(s) = (r(s), z(s))$$

be a unit speed curve defined for s in some interval (s_0, s_1) with $r(s) > 0$ and

$$M = \{(r(s) \cos \theta, r(s) \sin \theta, z(s)) \mid 0 \leq \theta \leq 2\pi, s \in (s_0, s_1)\}.$$

M is a surface and can be covered with two coordinate patches. Since $\boldsymbol{\alpha}$ is unit speed we know (Problem 3.1) that the metric matrix with respect to s and θ is

$$(g_{ij}) = \begin{pmatrix} 1 & 0 \\ 0 & r^2 \end{pmatrix}.$$

We also know (Problem 4.2) that the second fundamental form has matrix

$$(L_{ij}) = \begin{pmatrix} r'z'' - z'r'' & 0 \\ 0 & rz' \end{pmatrix}.$$

From these equations you can derive (Problem 11.1) the formula for the Gaussian curvature K of M:

$$K = -\frac{r''}{r}. \tag{11-1}$$

We shall integrate Equation (11-1) to find all surfaces of revolution of constant curvature, handling first the case of positive curvature, then zero, and finally negative curvature.

Case 1: $K = a^2$, $a > 0$.

In this situation Equation (11-1) becomes $r'' + a^2 r = 0$, which can be solved to yield $r(s) = a_1 \cos(as) + a_2 \sin(as)$ or by rearranging the constants of integration $r(s) = A \cos(as + b)$. By an appropriate choice of the point from which we measure arc length we may force $b = 0$. Since $r(s) > 0$ we must have $A > 0$ and $|s| < \pi/2a$. Since $\boldsymbol{\alpha}$ is unit speed, $z' = \pm\sqrt{1 - (r')^2}$ and we have

PROPOSITION 11.1. If M is the surface of revolution generated by the unit speed curve $\boldsymbol{\alpha}(s)$ and M has constant positive Gaussian curvature $K = a^2$, then $\boldsymbol{\alpha}(s) = (r(s), z(s))$ is given by

$$\begin{aligned} r(s) &= A\cos(as), \qquad |s| < \pi/2a \\ z(s) &= \pm\int_0^s \sqrt{1 - a^2A^2 \sin^2(at)}\, dt + C \end{aligned} \tag{11-2}$$

where $A > 0$ and C are constants.

Note that if $A > 1/a$, $z(s)$ will be defined only for

$$|s| < \frac{1}{a}\sin^{-1}\left(\frac{1}{aA}\right) < \frac{\pi}{2a}.$$

If $A \neq 1/a$, the integral giving $z(s)$ is nonelementary (it is an elliptic integral) and cannot be computed by elementary functions.

If $A = 1/a$, then $z(s) = \pm\int_0^s \cos(at)\, dt + C = \pm(1/a)\sin(as) + C$. Thus if $A = 1/a$, M is part of a sphere of radius $1/a$. The three cases $A < 1/a$, $A = 1/a$, $A > 1/a$ are pictured in Figures 4.27, 4.28, and 4.29 respectively. Note that for $A < 1/a$ the surface tends to a sharp point on the axis of revolution and that for $A > 1/a$ the limiting tangent plane for $s = \pm(1/a)\sin^{-1}(1/aA)$ is perpendicular to the axis.

We thus have a whole family of surfaces of revolution with constant curvature $a^2 > 0$. This family is parametrized by the integration constant A.

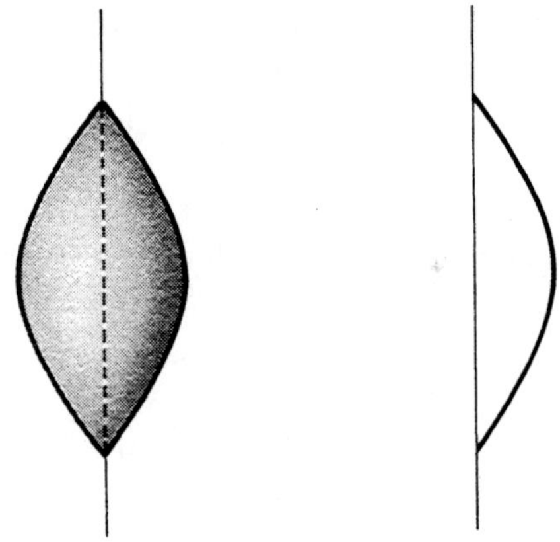

FIGURE 4.27

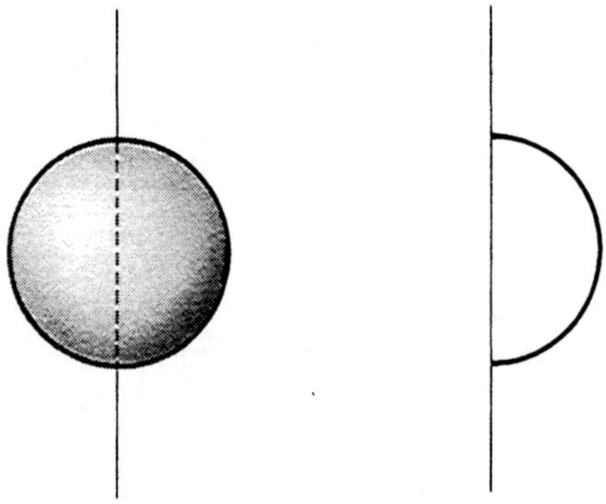

FIGURE 4.28

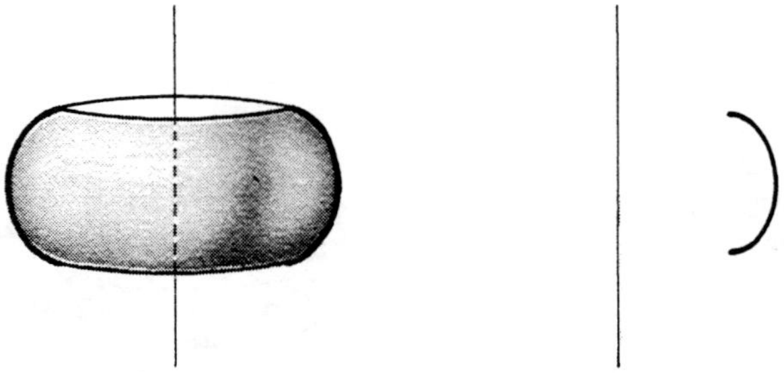

FIGURE 4.29

The extrinsic geometry varies with the parameter A because different values of A give geometrically distinct surfaces. The intrinsic geometry is independent of the parameter A (but not a).

PROPOSITION 11.2. Let M_1 and M_2 be two surfaces of revolution of unit speed curves $\boldsymbol{\alpha}_1$ and $\boldsymbol{\alpha}_2$. If both M_1 and M_2 have constant curvature $a^2 > 0$, then M_1 is locally isometric to M_2.

Proof: We shall show that there is a change of coordinates so that the metric matrix for any surface of revolution in the family (11-2) takes the form

$$\begin{pmatrix} 1 & 0 \\ 0 & \frac{1}{a^2}\cos^2(as) \end{pmatrix}.$$

The result will then follow from Proposition 10.5.

With respect to the coordinates (s, θ), the metric matrix is

$$\begin{pmatrix} 1 & 0 \\ 0 & A^2\cos^2(as) \end{pmatrix}.$$

Define new coordinates (s, ϕ) by $\phi = aA\theta$. Then the metric matrix with respect to (s, ϕ) is

$$\begin{pmatrix} 1 & 0 \\ 0 & \frac{1}{aA} \end{pmatrix}\begin{pmatrix} 1 & 0 \\ 0 & A^2\cos^2(as) \end{pmatrix}\begin{pmatrix} 1 & 0 \\ 0 & \frac{1}{aA} \end{pmatrix} = \begin{pmatrix} 1 & 0 \\ 0 & \frac{1}{a^2}\cos^2(as) \end{pmatrix}.$$

Thus locally, in terms of the coordinates (s, ϕ), all members of this family have the same metric matrix and so are locally isometric by Proposition 10.5. ∎

Thus all surfaces of revolution with constant curvature a^2 are locally isometric to the sphere of radius $1/a$. However, their global properties are different. For example, a surface of revolution given by Equation (11-2) with $A > 1/a$ cannot be extended past the boundary $|s| = (1/a)\sin^{-1}(1/aA)$ and remain a surface of revolution of curvature a^2. A path around the "equator" cannot be stretched and deformed to a single point without leaving the surface. For a sphere the equator can be so shrunk by passing it up over the north pole. This is a topological difference between the surfaces. See Figure 4.30.

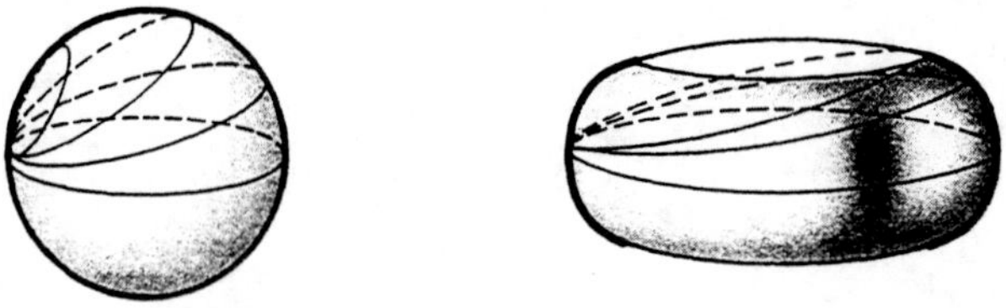

FIGURE 4.30

Case 2: $K \equiv 0$.

PROPOSITION 11.3. If M is a surface of revolution of a unit speed curve $\boldsymbol{\alpha}$ and M has constant curvature zero then M is either

(a) part of a circular cylinder;

(b) part of a plane; or

(c) part of a circular cone.

Furthermore, these surfaces are locally isometric. (See Figures 4.31, 4.32 and 4.33.)

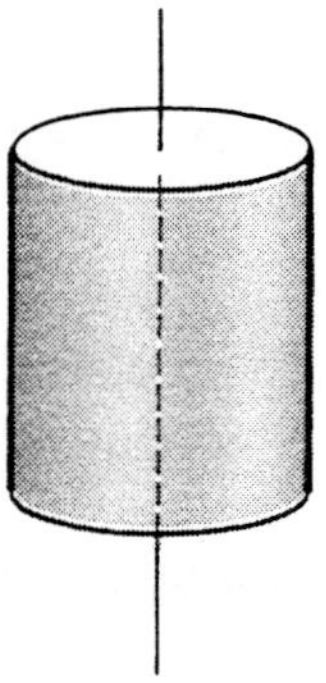

FIGURE 4.31

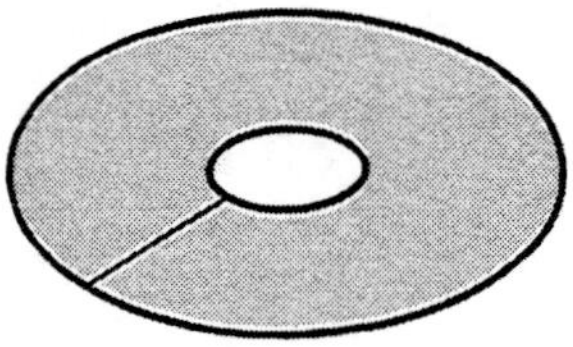

FIGURE 4.32

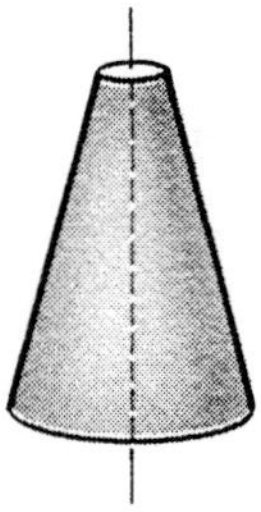

FIGURE 4.33

Proof: Equation (11-1) becomes $r'' = 0$ so that there are constants $a, b \in \mathbf{R}$ such that $r = as + b$. Since $z' = \pm\sqrt{1 - (r')^2}$, $z(s) = \pm\sqrt{1 - a^2}\, s + d$. Thus $\boldsymbol{\alpha}(s)$ is a straight line and is given by

$$\boldsymbol{\alpha}(s) = (as + b, cs + d), \quad \text{where } a^2 + c^2 = 1. \tag{11-3}$$

If $a = 0$ then M is a part of a circular cylinder, which is possibility (a). If $c = 0$, M is part of a plane (possibility (b)). If $ac \neq 0$ then $\boldsymbol{\alpha}$ is parallel to neither the r nor z axes and M is part of a circular cone (possibility (c)).

Again these surfaces are quite different extrinsically but not intrinsically. As before we could carefully describe a change of coordinates that would indicate the local intrinsic geometry is like that of the plane. However, the formulas become quite messy in the case of the cones, and we shall be content with the following heuristic description. If M is a piece of a cylinder or a cone, make a slit along one of the meridians. The surface may now be unrolled flat and laid on the plane. The placement gives rise to a coordinate patch for M with respect to which the metric matrix is the identity matrix. This is true because the process of unrolling the surface did not cause any distortions and hence did not change the length of any curve. ∎

Case 3: $K = -a^2$, $a > 0$.

Equation (11-1) becomes $r'' - a^2 r = 0$. This has the general solution (see Problem 11.5)

$$r(s) = c_1 \cosh(as) + c_2 \sinh(as). \tag{11-4}$$

This may be rewritten for some real numbers B, b or C, c as

$$r(s) = \begin{cases} B \cosh(as + b) & \text{if } c_1 > c_2 \\ A\, e^{as} & \text{if } c_1 = c_2 = A \\ C \sinh(as + c) & \text{if } c_1 < c_2. \end{cases} \tag{11-5}$$

By choosing the point from which arc length is measured we may force $b = 0$ and $c = 0$ in the above. Thus we get three basic kinds of surfaces.

$$(11\text{-}6)\qquad \begin{cases} r(s) = Ae^{as} \\ z(s) = \pm\displaystyle\int_0^s \sqrt{1 - a^2A^2e^{2at}}\,dt + D \end{cases}$$

$$(11\text{-}7)\qquad \begin{cases} r(s) = B\cosh(as) \\ z(s) = \pm\displaystyle\int_0^s \sqrt{1 - a^2B^2\sinh^2(at)}\,dt + D \end{cases}$$

$$(11\text{-}8)\qquad \begin{cases} r(s) = C\sinh(as) \\ z(s) = \pm\displaystyle\int_0^s \sqrt{1 - a^2C^2\cosh^2(at)}\,dt + D \end{cases}$$

In order that z be defined we must have $s < (1/a)\ln(1/aA)$ in Equation (11-6). By use of the substitution $aAe^{at} = \sin\phi$, the integral describing $z(s)$ may be computed. See Problem 11.4. Geometrically ϕ measures the angle between the tangent to $\boldsymbol{\alpha}$ and the z-axis. This curve is called a *tractrix* or *drag curve*. The terminology drag curve comes from the geometric description of the curve: walk along the z-axis dragging an object by a rope of length $1/a$, starting at the origin with the object at a distance $1/a$ out the r-axis. The resulting path of the object is the tractrix. The surface of revolution generated by the tractrix is called the *pseudo-sphere* of radius $1/a$. The limiting tangent plane as $s \longrightarrow (1/a)\ln(1/aA)$ is perpendicular to the axis of revolution. See Figure 4.34. The pseudo-sphere was one of the first models of non-Euclidean geometry and was discovered by E. Beltrami (1835–1900) in 1868.

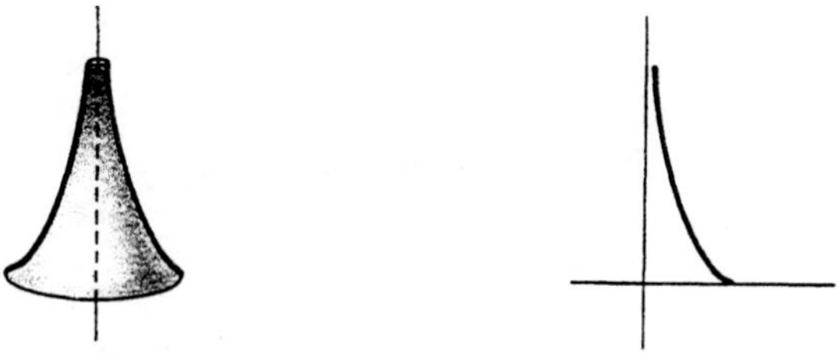

FIGURE 4.34

In Equation (11-7) we must have $|s| < (1/a)\sinh^{-1}(1/aB)$. The resulting surface is pictured in Figure 4.35. The limiting tangent planes as $s \longrightarrow \pm(1/a)\sinh^{-1}(1/aB)$ are perpendicular to the axis.

In Equation (11-8) we must have $0 < aC < 1$ and then

$$0 < s < \frac{1}{a}\cosh^{-1}\left(\frac{1}{aC}\right).$$

The resulting surface is pictured in Figure 4.36. Again the limiting tangent planes are perpendicular to the axis.

In analogy with the case of positive curvature, the surfaces of constant curvature $-a^2 < 0$ are locally isometric to the pseudo-sphere of radius $1/a$. This result will follow from a more general result to be proved in Section 6-2.

FIGURE 4.35

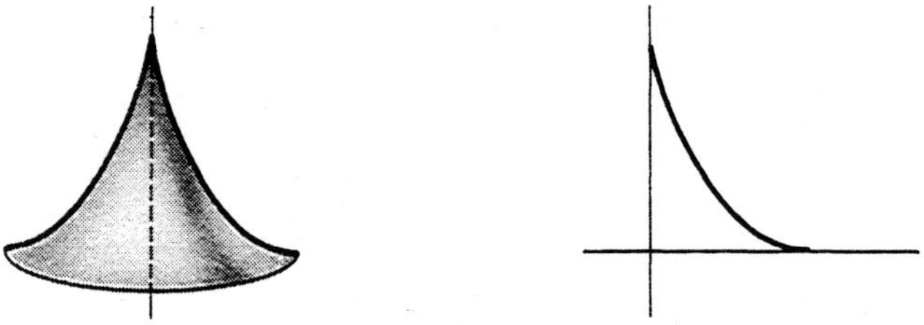

FIGURE 4.36

PROBLEMS

11.1. Let M be the surface of revolution generated by the unit speed curve $\boldsymbol{\alpha}(s) = (r(s), z(s))$ and show that $K = -r''/r$. (*Hint:* First show $K = z'(z''r' - r''z')/z$ and then use the relationship obtained by differentiating $\langle \boldsymbol{\alpha}', \boldsymbol{\alpha}' \rangle = 1$.)

11.2. In Case 1, if $A > 1/a$, verify that as $s \longrightarrow \pm(1/a)\sin^{-1}(1/aA)$ the tangent to $\boldsymbol{\alpha}$ becomes perpendicular to the z-axis. Use this to conclude that the limiting tangent planes are perpendicular to the axis as claimed.

11.3. In Case 1, if $A < 1/a$, determine the angle between $\boldsymbol{\alpha}'$ and the z-axis as $s \longrightarrow \pm\pi/2a$.

11.4. Use the trigonometric substitution $aAe^{at} = \sin\phi$ to compute $\int_0^s \sqrt{1 - a^2A^2e^{2at}}\,dt$ and then parametrize the tractrix as a function of ϕ.

11.5. Verify that if $K = -a^2$ then Equation (11-4) does give the general solution to $K = -r''/r$. (You must be able to solve linear differential equations with constant coefficients.)

11.6. Derive the basic forms of Equation (11-5) from Equation (11-4).

11.7. Verify that the limiting tangent planes for Case 3 behave as described. For the surface given by Equation (11-8) determine the angle between the meridian and the axis of revolution at the sharp point.

11.8. Prove that the surface $\mathbf{x}(s, t) = (\cos s, 2 \sin s, t)$ has constant Gaussian curvature but is not a surface of revolution. Hence the results of this section do not give all surfaces of constant curvature.

11.9. Compute the area of each of the surfaces in Cases 1 and 3 in terms of the arbitrary constants.

5

Global Theory of Space Curves

In this chapter we will prove analogues of the Rotation Index Theorem (Theorem 2.9 of Chapter 3) for space curves. For a quick overview you should reread Section 3-6 at this time. The notion of the tangent spherical image of a curve (Problems 4.15–4.21 of Chapter 2) is critical for this discussion because the total curvature of a curve is the length of the tangent spherical image of that curve. Anyone wishing to delve further into the material covered in this chapter should see Fenchel [1951], who gives a good survey of related questions.

5–1. FENCHEL'S THEOREM

DEFINITION. The *total curvature* of a regular curve $\boldsymbol{\alpha}: [0, L] \longrightarrow \mathbf{R}^3$ is $\int_0^L \kappa \, ds$.

Since $\kappa = |\mathbf{T}'|$, the total curvature of $\boldsymbol{\alpha}$ is the length of the tangent spherical image

$$\mathbf{T}: [0, L] \longrightarrow S^2 = \{(x, y, z) \in \mathbf{R}^3 \,|\, x^2 + y^2 + z^2 = 1\} = \{\mathbf{a} \in \mathbf{R}^3 \,||\, \mathbf{a}\,| = 1\}.$$

Thus it is not surprising that we need some results about sphere curves in order to prove Fenchel's Theorem on the total curvature of a space curve.

Definition. If $A, B \in S^2$, then $\widehat{AB}$ denotes the distance from A to B along any shortest geodesic (i.e., along a great circle through A and B).

Definition. The *open hemisphere with pole N* is $\{X \in S^2 \mid \widehat{NX} < \pi/2\}$. If S^2 is viewed as the set of unit vectors, the open hemisphere with pole $\mathbf{n}$ is

$$\left\{\mathbf{x} \in S^2 \mid \langle \mathbf{n}, \mathbf{x} \rangle > 0\right\} = \left\{\mathbf{x} \in S^2 \mid 0 \leq \sphericalangle(\mathbf{n}, \mathbf{x}) < \frac{\pi}{2}\right\}$$

where $\sphericalangle(\mathbf{a}, \mathbf{b})$ denotes the (positive) angle between $\mathbf{a}$ and $\mathbf{b}$.

The *closed hemisphere with pole N* is $\{X \in S^2 \mid \widehat{NX} \leq \pi/2\}$. In vector notation, the closed hemisphere with pole $\mathbf{n}$ is

$$\left\{\mathbf{x} \in S^2 \mid \langle \mathbf{n}, \mathbf{x} \rangle \geq 0\right\} = \left\{\mathbf{x} \in S^2 \mid 0 \leq \sphericalangle(\mathbf{n}, \mathbf{x}) \leq \frac{\pi}{2}\right\}.$$

Definition. If $A, B \in S^2$, then A and B are *antipodal points* if $\widehat{AB} = \pi$. In vector notation, $\mathbf{a}$ and $\mathbf{b}$ are antipodal if $\mathbf{a} = -\mathbf{b}$. (See Figure 5.1.)

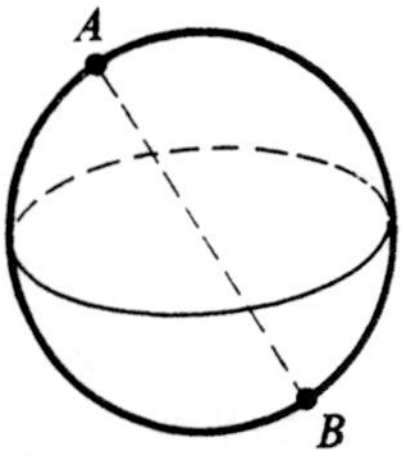

FIGURE 5.1

Lemma 1.1. If $\boldsymbol{\alpha}$ is a regular closed curve, then its tangent spherical image does not lie in any open hemisphere.

Proof: We use vector notation and give a proof by contradiction. Suppose that the tangent spherical image lies in the open hemisphere with pole $\mathbf{a}$, so that $\langle \mathbf{a}, \mathbf{T} \rangle > 0$. Then

$$0 < \int_0^L \langle \mathbf{a}, \mathbf{T} \rangle \, ds = \int_0^L \langle \mathbf{a}, \boldsymbol{\alpha}' \rangle \, ds = \int_0^L \langle \mathbf{a}, \boldsymbol{\alpha} \rangle' \, ds = \langle \mathbf{a}, \boldsymbol{\alpha} \rangle \Big|_0^L = 0$$

(since $\boldsymbol{\alpha}$ is closed), which is impossible. ∎

Corollary 1.2. The tangent spherical image of a regular closed curve does not lie in any closed hemisphere unless it lies in the great circle that bounds the hemisphere.

Proof: Problem 1.1. ∎

Note that the next lemma does *not* require that the curve $\boldsymbol{\gamma}$ be regular.

LEMMA 1.3. Let $\gamma(t)$ be a closed C^1 curve on the unit sphere S^2. The image $\mathcal{C}$ of γ is contained in an open hemisphere if either of the following two conditions hold:

(a) the length l of γ is less than 2π; or

(b) $l = 2\pi$ but the image of γ is not the union of two great semicircles.

Proof:

(a) Assume $l < 2\pi$. Let A and B be two points of $\mathcal{C}$ which divide γ into two segments of equal length $l/2$. We say A and B *bisect* γ. $\widehat{AB} \leq l/2 < \pi$ since geodesics (great circles) give the shortest distance between points. Thus A and B are not antipodal and there is a unique great circle through A and B (formed by the intersection of the sphere with the plane determined by A, B, and the center of the sphere). The shorter of the two segments of this circle has length $\widehat{AB}$. Let M be the midpoint of this segment. We will show that $\mathcal{C}$ lies in the open hemisphere with pole M. This means we must show that if X is a point on $\mathcal{C}$ then $\widehat{XM} < \pi/2$.

Since the spherical distance from X to M is a continuous function and since $\widehat{AM} < \pi/2$, if there is a point Y on $\mathcal{C}$ whose distance from M is not less than $\pi/2$, the Intermediate Value Theorem of calculus (see Fulks [1969]) implies that for each real number r with $\widehat{AM} < r < \widehat{YM}$ there is a point $Y_r \in \mathcal{C}$ with $\widehat{Y_rM} = r$. In the next paragraph we show that if $C \in \mathcal{C}$ with $\widehat{CM} < \pi/2$ then, in fact, $\widehat{CM} \leq l/4$. This says that there is no $Y_r \in \mathcal{C}$ with $\widehat{Y_rM} = r$ if $l/4 < r < \pi/2$. This contradiction of the Intermediate Value Theorem implies there is no $Y \in \mathcal{C}$ with $\widehat{YM} \geq \pi/2$.

Suppose $C \in \mathcal{C}$ with $\widehat{CM} < \pi/2$. Let $D \in S^2$ be chosen so that C, M, D all lie on the same great circle, $D \neq C$, and $\widehat{CM} = \widehat{MD}$. (See Figure 5.2.) D is called the *reflection* of C through M. Note that $\widehat{CD} = \widehat{CM} + \widehat{MD}$ since $\widehat{CM} < \pi/2$ and that $\widehat{AC} = \widehat{BD}$ since the spherical triangles AMC and BMD are congruent. Hence $2\widehat{CM} = \widehat{CM} + \widehat{MD} = \widehat{CD} \leq \widehat{CB} + \widehat{BD} = \widehat{CB} + \widehat{AC}$

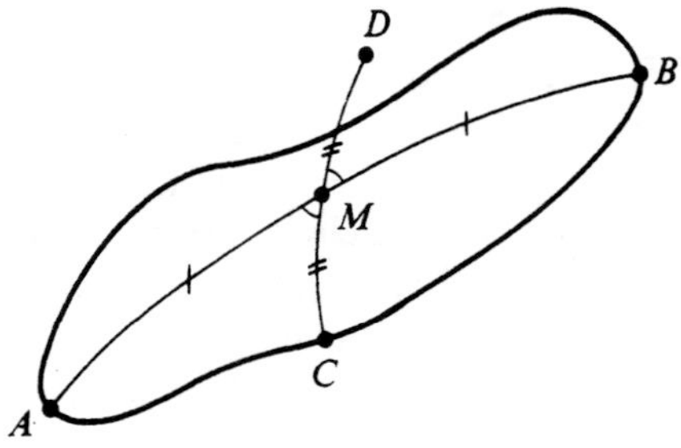

FIGURE 5.2

$\leq$ distance along γ from A to C to $B = l/2$. Thus $\widehat{CM} \leq l/4$. Hence there is no Y on the image of γ with $\widehat{YM} \geq \pi/2$ and (a) is proved.

(b) Now assume that $l = 2\pi$. This time we must choose A and B more carefully. We first show that there is no pair of antipodal points on $\mathcal{C}$. If P and Q are on $\mathcal{C}$ and are antipodal, then P and Q divide γ into two segments whose lengths are at least π since $\widehat{PQ} = \pi$. Since $l = 2\pi$, each of these segments must have length π and must give a curve of shortest length joining P to Q, and so must be a geodesic by Theorem 5.9 of Chapter 4, i.e., a great semicircle. But this means that $\mathcal{C}$ is a union of two great semicircles, which contradicts the hypothesis. Thus there are no antipodal points on $\mathcal{C}$.

Next we show that there is a pair of points A, B that bisect γ such that $\widehat{AF} + \widehat{FB} < \pi$ for all $F \in \mathcal{C}$. Suppose P, Q bisect γ. If P, Q have the desired property (i.e., if we can take $A = P$ and $B = Q$) we are done. Otherwise there is a point X on $\mathcal{C}$ with $\widehat{PX} + \widehat{XQ} = \pi$. Since the distance along γ from P to X to Q is also π, each of the two segments PX and XQ are great circle segments (geodesics). They are not part of the same great circle since P and Q are not antipodal. (See Figure 5.3.)

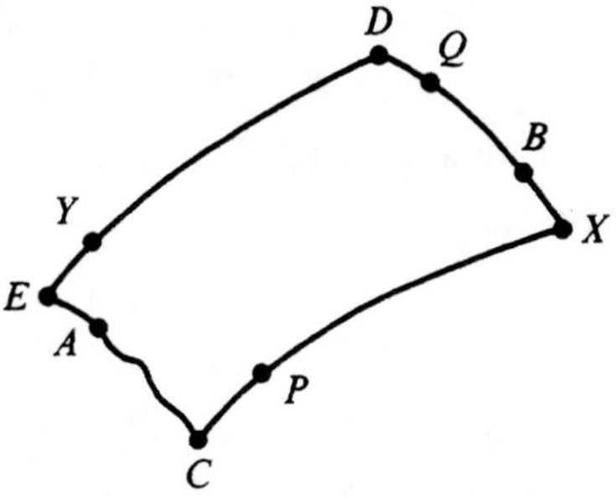

FIGURE 5.3

Consider the great circle segment XP. We know that it does not continue along $\mathcal{C}$ past X. However, it may continue along $\mathcal{C}$ past P. Let C be the point of the $\mathcal{C}$ beyond which this great circle segment does not continue in $\mathcal{C}$. Likewise the great circle segment XQ extends to a point D. $\widehat{CX} < \pi$ and $\widehat{XD} < \pi$. Let Y be chosen so that X and Y bisect γ. If X and Y serve as A and B, we are done. Otherwise there is a point R on $\mathcal{C}$ with $\widehat{XR} + \widehat{RY} = \pi$. Then the segment of γ from X to R to Y is the union of two great circle segments as before. Thus $R = C$ or $R = D$. Without the loss of generality we may assume that $R = D$. The great circle segment D to Y in $\mathcal{C}$ extends to a point E.

If $E = C$, then γ is a spherical triangle of perimeter 2π. Then the solid angle at the center O of the sphere has 2π as the sum of its face angles (see Figure 5.4). Hence it is flat and γ is a great circle, a contradiction. Thus $E \neq C$.

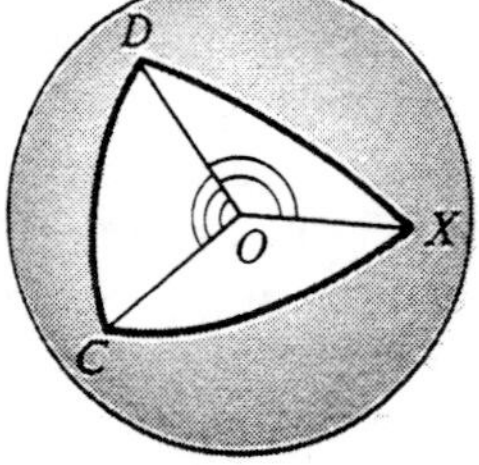

FIGURE 5.4

Let A be a point on $\mathcal{C}$ between C and E and let B be chosen so that A, B bisect γ. Note that B must be between D and X. Then no matter what point F on $\mathcal{C}$ is chosen, the segment of γ from A to F to B has either D and Y or C and X on it. Thus this segment is not the union of just two great circle segments. Hence $\widehat{AF} + \widehat{FB} < \pi$ and we have the desired A and B.

A and B are not antipodal. Let M be the midpoint of the unique great circle segment of length $\widehat{AB} < \pi$ joining A to B. If C is on $\mathcal{C}$ with $\widehat{CM} \leq \pi/2$, let D be its reflection through M as before. $2\widehat{CM} = \widehat{CM} + \widehat{MD} = \widehat{CD} \leq \widehat{CB} + \widehat{BD} = \widehat{CB} + \widehat{AC} < \pi$ by choice of A and B. Thus $\widehat{CM} < \pi/2$ and there are no points on $\mathcal{C}$ whose distance from M is $\pi/2$. Since $\widehat{AC} < \pi/2$, all points $X \in \mathcal{C}$ satisfy $\widehat{XM} < \pi/2$, or else the Intermediate Value Theorem implies there are points on $\mathcal{C}$ at a distance $\pi/2$ from M. Hence $\mathcal{C}$ lies in the open hemisphere with M as pole. ▮

THEOREM 1.4 (Fenchel, 1929). The total curvature of a closed space curve $\boldsymbol{\alpha}$ is at least 2π. It equals 2π if and only if $\boldsymbol{\alpha}$ is a plane convex curve.

Proof: Let $\boldsymbol{\gamma}$ be the tangent spherical image of $\boldsymbol{\alpha}$. Its length is $\int_0^L \kappa\, ds$. If this is less than 2π, Lemma 1.3 shows that the image of $\boldsymbol{\gamma}$ lies in an open hemisphere, which contradicts Lemma 1.1. Thus $\int_0^L \kappa\, ds \geq 2\pi$.

Now assume that $\int_0^L \kappa\, ds = 2\pi$. Then $\boldsymbol{\gamma}$ is the union of two great semicircles by Lemma 1.3. If $\boldsymbol{\gamma}$ is not a single great circle, the image of $\boldsymbol{\gamma}$ lies in a closed hemisphere but not entirely in the boundary, which contradicts Corollary 1.2. Thus $\boldsymbol{\gamma}$ is a single great circle. By Problem 4.20 of Chapter 2, $\boldsymbol{\alpha}$ is a plane curve. $\int_0^L \kappa\, ds$ measures the length of the parametrized curve $\boldsymbol{\gamma}$ and counts those segments traversed more than once with appropriate (positive) multiplicity. Since the image set of $\boldsymbol{\gamma}$ is a great circle whose length is 2π, no segment of this circle can be traversed twice, or else $\int_0^L \kappa\, ds > 2\pi$. If the plane curvature $k = d\theta/ds$ of $\boldsymbol{\alpha}$ takes on both positive and negative values, then some segment of the tangent spherical image $\boldsymbol{\gamma}$ is traversed at least twice, a contradiction. Hence k does not change sign and, by Theorem 3.2 of Chapter 3, $\boldsymbol{\alpha}$ is convex. ▮

Fenchel's Theorem is also true for piecewise smooth curves provided that the total curvature is defined as $\int \kappa\, ds + \sum \Delta\theta_i$, where the $\Delta\theta_i$ are the (positive) jump angles of $\mathbf{T}$ at the junction points. The tangent spherical image $\boldsymbol{\gamma}$ of such a curve consists of a collection of disconnected segments. These may be joined by great circle segments of lengths $\Delta\theta_i$ to get $\boldsymbol{\beta}$, whose length is the total curvature. Lemma 1.1 still holds with virtually the same proof. Lemma 1.3 holds for piecewise smooth curves, in particular $\boldsymbol{\beta}$. (A piecewise smooth curve may be parametrized to be C^1 but need not be regular or unit speed.) It is possible to have piecewise smooth convex curves, although we have not made a formal definition. The proof of Theorem 1.4 can be modified to handle this case. We have not done this because there is a tendency to get lost in the technicalities.

The proof of Fenchel's Theorem that we have given is due to B. Segre [1934] and to H. Rutishauser and H. Samelson [1948]. A proof due to K. Voss [1955] may be found in Laugwitz [1965]. This proof involves surrounding the curve with a tubular surface and relating $\int \kappa\, ds$ to $\iint_{K\geq 0} K\, dA$, where K is the Gaussian curvature of the surface. Both Fary [1949] and Milnor [1950] gave proofs of Fenchel's Theorem while studying the total curvature of knots. In Problem 2.3 you will prove the easy part ($\int \kappa\, ds \geq 2\pi$) by using Crofton's Formula.

Fenchel's Theorem is also true for curves in $\mathbf{R}^n$, as has been proved by Borsuk [1948]. In fact, it was Borsuk who first conjectured the Fary-Milnor Theorem.

EXAMPLE 1.5. We verify Fenchel's Theorem in the special case where $\mathcal{C}$ is the ellipse $\boldsymbol{\alpha}(t) = (2\cos t, \sin t, 0)$ for $0 \leq t \leq 2\pi$. If s is arc length on $\mathcal{C}$,

$$\int_{\mathcal{C}} \kappa\, ds = \int_0^{2\pi} \kappa(s(t)) \frac{ds}{dt}\, dt = 4 \int_0^{\pi/2} \kappa(s(t)) \frac{ds}{dt}\, dt$$

because of the symmetry of the ellipse. Now

$$\frac{ds}{dt} = |\dot{\boldsymbol{\alpha}}| = \sqrt{4\sin^2 t + \cos^2 t}.$$

By the results of Section 2-6, $\kappa(s(t)) = 2/\sqrt{4\sin^2 t + \cos^2 t}^{\,3}$. Hence

$$\int_{\mathcal{C}} \kappa\, ds = 4 \int_0^{\pi/2} \frac{2dt}{4\sin^2 t + \cos^2 t} = 4 \int_0^{\pi/2} \frac{2\sec^2 t\, dt}{1 + 4\tan^2 t}.$$

Let $u = 2\tan t$ so that $du = 2\sec^2 t\, dt$.

$$\int_{\mathcal{C}} \kappa\, ds = 4 \int_0^{\infty} \frac{du}{1+u^2} = 4\tan^{-1} u \Big|_0^{\infty} = 4\frac{\pi}{2} = 2\pi.$$

Note that $\mathcal{C}$ *is* plane and convex.

PROBLEMS

***1.1.** Prove Corollary 1.2.

1.2. Let $\boldsymbol{\alpha}(s)$ be a closed space curve. Suppose $0 \leq \kappa \leq 1/R$ for some real number $R > 0$. Prove that the length of $\boldsymbol{\alpha}$ is at least $2\pi R$.

1.3. Compute the tangent spherical image for the ellipse in Example 1.5. What does the proof of Fenchel's Theorem say about this image?

1.4. Consider the piecewise regular plane nonconvex curve $\boldsymbol{\alpha}$ in Figure 5.5 made up of three circular segments of radius 2. Compute the total curvature of $\boldsymbol{\alpha}$.

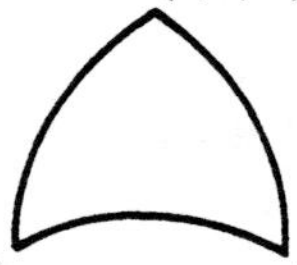

FIGURE 5.5

5–2. THE FARY-MILNOR THEOREM

Suppose that $\boldsymbol{\omega}$ is an *oriented* great circle on S^2. Then there is a unique point $W \in S^2$ associated with $\boldsymbol{\omega}$, namely the pole of the hemisphere to the left as $\boldsymbol{\omega}$ is traversed. (See Figure 5.6.) Conversely every point of S^2 is associated with some oriented great circle. Thus the set of oriented great circles is in one-to-one correspondence with the points of S^2. The orientation is what allows this to be one-to-one—it gives a well defined choice of a pole (from two possibilities) for each oriented great circle. If the north pole corresponds to the equator with one orientation, the south pole corresponds to the equator with the opposite orientation.

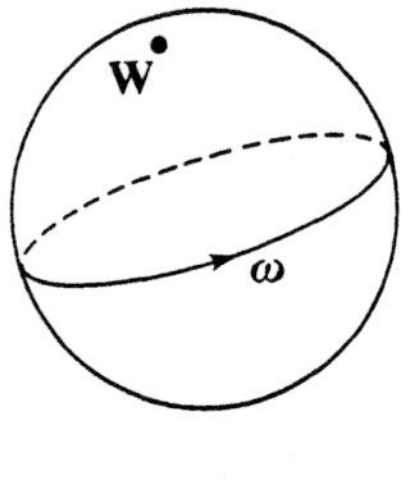

FIGURE 5.6

DEFINITION. The *measure* of a set of oriented great circles (counted with multiplicity) is the area of the corresponding subset of S^2 (counted with multiplicity).

If $W \in S^2$, let $W^\perp$ be the associated great circle. For a regular curve $\boldsymbol{\gamma}$ with image set $\mathcal{C}$, let $n_\gamma(W)$ be the number of points in $\mathcal{C} \cap W^\perp$ (which might be infinite). Note that $n_\gamma(W)$ does not depend upon the parametrization of $\boldsymbol{\gamma}$. We shall usually omit the subscript $\boldsymbol{\gamma}$ in $n_\gamma(W)$ if no confusion will result.

THEOREM 2.1 (Crofton's Formula). Let $\mathcal{C}$ be the image of a regular curve $\boldsymbol{\gamma}(t)$ on S^2 of length l. The measure of the set of oriented great circles that intersect $\mathcal{C}$, counted with multiplicity, is $4l$.

Proof: Since neither $n(W)$ nor l depends on the parametrization, we may assume that $\boldsymbol{\gamma}$ is unit speed. Let $S = \{W \in S^2 \mid W^\perp \cap \mathcal{C} \neq \varnothing\}$. The measure of the set of oriented great circles that intersect $\mathcal{C}$ is the area of S; that is, $\iint_S dA$. We wish to count the circles with multiplicity—if $\mathcal{C} \cap W^\perp$ has three points, we want to count $W^\perp$ three times. This measuring is done by the integral $\iint_S n(W)\, dA$. We are to prove $\iint_S n(W)\, dA = 4l$.

Let $\mathbf{a}(t) = \boldsymbol{\gamma}(t)$, $\mathbf{b}(t) = d\boldsymbol{\gamma}/dt$ and $\mathbf{c}(t) = \mathbf{a} \times \mathbf{b}$. Then analogously to the Frenet-Serret equations, for some function $\lambda(t)$ we have the equations

$$\begin{aligned} \mathbf{a}' &= \quad\quad \mathbf{b} \\ \mathbf{b}' &= -\mathbf{a} \quad\quad\quad + \lambda(t)\mathbf{c} \\ \mathbf{c}' &= \quad\quad - \lambda(t)\mathbf{b}. \end{aligned}$$

(See Problem 2.1.) $\mathbf{a}$, $\mathbf{b}$, and $\mathbf{c}$ give an orthonormal basis of $\mathbf{R}^3$ for each $t \in [0, l]$. If $W^\perp$ meets $\mathcal{C}$ at $\boldsymbol{\gamma}(t_0)$, then the unit vector $\mathbf{w}$ which represents the pole of $W^\perp$ is orthogonal to $\boldsymbol{\gamma}(t_0) = \mathbf{a}(t_0)$. Hence

$$\mathbf{w} = -\sin\phi\, \mathbf{b}(t_0) + \cos\phi\, \mathbf{c}(t_0),$$

where ϕ is the angle between the tangent vector $\mathbf{b}$ of $\boldsymbol{\gamma}$ and the tangent vector of $W^\perp$ at $\boldsymbol{\gamma}(t_0)$.

Define a map $\mathbf{w}\colon [0, l) \times [0, 2\pi) \longrightarrow S^2$ by

$$\mathbf{w}(t, \phi) = -\sin\phi\, \mathbf{b}(t) + \cos\phi\, \mathbf{c}(t).$$

The image of $\mathbf{w}$ is S. This makes S a parametrized surface. Despite the fact that $\mathbf{w}$ might not be regular (i.e., $\mathbf{w}_1 \times \mathbf{w}_2$ might be zero sometimes), the area of the parametrized surface S is given by

$$\int_0^{2\pi} \int_0^l \left| \frac{\partial \mathbf{w}}{\partial t} \times \frac{\partial \mathbf{w}}{\partial \phi} \right| dt\, d\phi,$$

just as for a regular surface (Equation (8-1) of Chapter 4). However, this is not the area of the *set* S, but rather the area counted with multiplicity ($\mathbf{w}$ need not be one-to-one either). Now, if $Y \in S$, precisely $n(Y)$ points in $[0, l) \times [0, 2\pi)$ are mapped to Y. Thus

$$\iint_{Y \in S} n(Y)\, dA = \int_0^{2\pi} \int_0^l \left| \frac{\partial \mathbf{w}}{\partial t} \times \frac{\partial \mathbf{w}}{\partial \phi} \right| dt\, d\phi.$$

This last equality can also be derived from Jacobi's rule for change of variables (Fulks [1969]). By Problem 2.2 we have

$$\left|\frac{\partial \mathbf{w}}{\partial t} \times \frac{\partial \mathbf{w}}{\partial \phi}\right| = |\sin \phi|.$$

Thus

$$\iint_S n(Y)\, dA = \int_0^{2\pi} \int_0^l |\sin \phi|\, dt\, d\phi = l \int_0^{2\pi} |\sin \phi|\, d\phi$$

$$= 4l \int_0^{\pi/2} \sin \phi\, d\phi = 4l. \quad ∎$$

COROLLARY 2.2. Crofton's Formula holds for piecewise smooth curves.

Proof: Use Theorem 2.1 on each segment of the curve and add. ∎

DEFINITION. $D^2 = \{(x, y) \in R^2 \mid x^2 + y^2 \leq 1\}$ is the *closed unit disk*. $S^1 = \{(x, y) \in R^2 \mid x^2 + y^2 = 1\}$ is the *unit circle*.

DEFINITION. A simple closed curve $\boldsymbol{\alpha}$ is *unknotted* if there is a one-to-one continuous function $\mathbf{g}\colon D^2 \longrightarrow \mathbf{R}^3$ that sends S^1 onto the image of $\boldsymbol{\alpha}$. Otherwise $\boldsymbol{\alpha}$ is *knotted*.

THEOREM 2.3 (Fary-Milnor). If $\boldsymbol{\alpha}$ is a simple knotted regular curve, then the total curvature of $\boldsymbol{\alpha}$ is at least 4π.

Proof: We shall prove the contrapositive: $\int \kappa\, ds < 4\pi$ implies $\boldsymbol{\alpha}$ is unknotted. Assume $\int \kappa\, ds < 4\pi$. Let $\boldsymbol{\gamma}$ be the tangent spherical image of $\boldsymbol{\alpha}$. Then $\frac{1}{4}\iint_{S^2} n_\gamma(W)\, dA = l = \int \kappa\, ds < 4\pi$. Then for some point $Y \in S^2$, $n(Y) < 4$, otherwise $\frac{1}{4}\iint n(W)\, dA \geq \frac{1}{4}4 \cdot 4\pi = 4\pi$. Let $\mathbf{y}$ be the unit vector representing Y and set $f(s) = \langle \mathbf{y}, \boldsymbol{\alpha}(s)\rangle$. Then $f'(s) = \langle \mathbf{y}, \mathbf{T}(s)\rangle = \langle \mathbf{y}, \boldsymbol{\gamma}(s)\rangle$ and $f'(s) = 0$ if and only if $Y^\perp$ intersects $\mathcal{C}$ at $\boldsymbol{\gamma}(s)$. Thus $f(s)$ has exactly $n(Y) < 4$ critical points. There is an absolute maximum, an absolute minimum, and possibly an inflection point (if $n(Y) = 3$). There are *no* relative extrema other than the absolute extrema.

We may assume that coordinates for $\mathbf{R}^3$ have been chosen so that $\mathbf{y} = (0, 0, 1)$. If M and m are the maximum and minimum values of f and $m < r < M$, the plane $z = r$ intersects $\boldsymbol{\alpha}$ in exactly two points, or else the Mean Value Theorem of calculus would imply some more relative extrema. Join these two points with a straight line segment. The totality of these segments gives the image of a one-to-one continuous function $\mathbf{g}\colon D^2 \longrightarrow \mathbf{R}^3$ making $\boldsymbol{\alpha}$ unknotted. See Figure 5.7. ∎

The above proof is essentially that outlined by Fenchel [1951]. The theorem holds for piecewise regular curves, but this proof does not work. (Why?) Milnor [1950] proved the theorem by approximating the curve with

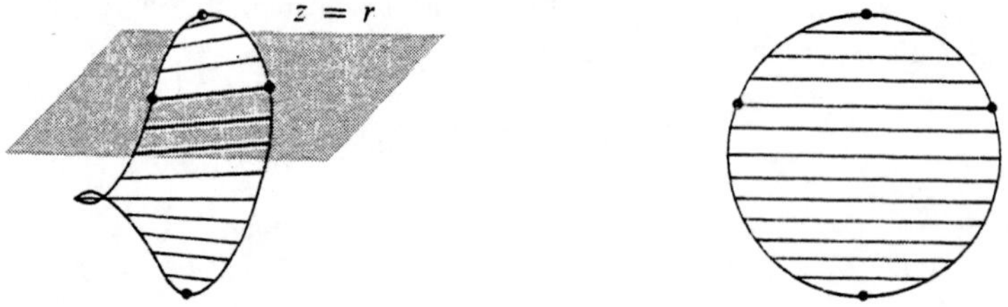

FIGURE 5.7

piecewise regular curves made up of straight segments. He used very geometric methods to compute the total curvature and then took the necessary limits. Fary [1949] also approximated $\boldsymbol{\alpha}$ by piecewise linear curves. One important step in his proof was the result that the total curvature of $\boldsymbol{\alpha}$ is the average of the total curvature of the projections of $\boldsymbol{\alpha}$ into the planes through a fixed point. Voss [1955] also outlines a proof.

PROBLEMS

***2.1.** Prove that $\mathbf{a}' = \mathbf{b}$, $\mathbf{b}' = -\mathbf{a} + \lambda\mathbf{c}$, $\mathbf{c}' = -\lambda\mathbf{b}$, as claimed in the proof of Theorem 2.1.

***2.2.** Prove that $|(\partial\mathbf{w}/\partial t) \times (\partial\mathbf{w}/\partial\phi)| = |\sin\phi|$ as claimed in the proof of Theorem 2.1.

2.3. Use Crofton's Formula to prove that $\int \kappa\, ds \geq 2\pi$ for all closed regular curves.

2.4. Let $\boldsymbol{\alpha}(s)$ be the curve on S^2 given by $\boldsymbol{\alpha}(s) = (\cos s, \sin s, 0)$, $0 \leq s \leq 2\pi$. Verify that Crofton's Formula is true for this curve.

5–3. TOTAL TORSION

After considering the total curvature of a space curve it would seem natural to investigate the integral $\int \tau\, ds$. Results here are not as strong as for $\int \kappa\, ds$. In fact, if $r \in R$, there is a closed curve $\boldsymbol{\alpha}$ with $\int \tau\, ds = r$. To see this, one need consider curves made up of a circular helix with the ends joined, as in Figure 5.8. By varying the pitch, the number of coils, and the right- or left-handedness of the helix, any desired total torsion may be found.

Despite this, there are some results on total torsion worth studying. Recall that $\rho = 1/\kappa$ is the radius of curvature and that $\sigma = 1/\tau$ is the radius of torsion of the curve $\boldsymbol{\alpha}$.

THEOREM 3.1. The *total torsion*, $\int \tau\, ds$, of a closed unit speed curve $\boldsymbol{\alpha}(s)$ on S^2 is zero.

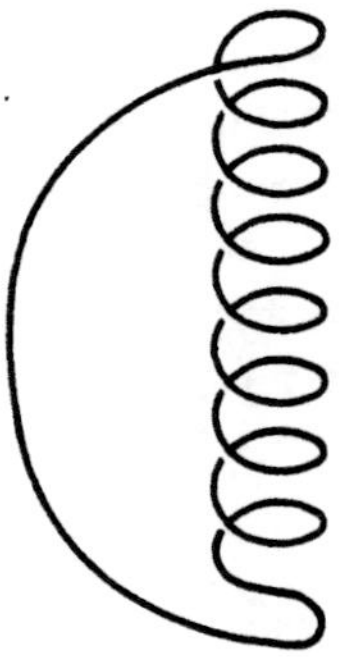

FIGURE 5.8

Proof: $\boldsymbol{\alpha}(s) = \lambda\mathbf{T} + \mu\mathbf{N} + \nu\mathbf{B}$. Since $\boldsymbol{\alpha}$ is a unit vector,

$$\lambda = \langle \boldsymbol{\alpha}, \mathbf{T} \rangle = \tfrac{1}{2}\langle \boldsymbol{\alpha}, \boldsymbol{\alpha} \rangle' = 0.$$

Thus $\boldsymbol{\alpha} = \mu\mathbf{N} + \nu\mathbf{B}$. $\mathbf{T} = \boldsymbol{\alpha}' = \mu'\mathbf{N} + \mu(-\kappa\mathbf{T} + \tau\mathbf{B}) + \nu'\mathbf{B} + \nu(-\tau\mathbf{N})$. Thus $\mathbf{T} = -\mu\kappa\mathbf{T} + (\mu' - \tau\nu)\mathbf{N} + (\mu\tau + \nu')\mathbf{B}$. Hence $1 = -\mu\kappa$ or $\mu = -1/\kappa = -\rho$, and $\mu' - \tau\nu = 0$. Since $\langle \boldsymbol{\alpha}, \boldsymbol{\alpha} \rangle = 1$, $\mu^2 + \nu^2 = 1$ and

$$\nu = \pm\sqrt{1 - \mu^2} = \pm\sqrt{1 - \rho^2}.$$

Hence $-\rho' = \pm\tau\sqrt{1 - \rho^2}$ and $\tau = \pm\rho'/\sqrt{1 - \rho^2}$ (when $\rho \neq 1$). If $\rho < 1$ always so that $\rho'/\sqrt{1 - \rho^2}$ always makes sense, and if a constant choice of sign in $\tau = \pm\rho'/\sqrt{1 - \rho^2}$ could be made, then $\tau = \pm\rho'/\sqrt{1 - \rho^2}$ everywhere. In this case

$$\int_0^L \tau\, ds = \pm\int_0^L \frac{\rho'}{\sqrt{1 - \rho^2}}\, ds = \pm\sin^{-1} \rho(s)\Big|_0^L = 0$$

since $\rho(L) = \rho(0)$. However, this may not be true. In fact, we shall see below that $\rho = 1$ whenever the geodesic curvature of $\boldsymbol{\alpha}$ is zero, and the choice of sign depends on the sign of the geodesic curvature.

We have $\boldsymbol{\alpha}'' = \kappa_n\mathbf{n} + \kappa_g\mathbf{S}$ where $\mathbf{S} = \mathbf{n} \times \mathbf{T}$. If we choose the inward pointing normal to S^2 as $\mathbf{n}$, then $\mathbf{n} = -\boldsymbol{\alpha}$ and $\kappa_n = +1$.

$$\begin{aligned}\kappa_g &= \langle \mathbf{S}, \boldsymbol{\alpha}'' \rangle = \langle \mathbf{n} \times \mathbf{T}, \boldsymbol{\alpha}'' \rangle = -\langle \boldsymbol{\alpha} \times \mathbf{T}, \mathbf{T}' \rangle \\ &= -\langle \boldsymbol{\alpha}, \mathbf{T} \times \kappa\mathbf{N} \rangle = -\kappa\langle \boldsymbol{\alpha}, \mathbf{B} \rangle.\end{aligned}$$

On the other hand, $\langle \boldsymbol{\alpha}, \mathbf{B} \rangle = -\rho'\sigma$ so that $\kappa_g = \kappa\rho'\sigma$ or $\tau = \kappa\rho'/\kappa_g$ if $\tau \neq 0$. Since $\kappa^2 = \kappa_n{}^2 + \kappa_g{}^2 = 1 + \kappa_g{}^2$, we have $\kappa_g = \pm\sqrt{\kappa^2 - 1}$ and $\kappa_g/\kappa = \pm\sqrt{1 - \rho^2}$. Thus $\tau = \kappa\rho'/\kappa_g = \rho'/(\kappa_g/\kappa) = \pm\rho'/\sqrt{1 - \rho^2}$ and the sign $\pm$ is that of κ_g. Note also that $\rho = 1$ if and only if $\kappa = 1$, which is true if and only if $\kappa_g = 0$. Thus $\tau = \pm\rho'/\sqrt{1 - \rho^2}$ whenever $\kappa_g \neq 0$ and the sign is that of κ_g.

To simplify the proof we shall assume there are a finite number of points $0 = s_0 < s_1 < \ldots < s_n = L$ where κ_g is zero and such that on (s_{i-1}, s_i) κ_g is either strictly positive, strictly negative, or identically zero. (This finiteness need not be true. In order to handle the general situation one needs some topology and a more sophisticated concept of integration. However, the basic idea is the same.)

$\int_0^L \tau\, ds = \sum \int_{s_{i-1}}^{s_i} \tau\, ds$. At each s_i, κ_g is zero, $\kappa = 1$ and thus $\rho(s_j) = 1$. If $\kappa_g > 0$ on (s_{i-1}, s_i), then $\tau = \rho'/\sqrt{1-\rho^2}$ and $\int_{s_{i-1}}^{s_i} \tau\, ds$ is the improper integral

$$\int_{s_{i-1}}^{s_i} \frac{\rho'}{\sqrt{1-\rho^2}}\, ds = \sin^{-1}(\rho(s))\Big|_{s_{i-1}}^{s_i} = \left(\frac{\pi}{2} - \frac{\pi}{2}\right) = 0.$$

Likewise if $\kappa_g < 0$ on (s_{i-1}, s_i), then $\int_{s_{i-1}}^{s_i} \tau\, ds = 0$. If $\kappa_g \equiv 0$ on (s_{i-1}, s_i) then $\kappa \equiv 1$ on (s_{i-1}, s_i). Problem 5.3 of Chapter 2 implies that this segment of α is part of a circle and hence is a plane curve. Thus $\tau \equiv 0$ on $[s_{i-1}, s_i]$ and $\int_{s_{i-1}}^{s_i} \tau\, ds = 0$. Therefore $\int_0^L \tau\, ds = 0$. ∎

W. Scherrer [1940] has proved the converse of this theorem: if M is a surface in $\mathbf{R}^3$ such that $\int \tau\, ds = 0$ for *all* closed curves in M, then M is part of a plane or part of a sphere. B. Segre [1947*b*] has studied the total absolute torsion $\int |\tau|\, ds$ and has obtained some results on lower bounds on the integral.

PROBLEM

3.1. Prove that $\int (\tau/\kappa)\, ds = 0$ for any closed unit speed sphere curve. (This result and its converse is due to B. Segre [1947*a*].)

6

Global Theory of Surfaces

After the brief hiatus of Chapter 5 we return once again to surfaces. We now study results about surfaces which are of a global nature—that is, they do not just depend on the behavior of a surface in a neighborhood of a point but rather on the surface as a whole. These results are very much in the spirit of modern research in differential geometry. We get quite unexpected relationships between quantities that seem totally different (e.g., Euler characteristic and total curvature), especially mixing results of a topological nature with those of a geometric one. One word of caution: the concepts in this chapter are more difficult than those of Chapter 4 because we are dealing with global concepts. Note how much more difficult Chapters 3 and 5 were than was Chapter 2.

In the first section we define basic topological concepts and prove some simple results to give the reader the flavor of the kinds of theorems that we are aiming at. Sections 6-2 and 6-3 are rather technical, giving some results (geodesic coordinate patches) and introducing concepts (orientability and angular variation) which will be needed to prove the Gauss-Bonnet Formula in Section 6-4. This formula is truly beautiful because it relates the total geodesic curvature of a curve to the sum of the jump angles of the curve and the total Gaussian curvature of the area it encloses—seemingly unrelated things! The last three sections prove corollaries of the Gauss-Bonnet Formula, all of which are important theorems in global differential geometry (the Gauss-Bonnet Theorem, Hadamard's Theorem, Jacobi's Theorem, and the Poincaré-Brouwer Theorem). It is startling relationships of this sort between unexpectedly related quantities like these which give global differential geometry its fascination.

6–1. SIMPLE CURVATURE RESULTS

In this section we will be concerned with some simple results relating the curvature of the surface to its "topology." The first theorem (Theorem 1.2) shows how a topological assumption (compactness) yields a geometric conclusion (that the curvature must be positive at some point). The second theorem (Theorem 1.3) shows how a geometric assumption (all points are umbilics) can yield an extremely strong conclusion (M is a sphere).

DEFINITION. If $P \in \mathbf{R}^3$, then $|P|$ denotes the distance from P to the origin O of $\mathbf{R}^3$. (This agrees with the vector notation: if $\mathbf{p}$ is the vector from O to P, $|P| = |\mathbf{p}|$.)

DEFINITION. Let $\bar{B}_r = \{P \in \mathbf{R}^3 \mid |P| \leq r\}$. $M \subset \mathbf{R}^3$ is *bounded* if there is an $r > 0$ such that $M \subset \bar{B}_r$. M is *closed* if for each sequence $\{P_n\}$ such that $P_n \in M$ and $\lim_{n\to\infty} P_n = P$ exists (in $\mathbf{R}^3$), P is actually in M.

The ball $B = \{P \in \mathbf{R}^3 \mid |P| < 1\}$ is bounded but not closed. The (x, y) plane in $\mathbf{R}^3$ is closed but not bounded. Observe that if M is bounded, then (1) there is a minimum r such that $M \subset \bar{B}_r$, and (2) if $\{P_n\}$ is a sequence in M, then $\{P_n\}$ must contain a subsequence which converges in $\mathbf{R}^3$.

DEFINITION. If $M \subset \mathbf{R}^3$ is closed and bounded, then M is *compact*.

LEMMA 1.1. Let M be a compact surface in $\mathbf{R}^3$ and r the minimum number such that $M \subset \bar{B}_r$. Then there is a point $P \in M$ with $|P| = r$, i.e., $M \cap S_r \neq \varnothing$, where S_r is the sphere of radius r.

Proof: Let $r_n = r - 1/n$ for $n > 0$. By the choice of r we have $M - \bar{B}_{r_n} \neq \varnothing$. Let $P_n \in M - \bar{B}_{r_n}$. Since M is bounded and $P_n \in M$, there is a subsequence (which we also name P_n) which converges to $P \in \mathbf{R}^3$. Since M is closed, $P \in M$. Clearly, $\lim |P_n| = r$ so that $|P| = r$. ▮

THEOREM 1.2. Let M be a compact surface in $\mathbf{R}^3$. Then there is a point $P \in M$ such that the Gaussian curvature at P is positive.

Proof: Let r be the minimum number such that $M \subset \bar{B}_r$. By Lemma 1.1 we know that the sphere S_r of radius r and center O satisfies $S_r \cap M \neq \varnothing$. We will show that M has positive curvature at all points of $S_r \cap M$. Let $P \in S_r \cap M$. M and S_r have the same unit normal $\mathbf{n}$ at P, or else there is a point of M outside S_r. (This argument is similar to that about parallel lines in the proof of Proposition 3.2 of Chapter 3.) Hence M and S_r have the same tangent plane at P also. Let $\mathbf{X}$ be tangent to both M and S_r at P. Let Π be the plane spanned by $\mathbf{n}$ and $\mathbf{X}$. The normal curvature κ_n of M in the direction $\mathbf{X}$ is the curvature (with respect to $\mathbf{n}$) of the curve $\Pi \cap M$. This has the same sign as the normal curvature of S_r at P, which is $\pm 1/r$. Clearly $|\kappa_n| \geq 1/r$. If κ_1, κ_2 are the principal curvatures of M at P we have $K = \kappa_1\kappa_2 \geq 1/r^2 > 0$. See Figure 6.1. ∎

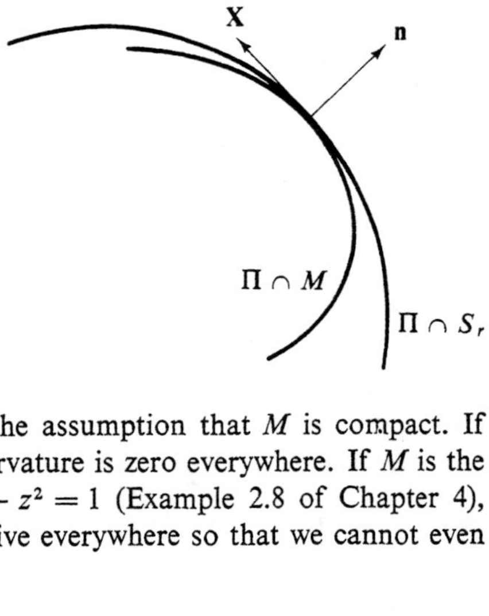

FIGURE 6.1

This theorem is false without the assumption that M is compact. If $M = \mathbf{R}^2 \subset \mathbf{R}^3$, then the Gaussian curvature is zero everywhere. If M is the hyperboloid of revolution $x^2 + y^2 - z^2 = 1$ (Example 2.8 of Chapter 4), then the Gaussian curvature is negative everywhere so that we cannot even hope for $K = 0$ at some point.

THEOREM 1.3 (Meusnier, 1785). Let M be a compact connected surface all of whose points are umbilics. Then M is a sphere.

Proof: We first prove that the curvature is constant in any coordinate patch. Let $\mathbf{x}\colon \mathcal{U} \to \mathbf{R}^3$ be a coordinate patch of M. Then since every direction is principal at an umbilic, $(L^i{}_j) = \begin{pmatrix} \kappa_1 & 0 \\ 0 & \kappa_1 \end{pmatrix}$. Weingarten's equations ((7-6) of Chapter 4) are thus $\mathbf{n}_i = -\kappa_1 \mathbf{x}_i$. Hence $\mathbf{n}_{ij} = -(\partial\kappa_1/\partial u^j)\mathbf{x}_i - \kappa_1 \mathbf{x}_{ij}$ and $\mathbf{0} = \mathbf{n}_{12} - \mathbf{n}_{21} = (\partial\kappa_1/\partial u^1)\mathbf{x}_2 - (\partial\kappa_1/\partial u^2)\mathbf{x}_1$. Since $\mathbf{x}_1$ and $\mathbf{x}_2$ are independent, $\partial\kappa_1/\partial u^i = 0$ and κ_1 is constant. Hence $K = \kappa_1\kappa_1$ is constant.

In overlapping patches K must have the same (constant) value. Thus K is constant on M. By Theorem 1.2, $K > 0$ and $\kappa_1 \neq 0$. In a patch $\mathbf{x}$ we have $\mathbf{n} = \int \mathbf{n}_1\, du^1 = -\kappa_1 \int \mathbf{x}_1\, du^1 = -\kappa_1(\mathbf{x} - \mathbf{c})$. Take the length of each side of this equation to obtain $1 = K\langle \mathbf{x} - \mathbf{c}, \mathbf{x} - \mathbf{c}\rangle$ so that the patch is contained in a sphere of radius $1/\sqrt{K}$ centered at $\mathbf{c}$. An overlapping patch must be contained in a sphere of radius $1/\sqrt{K}$ and centered at some point $\mathbf{d}$. The overlap of the two patches is a simple surface and lies in both spheres. Hence $\mathbf{c} = \mathbf{d}$. Thus M is contained in a sphere. Using compactness, we see that M is the entire sphere. ▮

PROBLEMS

1.1. If M is a surface in $\mathbf{R}^3$ each of whose points is an umbilic, prove that each point of M has a coordinate patch whose image is either an open subset of a plane or an open subset of a sphere. (*Hint:* Examine the proof of Theorem 1.3 carefully.) What invariant differentiates between these two cases?

1.2. Parametrize the level surface $x^2 + y^2 - z^2 = 1$ as a surface of revolution. Compute its Gaussian curvature and show that it is negative at each point.

6–2. GEODESIC COORDINATE PATCHES

In this section we will show that any point of a surface M is contained in a special kind of coordinate patch, called a geodesic coordinate patch. This kind of patch is very useful for the theoretical computations involved in proving the Gauss-Bonnet Theorem. (See Sections 6-4 and 6-5.)

DEFINITION. If $\mathbf{x}: \mathcal{U} \longrightarrow \mathbf{R}^3$ is a coordinate patch such that $g_{11} \equiv 1$ and $g_{12} \equiv 0$ (hence $g_{21} \equiv 0$) in $\mathcal{U}$, then $\mathbf{x}$ is called a *geodesic coordinate patch*. If, in addition, there is a curve $\boldsymbol{\gamma}$ on M defined on $[a, b]$ such that $\boldsymbol{\gamma}([a, b]) \subset \mathbf{x}(\mathcal{U})$ and such that the u^2-curve through a point of the image of $\boldsymbol{\gamma}$ is $\boldsymbol{\gamma}$ itself, then $\mathbf{x}$ is called a *geodesic coordinate patch along* $\boldsymbol{\gamma}$.

In Problem 5.5 of Chapter 4 we saw that the u^1-curves in a geodesic coordinate patch are geodesics. We will obtain a geodesic coordinate patch by starting with any nonclosed curve $\boldsymbol{\alpha}$ and constructing geodesics perpendicular to $\boldsymbol{\alpha}$ which fill out a neighborhood. (See Figure 6.2.) This will mean that the u^1-curves are geodesics and the u^2-curve through any point of $\boldsymbol{\alpha}$ is $\boldsymbol{\alpha}$.

PROPOSITION 2.1. Let M be a surface and let $\boldsymbol{\alpha}: [a, b] \longrightarrow M$ be a simple regular curve which is not closed. Then there is a geodesic coordinate patch $\mathbf{x}: \mathcal{U} \longrightarrow \mathbf{R}^3$ of M along $\boldsymbol{\alpha}$.

Proof: We may assume that $\boldsymbol{\alpha}$ is defined on (c, d) with $c < a \leq b < d$. Choose $e, f \in \mathbf{R}$ so that $c < e < a \leq b < f < d$. For each t, let $\mathbf{X}(t) = -\mathbf{S},$

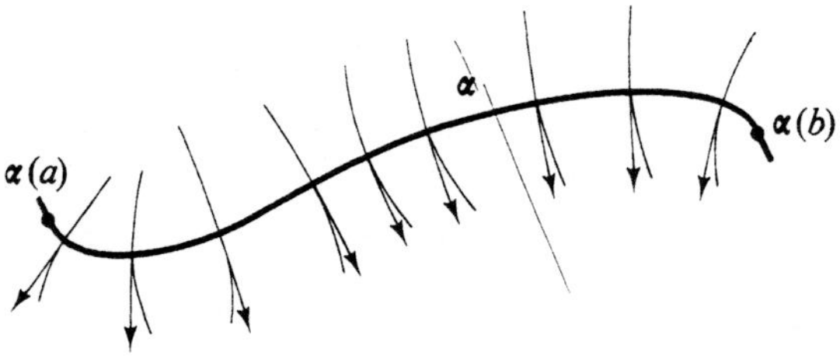

FIGURE 6.2

where $\mathbf{S}$ is the intrinsic normal $\mathbf{n} \times \dot{\boldsymbol{\alpha}}/|\dot{\boldsymbol{\alpha}}|$ and $\mathbf{n}$ is a continuous choice of a unit normal to M along $\boldsymbol{\alpha}$. (Such a choice of $\mathbf{n}$ can be made because $\boldsymbol{\alpha}$ is not closed.) For each $t \in [e, f]$ let $\boldsymbol{\alpha}_t(r)$ be the unique geodesic satisfying $\boldsymbol{\alpha}_t(0) = \boldsymbol{\alpha}(t)$ and $(d\boldsymbol{\alpha}_t/dr)(0) = \mathbf{X}(t)$. Let $\mathbf{x}(r, t) = \boldsymbol{\alpha}_t(r)$. $\mathbf{x}$ is of class C^3 because of the smooth dependence of the solution of differential equations upon a smoothly varying parameter t. (Recall that a geodesic satisfies a differential equation: Equation (5-1) of Chapter 4.) Because $[e, f]$ is a closed interval, there is a $\delta > 0$ such that $\mathbf{x}$ is one-to-one on $[-\delta, \delta] \times [e, f]$.

$$\mathbf{x}_1(0, t) = \mathbf{X}(t), \qquad \mathbf{x}_2(0, t) = d\boldsymbol{\alpha}/dt.$$

$\mathbf{x}_1 \times \mathbf{x}_2(0, t) \neq \mathbf{0}$ since $\mathbf{X}$ and $d\boldsymbol{\alpha}/dt$ are orthogonal and nonzero. Thus there is an $\epsilon > 0$ with $\epsilon \leq \delta$ such that $\mathbf{x}_1 \times \mathbf{x}_2 \neq \mathbf{0}$ on $[-\epsilon, \epsilon] \times [e, f]$. Then $\mathbf{x}\colon (-\epsilon, \epsilon) \times (e, f) \longrightarrow \mathbf{R}^3$ is a C^3 coordinate patch containing $\boldsymbol{\alpha}([a, b])$.

$g_{11} \equiv 1$ since $\mathbf{x}_1 = \partial\boldsymbol{\alpha}_t/\partial r$ and $\boldsymbol{\alpha}_t(r)$ is a geodesic with r arc length.

$$\frac{\partial g_{12}}{\partial r} = \frac{\partial}{\partial r}\langle \mathbf{x}_1, \mathbf{x}_2 \rangle = \langle \mathbf{x}_{11}, \mathbf{x}_2 \rangle + \langle \mathbf{x}_1, \mathbf{x}_{21} \rangle$$

or

$$\text{(2-1)} \qquad \frac{\partial g_{12}}{\partial r} = \langle \mathbf{x}_{11}, \mathbf{x}_2 \rangle + \langle \mathbf{x}_1, \mathbf{x}_{12} \rangle.$$

Now $\mathbf{x}_{11}$ equals $\partial^2\boldsymbol{\alpha}_t/\partial r^2$, which is normal to M since $\boldsymbol{\alpha}_t$ is a geodesic. Thus $\langle \mathbf{x}_{11}, \mathbf{x}_2 \rangle = 0$. Since $\langle \mathbf{x}_1, \mathbf{x}_1 \rangle = 1$, we have $2\langle \mathbf{x}_{12}, \mathbf{x}_1 \rangle = 0$ and hence, by Equation (2-1), $\partial g_{12}/\partial r = 0$. This means that g_{12} is constant with respect to r so that $g_{12}(r, t) = g_{12}(0, t) = \langle \dot{\boldsymbol{\alpha}}, \mathbf{X} \rangle = 0$. Therefore $g_{12} \equiv 0$ and $\mathbf{x}$ is a geodesic coordinate patch (with $u^1 = r$ and $u^2 = t$). The u^2-curve through $\boldsymbol{\alpha}(a) = \boldsymbol{\alpha}_a(0) = \mathbf{x}(0, a)$ is $\mathbf{x}(0, a + t) = \boldsymbol{\alpha}_{a+t}(0) = \boldsymbol{\alpha}(a + t)$, i.e., is $\boldsymbol{\alpha}$. ∎

We note that the first fundamental form in the coordinate patch has as its matrix $\begin{pmatrix} 1 & 0 \\ 0 & h^2 \end{pmatrix}$, where $0 < h = |\mathbf{x}_2|$.

In the special case where $\boldsymbol{\alpha}$ in Proposition 2.1 is a (unit speed) geodesic, the function $h = |\mathbf{x}_2|$ has two important properties. In the first place, $h(0, u^2) = |\mathbf{x}_2(0, u^2)| = |\boldsymbol{\alpha}'(u^2)| = 1$. Secondly, since $\boldsymbol{\alpha}$ is a geodesic,

$\mathbf{x}_{22}(0, u^2) = \boldsymbol{\alpha}''(u^2)$ is normal to the surface. Thus

$$\frac{\partial(h^2)}{\partial u^1} = \frac{\partial}{\partial u^1}\langle \mathbf{x}_2, \mathbf{x}_2\rangle = 2\langle \mathbf{x}_{21}, \mathbf{x}_2\rangle = 2\langle \mathbf{x}_{12}, \mathbf{x}_2\rangle$$

$$= 2\left(\frac{\partial}{\partial u^2}\langle \mathbf{x}_1, \mathbf{x}_2\rangle - \langle \mathbf{x}_1, \mathbf{x}_{22}\rangle\right) = -2\langle \mathbf{x}_1, \mathbf{x}_{22}\rangle.$$

Hence at $u^1 = 0$ we have $(\partial(h^2)/\partial u^1)(0, u^2) = 0$ and hence $(\partial h/\partial u^1)(0, u^2) = 0$.

These results, that $h(0, u^2) = 1$ and $(\partial h/\partial u^1)(0, u^2) = 0$ if $\boldsymbol{\alpha}$ is a geodesic, are very useful in proving the next theorem, which gives a nice application of geodesic coordinate patches.

THEOREM 2.2. Any two surfaces of the same constant Gaussian curvature $K \equiv c$ are locally isometric.

Proof: Suppose M and N both satisfy $K \equiv c$. Let $P \in M$, $\mathbf{X} \in T_PM$, $Q \in N$, and $\mathbf{Y} \in T_QM$, where $\mathbf{X}$ and $\mathbf{Y}$ are unit vectors. Let $\boldsymbol{\alpha}\colon (-\eta, \eta) \longrightarrow M$ and $\boldsymbol{\gamma}\colon (-\eta, \eta) \longrightarrow N$ be the unique geodesics such that $\boldsymbol{\alpha}(0) = P$, $\boldsymbol{\alpha}'(0) = \mathbf{X}$, $\boldsymbol{\gamma}(0) = Q$, and $\boldsymbol{\gamma}'(0) = \mathbf{Y}$. Construct geodesic coordinate patches $\mathbf{x}\colon(-\epsilon, \epsilon) \times (-\eta, \eta) \longrightarrow M$ and $\mathbf{y}\colon(-\epsilon, \epsilon) \times (-\eta, \eta) \longrightarrow N$ along $\boldsymbol{\alpha}$ and $\boldsymbol{\gamma}$ respectively, denoting the parameter in $(-\epsilon, \epsilon)$ by t and that in $(-\eta, \eta)$ by s.

The metric matrices for $\mathbf{x}$ and $\mathbf{y}$ are

$$(g_{ij}) = \begin{pmatrix} 1 & 0 \\ 0 & e^2 \end{pmatrix} \quad \text{and} \quad (h_{ij}) = \begin{pmatrix} 1 & 0 \\ 0 & f^2 \end{pmatrix},$$

where e and f satisfy

$$e(0, s) = f(0, s) = 1 \quad \text{and} \quad \frac{\partial e}{\partial t}(0, s) = \frac{\partial f}{\partial t}(0, s) = 0.$$

According to Problem 2.3, the Gaussian curvature of $\mathbf{x}$ and $\mathbf{y}$ are given by $K = (-\partial^2 e/\partial t^2)/e$ and $K = (-\partial^2 f/\partial t^2)/f$, respectively. Hence, for each fixed value of s, e and f both solve the initial value problem

$$\text{(2-2)} \qquad \begin{gathered} \ddot{u} = -cu, \\ u(0) = 1, \quad \dot{u}(0) = 0. \end{gathered}$$

By Picard's Theorem, Equation (2-2) has a unique solution. In fact, it is given by

$$\text{(2-3)} \qquad u(t) = \begin{cases} \cos(\sqrt{c}\, t) & \text{if } c > 0 \\ 1 & \text{if } c = 0 \\ \cosh(\sqrt{-c}\, t) & \text{if } c < 0. \end{cases}$$

In particular, $e = f$ and there exist coordinate patches $\mathbf{x}\colon \mathcal{U} \longrightarrow M$, $\mathbf{y}\colon \mathcal{U} \longrightarrow N$ such that the metric matrices are the same. Hence, M and N are locally isometric by Proposition 10.5 of Chapter 4. ∎

It should be noted that the above proof gives more than is required. In fact, it shows that for surfaces of the same constant curvature if there are given any two points P and Q in M and N and geodesics $\boldsymbol{\alpha}$ and $\boldsymbol{\gamma}$ through P and Q, then there is a local isometry f of a neighborhood $\mathcal{U}'$ of P with a neighborhood $\mathcal{V}'$ of Q such that $f(P) = Q$ and $\boldsymbol{\gamma} = f \circ \boldsymbol{\alpha}$. In particular, if $M = N$ there is a local isometry mapping P to Q (a generalized "translation") and if $P = Q$ there is a local isometry sending a given geodesic through P to a second given geodesic through P (a generalized "rotation").

EXAMPLE 2.3. The standard parametrization of a surface of revolution is a geodesic coordinate patch along a circle of latitude.

PROBLEMS

***2.1.** Prove the assertion in Example 2.3.

2.2. Suppose the curve $\boldsymbol{\alpha}$ used in Proposition 2.1 is a geodesic. Prove that along this curve $(g_{ij}) = (\delta_{ij})$. Are the other u^2-curves geodesics, or even unit speed, in general?

***2.3.** Let $\mathbf{x}$ be a geodesic coordinate patch with $(g_{ij}) = \begin{pmatrix} 1 & 0 \\ 0 & h^2 \end{pmatrix}$ and $h > 0$. Show that $\Gamma_{12}{}^2 = \Gamma_{21}{}^2 = h_1/h$, $\Gamma_{22}{}^1 = -hh_1$, $\Gamma_{22}{}^2 = h_2/h$ and all other $\Gamma_{ij}{}^k$ are zero, where $h_i = \partial h/\partial u^i$. Show that $K = -h_{11}/h$.

2.4. (Hyperbolic Half Plane.) Suppose there is a simple surface $\mathbf{x}\colon \mathcal{U} \longrightarrow \mathbf{R}^3$ with

$$\mathcal{U} = \{(x, y) \in \mathbf{R}^2 \mid y > 0\} \quad \text{and} \quad (g_{ij}) = \begin{pmatrix} \dfrac{1}{y^2} & 0 \\ 0 & \dfrac{1}{y^2} \end{pmatrix}.$$

Show that $\Gamma_{11}{}^2 = -\Gamma_{12}{}^1 = -\Gamma_{21}{}^1 = -\Gamma_{22}{}^2 = 1/y$ and all other $\Gamma_{ij}{}^k$ are zero. Compute K. (See the comment below.)

2.5. (Poincaré Disk.) Suppose there is a simple surface $\mathbf{x}\colon \mathcal{U} \longrightarrow \mathbf{R}^3$ with

$$\mathcal{U} = \{(x, y) \in \mathbf{R}^2 \mid x^2 + y^2 < 1\}$$

and

$$(g_{ij}) = \begin{pmatrix} \dfrac{4}{(1 - x^2 - y^2)^2} & 0 \\ 0 & \dfrac{4}{(1 - x^2 - y^2)^2} \end{pmatrix}.$$

Show that

$$\Gamma_{22}{}^2 = \Gamma_{12}{}^1 = \Gamma_{21}{}^1 = -\Gamma_{11}{}^2 = \frac{2y}{1 - x^2 - y^2}$$

and

$$\Gamma_{11}{}^1 = \Gamma_{12}{}^2 = \Gamma_{21}{}^2 = -\Gamma_{22}{}^1 = \frac{2x}{1 - x^2 - y^2}.$$

Compute K.

Comment: These last two exercises are a bit misleading because there is no function $\mathbf{x}$ giving the desired properties. They are examples of abstract surfaces. Both are models of hyperbolic geometry, just as the sphere is a model of elliptic geometry. We shall return to them below. See also Chapter 7.

6–3. ORIENTABILITY AND ANGULAR VARIATION

Let $\gamma(t)$ be a piecewise regular simple closed curve in a surface M with period L. Let $\mathbf{Z}(t)$ be a continuous vector field along γ which is differentiable along the regular segments of $\boldsymbol{\gamma}$. In general $\mathbf{Z}(0) \neq \mathbf{Z}(L)$. We shall be interested in the angle between these two vectors. In the plane we measured angles counterclockwise from the x-axis. In our more general setting we do not have an "x-axis" to use as a reference. We must therefore find some reference "line" from which to compute angles as well as to define angles themselves. Given any two vectors, the cosine of the angle between them is well defined, but the angle itself is not. (Recall the problems we had in Chapter 3 in defining the rotation index.) We need a concept analogous to "counterclockwise" in the plane. We will do that first by defining orientation.

DEFINITION. A surface M is *orientable* if there is a continuous function $\nu: M \longrightarrow S^2$ with $\nu(P)$ normal to M at P for all $P \in M$.

Thus for an orientable surface the Gauss map is globally defined.

EXAMPLE 3.1. S^2 is orientable; let $\nu(P)$ be the outward pointing normal at P. $\nu(P) = \mathbf{p}$, where $\mathbf{p}$ is the vector from the origin to P.

EXAMPLE 3.2. The torus T^2 (see Problem 1.1 of Chapter 4) is orientable; let $\nu(P)$ be the outward pointing normal at P. See Figure 6.3.

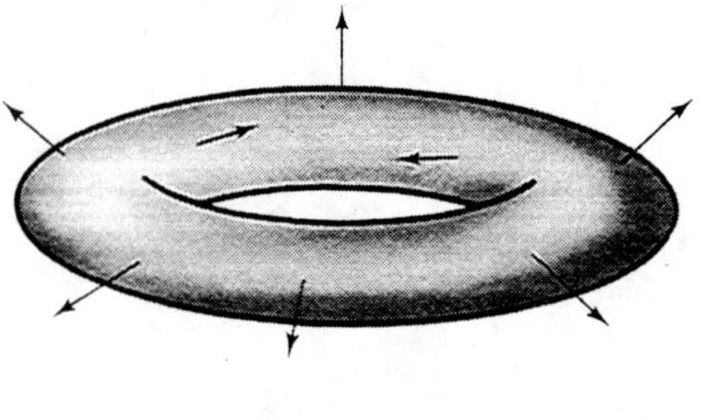

FIGURE 6.3

EXAMPLE 3.3. The Möbius band is not orientable. See Figure 6.4 and Problem 2.5 of Chapter 4.

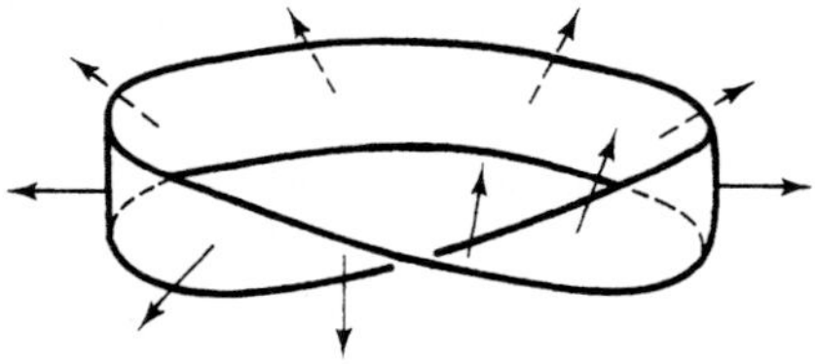

FIGURE 6.4

EXAMPLE 3.4. Recall the example of a surface mentioned between Examples 2.4 and 2.5 in Chapter 4. Let $f: \mathbf{R}^3 \to \mathbf{R}$ be a differentiable function and $(\nabla f)(P) = (f_x, f_y, f_z)$ be the gradient of f (where the derivatives are all evaluated at P). We mentioned that $M = \{P \in \mathbf{R}^3 \mid f(P) = 0\}$ is a surface if $(\nabla f)(P) \neq \mathbf{0}$ for all $P \in M$. M is called a *level surface of* f. A level surface of f is always orientable. (Just take $\nu(P) = \nabla f(P)/|\nabla f(P)|$.) This example subsumes most examples we will run into in practice. Level surfaces are treated in depth in Chapter 7.

The following is a very deep topological result and is beyond the scope of this book. A recent proof may be found in Samelson [1969].

THEOREM 3.5. Every compact surface in $\mathbf{R}^3$ is orientable.

An orientation (choice of $\mathbf{n}$ globally) aids in measuring angles. If $\mathbf{X}$ and $\mathbf{Y}$ are tangent to M at P, the angle from $\mathbf{X}$ to $\mathbf{Y}$ is defined up to sign by $\cos\theta = \langle \mathbf{X}, \mathbf{Y}\rangle/|\mathbf{X}||\mathbf{Y}|$. The sign of θ is that of $[\mathbf{X}, \mathbf{Y}, \mathbf{n}]$. This corresponds to counterclockwise, or the right-hand rule. We write $\sphericalangle(\mathbf{X}, \mathbf{Y})$ for the angle θ from $\mathbf{X}$ to $\mathbf{Y}$.

DEFINITION. A subset $\mathcal{R}$ of a surface M is a *region* in M if $\mathcal{R}$ is open and if any two points of $\mathcal{R}$ may be joined by a curve in $\mathcal{R}$. If $\mathcal{R}$ is a region in M, then the *boundary* of $\mathcal{R}$, bd $\mathcal{R}$, is the set $\{P \in M \mid P \notin \mathcal{R}$ and there is a sequence $\{P_n\}$ in $\mathcal{R}$ with $\lim P_n = P\}$. A curve $\boldsymbol{\gamma}$ *bounds* a region $\mathcal{R}$ if the image of $\boldsymbol{\gamma}$ is the boundary of $\mathcal{R}$ and $\mathbf{S}$ points into $\mathcal{R}$ at all points of $\boldsymbol{\gamma}$ while $-\mathbf{S}$ points out, where $\mathbf{S}$ is the intrinsic normal of $\boldsymbol{\gamma}$.

EXAMPLE 3.6. Let $M = \mathbf{R}^2$. $\mathcal{R} = \{(x, y, 0) \in \mathbf{R}^2 \mid x^2 + y^2 < 1\}$ is a region. The curve $\boldsymbol{\alpha}(\theta) = (\cos\theta, \sin\theta, 0)$ bounds $\mathcal{R}$.

EXAMPLE 3.7. Let M be the torus T^2 and let $\mathcal{R}$ be all the points of T^2 except those on the image of $\boldsymbol{\gamma}$, as in Figure 6.5. $\mathcal{R}$ is a region and the boundary of $\mathcal{R}$ is the image of $\boldsymbol{\gamma}$ but $\boldsymbol{\gamma}$ does not bound $\mathcal{R}$ since both $\mathbf{S}$ and $-\mathbf{S}$ point into $\mathcal{R}$.

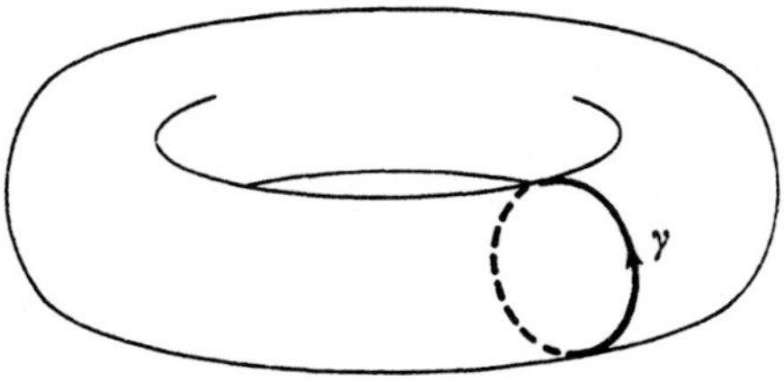

FIGURE 6.5

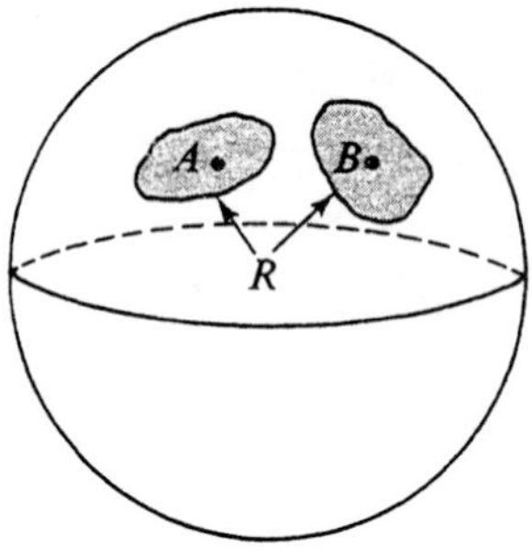

FIGURE 6.6

EXAMPLE 3.8. Let M be S^2 and let $\mathcal{R}$ be the set pictured in Figure 6.6. $\mathcal{R}$ is not a region because there is no curve in $\mathcal{R}$ joining A to B.

Now in order to measure the total angular change of $\mathbf{Z}$ along $\boldsymbol{\gamma}$ we need to replace the x-axis. Suppose $\boldsymbol{\gamma}$ bounds a region $\mathcal{R}$. It can be shown that there is a region $\mathcal{S}$ in M containing $\mathcal{R}$ and the image of $\boldsymbol{\gamma}$ and a field of unit vectors in $\mathcal{S}$, i.e., an assignment of a unit tangent vector $\mathbf{V}(P)$ to each point $P \in \mathcal{S}$, which is differentiable. (See also Problem 3.1.)

Let $\alpha(t) = \sphericalangle(\mathbf{V}(\boldsymbol{\gamma}(t)), \mathbf{Z}(t))$ with α continuous, and hence differentiable. Once $\mathbf{V}$ is fixed, α is unique up to an integral multiple of 2π and so $d\alpha/dt$ depends only on $\mathbf{Z}$.

DEFINITION. The *total angular variation* of $\mathbf{Z}$ along $\boldsymbol{\gamma}$ with respect to $\mathbf{V}$ is $\delta_{\mathbf{V}}\alpha = \int_0^L (d\alpha/dt)\, dt$.

In general the total angular variation depends on $\mathbf{V}$. However, if the curve $\boldsymbol{\gamma}$ can be continuously shrunk in $\mathcal{R}$ to a point, we will show that $\delta_{\mathbf{V}}\alpha$ is independent of $\mathbf{V}$. In this case we shall write $\delta\alpha$ for $\delta\sphericalangle(\mathbf{V}(\boldsymbol{\gamma}(t)), \mathbf{Z}(t))$ for any choice of the unit vector field $\mathbf{V}$. We shall now make the idea of "shrinking" precise.

DEFINITION. Let $\boldsymbol{\gamma}$ be a closed curve which bounds a region $\mathcal{R}$. Let $\boldsymbol{\sigma}$ be any closed curve of period L which is either $\boldsymbol{\gamma}$ or lies in $\mathcal{R}$. Let $\boldsymbol{\sigma}(0) = \mathbf{x}_0$. $\boldsymbol{\sigma}$ is *null-homotopic* in $\mathcal{R}$ if for each $s \in [0, 1]$ there is a closed curve $\boldsymbol{\sigma}_s$ in M such that

(a) $\boldsymbol{\sigma}_s(0) = \mathbf{x}_0$;
(b) $\boldsymbol{\sigma}_0(t) = \boldsymbol{\sigma}(t)$ and $\boldsymbol{\sigma}_1(t) = \mathbf{x}_0$ (i.e., $\boldsymbol{\sigma}_1$ is the constant curve);
(c) $\boldsymbol{\sigma}_s(t) \in \mathcal{R}$ for all $0 < s \leq 1$ and $t \in (0, L)$; and
(d) the function $\boldsymbol{\Gamma}\colon [0, L] \times [0, 1] \longrightarrow M$ given by $\boldsymbol{\Gamma}(t, s) = \boldsymbol{\sigma}_s(t)$ is continuous.

The $\boldsymbol{\sigma}_s$ represent the shrinking and the fourth condition is a precise statement that the shrinking is continuous. Note that the third condition says that each of the curves $\boldsymbol{\sigma}_s$ must lie in $\mathcal{R}$ if $s > 0$ (except possibly for the end point $\mathbf{x}_0$). The notion of homotopy is discussed more fully in Singer and Thorpe [1967].

EXAMPLE 3.9. Let $M = \mathbf{R}^2$ and $\boldsymbol{\gamma}(t) = (\cos t, \sin t)$, $0 \leq t \leq 2\pi$. Then $\mathcal{R} = \{(x, y) \mid x^2 + y^2 < 1\}$. That $\boldsymbol{\gamma}$ is null-homotopic is clear from Figure 6.7. More formally, if $\boldsymbol{\sigma}_s(t) = (1 - s)\boldsymbol{\gamma}(t) + (s, 0)$, then $\boldsymbol{\sigma}_s$ satisfies the definition above.

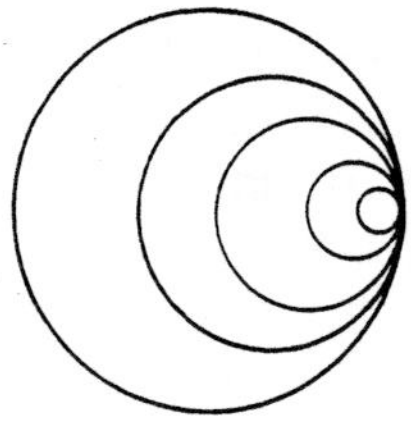

FIGURE 6.7

EXAMPLE 3.10. Let $\boldsymbol{\gamma}(t)$ be as above with $\mathcal{R} = \{(x, y) \mid 0 < x^2 + y^2 < 1\}$ and $M = \mathbf{R}^2 - \{(0, 0)\}$. Then $\boldsymbol{\gamma}(t)$ is not null-homotopic. This is intuitively clear because $\boldsymbol{\gamma}(t)$ goes around the hole (i.e., the origin) and so cannot be shrunk through it. Notice that an attempt to mimic the procedure of Example 3.9 fails because $\boldsymbol{\sigma}_{1/2}(\pi) = (0, 0)$ *which is not in* $\mathcal{R}$ and so violates condition (*c*).

Because the proof of the next lemma is very technical we shall only sketch it here.

LEMMA 3.11. If $\boldsymbol{\gamma}$ bounds the region $\mathcal{R}$ and is null-homotopic in $\mathcal{R}$, then $\delta_{\mathbf{V}}\alpha$ does not depend on the choice of $\mathbf{V}$.

Proof: Let $\mathbf{W}$ be another field of unit vectors.

$$\alpha = \measuredangle(\mathbf{V}, \mathbf{Z}) = \measuredangle(\mathbf{V}, \mathbf{W}) + \measuredangle(\mathbf{W}, \mathbf{Z}) + 2\pi n = \theta + \beta + 2\pi n.$$

Hence $d\alpha/dt = d\theta/dt + d\beta/dt$. We need only show that

$$\int_0^L \frac{d\theta}{dt}\, dt = \delta\theta = 0.$$

Since $\mathbf{V}(0) = \mathbf{V}(L)$ and $\mathbf{W}(0) = \mathbf{W}(L)$, $\delta\theta = 2\pi r$ for some integer r.

To say that $\boldsymbol{\gamma}$ is null-homotopic means we have a family of curves $\boldsymbol{\sigma}_s(t)$ for $0 \leq s \leq 1$ with $\boldsymbol{\sigma}_0 = \boldsymbol{\gamma}$, $\boldsymbol{\sigma}_{1-\epsilon}$ = a very small circle, $\boldsymbol{\sigma}_s(t)$ continuous in s and t, and $\boldsymbol{\sigma}_s(t)$ piecewise differentiable in t. Then $\delta\theta_s = \int_{\boldsymbol{\sigma}_s} (d\theta/dt)\, dt$ is defined and continuous in s. Since it is always an integral multiple of 2π it must be constant. But along a very small circle **V** and **W** can't change much relative to each other and so $\delta\theta_{1-\epsilon} = 0$. Then $\delta\theta = \delta\theta_0 = \delta\theta_{1-\epsilon} = 0$ and $\delta_{\mathbf{V}}\alpha = \delta_{\mathbf{W}}\beta$. ∎

EXAMPLE 3.12. Let M be S^2 and take $\boldsymbol{\gamma}$ to be the circle of latitude $\phi = \phi_0$ so that $\boldsymbol{\gamma}(\theta) = (\sin\phi_0 \cos\theta, \sin\phi_0 \sin\theta, \cos\phi_0)$. We work in the coordinate patch $\mathbf{x}(\phi, \theta) = (\sin\phi\cos\theta, \sin\phi\sin\theta, \cos\phi)$. Let

$$\mathbf{Z}(\theta) = \cos((\cos\phi_0)\theta)\mathbf{x}_1 - \frac{\sin((\cos\phi_0)\theta)}{\sin\phi_0}\mathbf{x}_2$$

as in Example 6.8 of Chapter 4. Recall that $\mathbf{Z}(\theta)$ is parallel along $\boldsymbol{\gamma}$ with $\mathbf{Z}(0) = \mathbf{x}_1$. Let **V** be a field of unit vectors on S^2 defined everywhere except at the south pole S (see Problem 3.2). We will compute $\delta\alpha = \delta\sphericalangle(\mathbf{V}, \mathbf{Z})$ and show that $\delta\alpha = \iint_{\mathcal{R}} K\,dA$ where $\mathcal{R}$ is the cap of the sphere bounded by $\boldsymbol{\gamma}$ and K is the Gaussian curvature. The fact that $\delta\alpha = \iint_{\mathcal{R}} K\,dA$ is a special case of the Gauss-Bonnet Formula, which we prove in Section 6-4.

Note that

$$\delta\alpha = \delta\sphericalangle(\mathbf{V}, \dot{\boldsymbol{\gamma}}) + \delta\sphericalangle(\dot{\boldsymbol{\gamma}}, \mathbf{Z}). \tag{3-1}$$

We first compute $\delta\sphericalangle(\dot{\boldsymbol{\gamma}}, \mathbf{Z})$. Now $\dot{\boldsymbol{\gamma}} = \mathbf{x}_2(\phi_0, \theta)$ and $g_{22} = \sin^2\phi$ so that

$$\cos(\sphericalangle(\dot{\boldsymbol{\gamma}}, \mathbf{Z})) = \frac{\langle\dot{\boldsymbol{\gamma}}, \mathbf{Z}\rangle}{|\dot{\boldsymbol{\gamma}}||\mathbf{Z}|} = \frac{-\sin((\cos\phi_0)\theta)g_{22}}{\sin\phi_0\sqrt{g_{22}}}$$

$$= -\sin((\cos\phi_0)\theta) = -\cos\left(\frac{\pi}{2} - (\cos\phi_0)\theta\right).$$

Thus

$$\sphericalangle(\dot{\boldsymbol{\gamma}}, \mathbf{Z}) = \frac{3\pi}{2} - (\cos\phi_0)\theta$$

and

$$\delta\sphericalangle(\dot{\boldsymbol{\gamma}}, \mathbf{Z}) = \int_0^{2\pi} \frac{d}{d\theta}\left(\frac{3\pi}{2} - (\cos\phi_0)\theta\right) d\theta = -2\pi\cos\phi_0.$$

Clearly $\delta\sphericalangle(\mathbf{V}, \dot{\boldsymbol{\gamma}}) = 2\pi n$ for some integer n. $\boldsymbol{\gamma}$ is null-homotopic in $\mathcal{R}$, and for a very small circle $\boldsymbol{\beta}$ it is clear that $\delta\sphericalangle(\mathbf{V}, \dot{\boldsymbol{\beta}}) = 2\pi$. Hence $n = 1$ and $\delta\sphericalangle(\mathbf{V}, \dot{\boldsymbol{\gamma}}) = 2\pi$. By Equation (3-1)

$$\delta\alpha = 2\pi - 2\pi\cos\phi_0 = 2\pi(1 - \cos\phi_0) = 2\pi\int_0^{\phi_0} \sin\phi\,d\phi$$

$$= \int_0^{2\pi}\int_0^{\phi_0} \sin\phi\,d\phi\,d\theta = \iint_{\mathcal{R}} dA = \iint_{\mathcal{R}} K\,dA$$

since $K \equiv 1$ on S^2.

PROBLEMS

***3.1.** Prove that if γ bounds a region $\mathcal{R}$ that is entirely contained in a single coordinate patch, then a field of unit vectors exists in $\mathcal{R}$.

***3.2.** Prove there is a field of unit vectors defined on all of the sphere except the south pole. (*Hint:* Problem 1.10 of Chapter 4.)

3.3. In Example 6.12 of Chapter 4 we considered parallel translating a vector along a certain piecewise regular curve. Show that the angular variation equals the area of the region surrounded in that case.

6–4. THE GAUSS-BONNET FORMULA

DEFINITION. A region of a surface is *simply connected* if every closed curve in that region is null-homotopic.

EXAMPLE 4.1. Any coordinate patch which has $\mathcal{U}$ equal to a disk or a rectangle is simply connected.

EXAMPLE 4.2. The entire sphere S^2 is simply connected. This is easy to visualize but difficult to prove. See Figure 6.8.

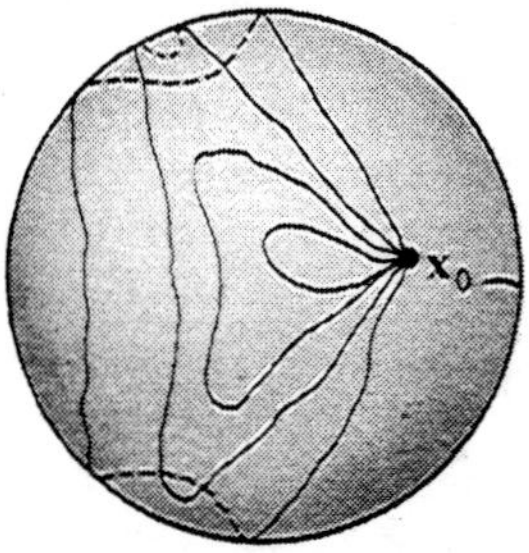

FIGURE 6.8

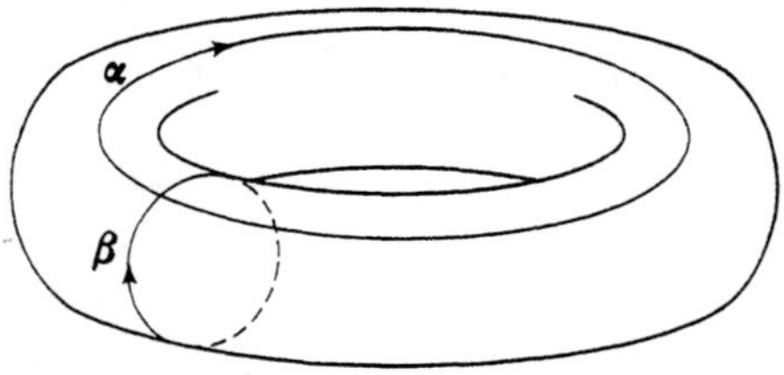

FIGURE 6.9

EXAMPLE 4.3. The entire torus, T^2, is not simply connected—a circle going around either "hole" cannot be shrunk to a point while staying in T^2. Neither of the curves $\boldsymbol{\alpha}$ and $\boldsymbol{\beta}$ in Figure 6.9 is null-homotopic.

THEOREM 4.4 (Gauss-Bonnet Formula). Let γ be a piecewise regular curve contained within a simply connected geodesic coordinate patch and bounding a region $\mathcal{R}$ in the patch. Let the jump angles at the junctions be $\alpha_1, \ldots, \alpha_n$. Then $\iint_{\mathcal{R}} K\,dA + \int_\gamma \kappa_g\,ds + \sum \alpha_i = 2\pi$.

Proof: We first point out that the geodesic coordinate patch referred to in the hypothesis is not a geodesic coordinate patch along $\boldsymbol{\gamma}$. It can't be because $\boldsymbol{\gamma}$ is closed.

Recall that for a geodesic coordinate patch, the metric tensor takes the form $\begin{pmatrix} 1 & 0 \\ 0 & h^2 \end{pmatrix}$, where $h > 0$ is a function of u^1 and u^2. By Problem 2.3,

$$\Gamma_{12}{}^2 = \Gamma_{21}{}^2 = \frac{h_1}{h}, \quad \Gamma_{22}{}^1 = -hh_1, \quad \Gamma_{22}{}^2 = \frac{h_2}{h},$$
$$\Gamma_{11}{}^1 = \Gamma_{12}{}^1 = \Gamma_{21}{}^1 = \Gamma_{11}{}^2 = 0 \quad \text{and} \quad K = \frac{-h_{11}}{h},$$

where $h_i = \partial h/\partial u^i$.

Since all the quantities in the conclusion are independent of parametrization, we may assume that $\boldsymbol{\gamma}$ is parametrized by arc length. We compute the angular variation of a vector field parallel along $\boldsymbol{\gamma}$.

Let $\mathbf{T}$ be the unit tangent to $\boldsymbol{\gamma}$ (where it exists) and let $\mathbf{P}$ be a unit vector field parallel along $\boldsymbol{\gamma}$, starting at a junction point. (Such a $\mathbf{P}$ exists by Theorem 6.7 of Chapter 4.) Note that $\mathbf{x}_1$ is a field of unit vectors in $\mathcal{R}$. Let $\alpha = \measuredangle(\mathbf{x}_1, \mathbf{T})$, $\phi = \measuredangle(\mathbf{x}_1, \mathbf{P})$ and $\theta = \measuredangle(\mathbf{P}, \mathbf{T})$. $d\alpha/ds = d\phi/ds + d\theta/ds$.

First we write $\boldsymbol{\gamma}(s) = \mathbf{x}(\gamma^1(s), \gamma^2(s))$. Note that $\cos\phi = \langle \mathbf{x}_1, \mathbf{P}\rangle$ so that $-(\sin\phi)\phi' = \langle \mathbf{x}_1{}', \mathbf{P}\rangle + \langle \mathbf{x}_1, \mathbf{P}'\rangle = \langle \mathbf{x}_1{}', \mathbf{P}\rangle$ since $\mathbf{P}$ is parallel along $\boldsymbol{\gamma}$. However, $\mathbf{x}_1{}' = \gamma^{1\prime}\mathbf{x}_{11} + \gamma^{2\prime}\mathbf{x}_{12}$ so that Gauss's formulas (Equations (4-8) of Chapter 4) yield

$$\begin{aligned} -(\sin\phi)\phi' &= \langle \gamma^{1\prime}\mathbf{x}_{11} + \gamma^{2\prime}\mathbf{x}_{12}, \mathbf{P}\rangle \\ &= \langle (\gamma^{1\prime}\Gamma_{11}{}^1 + \gamma^{2\prime}\Gamma_{12}{}^1)\mathbf{x}_1 + (\gamma^{1\prime}\Gamma_{11}{}^2 + \gamma^{2\prime}\Gamma_{12}{}^2)\mathbf{x}_2, \mathbf{P}\rangle. \end{aligned}$$

The specific form of the $\Gamma_{ij}{}^k$ allows us to conclude

$$\text{(4-1)} \qquad -(\sin\phi)\phi' = \gamma^{2\prime}\frac{h_1\langle \mathbf{x}_2, \mathbf{P}\rangle}{h}.$$

Because $g_{12} = 0$, $\{\mathbf{x}_1, \mathbf{x}_2/|\mathbf{x}_2|\}$ is an orthonormal basis of T_QM for each $Q \in \mathbf{x}(\mathcal{U})$; hence by Proposition 4.1 of Chapter 2,

$$\mathbf{P} = \langle \mathbf{P}, \mathbf{x}_1\rangle\mathbf{x}_1 + \frac{\langle \mathbf{P}, \mathbf{x}_2/|\mathbf{x}_2|\rangle\mathbf{x}_2}{|\mathbf{x}_2|}.$$

Since $\mathbf{P}$ is a unit vector and $\langle \mathbf{P}, \mathbf{x}_1\rangle = \cos\phi$, we have $\langle \mathbf{P}, \mathbf{x}_2\rangle/h = \sin\phi$. Hence Equation (4-1) becomes $\phi' = -h_1\gamma^{2\prime}$ so that

$$\text{(4-2)} \qquad \delta\phi = -\int_\gamma h_1\gamma^{2\prime}\,ds = -\int_\gamma h_1\,d\gamma^2 = -\int_\gamma h_1\,du^2.$$

We now show that

$$\text{(4-3)} \qquad \int_\gamma \kappa_g\,ds = \int_\gamma \theta'\,ds.$$

Note $\cos\theta = \langle \mathbf{T}, \mathbf{P}\rangle$ so that $-(\sin\theta)\theta' = \langle \mathbf{T}', \mathbf{P}\rangle + \langle \mathbf{T}, \mathbf{P}'\rangle = \langle \mathbf{T}', \mathbf{P}\rangle$.

$$\kappa_g = \langle \mathbf{n} \times \mathbf{T}, \mathbf{T}' \rangle = \langle \mathbf{n}, \mathbf{T} \times \mathbf{T}' \rangle = \frac{\langle \mathbf{P} \times \mathbf{T}, \mathbf{T} \times \mathbf{T}' \rangle}{\sin\theta}$$
$$= \frac{\langle \mathbf{P}, \mathbf{T} \times (\mathbf{T} \times \mathbf{T}') \rangle}{\sin\theta} = \frac{\langle \mathbf{P}, -\mathbf{T}' \rangle}{\sin\theta} = \theta'.$$

Hence $\int_\gamma \kappa_g\, ds = \int_\gamma \theta'\, ds$, which is Equation (4-3).

Because $\int_\gamma \alpha'\, ds = \int_\gamma \phi'\, ds + \int_\gamma \theta'\, ds$, Equations (4-2) and (4-3) imply $\int_\gamma \alpha'\, ds + \sum \alpha_i = -\int_\gamma h_1\, du^2 + \int_\gamma \kappa_g\, ds + \sum \alpha_i$. $\int_\gamma \alpha'\, ds + \sum \alpha_i$ is precisely the angular variation of $\mathbf{T}$ along γ taking into account the jump angles at the junctions. Since γ bounds $\mathcal{R}$, $\delta\alpha = \int_\gamma \alpha'\, ds + \sum \alpha_i = 2\pi$. (This is essentially the Rotation Index Theorem.)

Finally, Green's (Theorem 1.3 of Chapter 3) states that

$$-\int_\gamma h_1\, du^2 = -\iint_{\mathcal{R}} h_{11}\, du^1\, du^2 = -\iint_{\mathcal{R}} \frac{h_{11}}{h} h\, du^1\, du^2 = +\iint_{\mathcal{R}} K\, dA.$$

Thus $2\pi = \iint_{\mathcal{R}} K\, dA + \int_\gamma \kappa_g\, ds + \sum \alpha_i$. ∎

EXAMPLE 4.5. On S^2 let γ be a triangle whose sides are geodesics and whose interior angles are $\beta_1, \beta_2, \beta_3$. Then $\alpha_i = \pi - \beta_i$.

$$2\pi = \iint_{\mathcal{R}} K\, dA + \int \kappa_g\, ds + \sum \alpha_i = \text{area}(\mathcal{R}) + 0 + 3\pi - \sum \beta_i.$$

Thus $(\sum \beta_i) - \pi = \text{area}(\mathcal{R})$. The number $(\sum \beta_i) - \pi$ is called the *angular excess* of γ. The fact that the angular excess is equal numerically to area is a standard result of spherical geometry due to Legendre.

EXAMPLE 4.6. Let γ be a circle of latitude $\phi = \phi_0$ on S^2. The geodesic curvature of γ is $\kappa \cos\psi$, where $\kappa = 1/\sin\phi_0$ is the curvature of γ and ψ is the angle between the tangent plane of S^2 and the osculating plane of γ (this is the extrinsic formula for κ_g). This angle is also ϕ_0. Hence $\kappa_g = \cot\phi_0$. $\int \kappa_g\, ds = \kappa_g\,(\text{length of } \gamma) = \kappa_g\,(2\pi \sin\phi_0) = 2\pi\cos\phi_0$. Thus

$$2\pi = \iint_{\mathcal{R}} K\, ds + \int_\gamma \kappa_g\, ds + \sum \alpha_i = \text{area}(\mathcal{R}) + 2\pi\cos\phi_0$$

or

$$2\pi(1 - \cos\phi_0) = \text{area}(\mathcal{R})$$

as we found in Example 3.12.

PROBLEMS

4.1. Use the Gauss-Bonnet Formula to prove that the sum of the exterior angles of a Euclidean polygon is 2π.

4.2. In Problems 2.4 and 2.5 we had "surfaces" with $K \equiv -1$. Suppose γ is a geodesic triangle in either of these cases. What can you say about the sum of the interior angles? (*Hint:* Example 4.5.)

4.3. (Hyperbolic Half Plane, see Problem 2.4.) Prove $x =$ constant gives a geodesic. Prove $x = a + r\cos\theta$, $y = r\sin\theta$ gives a geodesic with a and r constant and $0 < \theta < \pi$. (Use Problem 5.8 of Chapter 4.) Note that in $\mathcal{U}$ these curves are vertical straight lines and circles which meet the x-axis at right angles. They give all the geodesics.

4.4. (Poincaré Disk, Problem 2.5.) Prove that $y = 0$ gives a geodesic. Prove that $x = a + r\cos\theta$, $y = r\sin\theta$, where a and r are constants related by $a^2 = 1 + r^2$, gives a geodesic (for a suitable range of θ). Note that in $\mathcal{U}$ this curve is a circle that intersects $x^2 + y^2 = 1$ at right angles. Due to the rotational symmetry of the matrix (g_{ij}) (it depends on $x^2 + y^2$, not really on x or y) any such circle (or straight line through the origin) gives a geodesic. These give all the geodesics.

4.5. In Problems 4.3 and 4.4 let γ be a geodesic and let P be a point not on the geodesic. Show that there are at least two geodesics through P that never meet γ. (Proof by reasonable picture is sufficient.) These "surfaces" give examples of geometries where Euclid's parallel axiom fails, if by straight line you mean geodesic and by parallel you mean never intersecting. These models of hyperbolic geometry are discussed more fully in Chapter 7.

6–5. THE GAUSS-BONNET THEOREM AND THE EULER CHARACTERISTIC

DEFINITION. A *polygon* on a surface M is a piecewise regular curve γ whose segments are geodesics and which bounds a simply connected region $\mathcal{R}$.

Let M be a compact surface in $\mathbf{R}^3$. Suppose M can be broken into regions bounded by polygons, each region contained in a simply connected geodesic coordinate patch. Let V equal the number of vertices, E the number of edges, and F the number of faces (i.e., the number of polygonally bounded regions). Let the Euler characteristic of M (with respect to this decomposition) be the integer $\chi = F - E + V$.

The next theorem is a truly amazing result. It relates two seemingly unrelated quantities, curvature (which is a geometric quantity) with the Euler characteristic χ (which is a topological or combinatorial quantity).

THEOREM 5.1 (Gauss-Bonnet). If M is compact then $\iint_M K\,dA = 2\pi\chi$.

Proof: Let the polygonally bounded regions be $\{\mathcal{R}_i \mid 1 \leq i \leq F\}$, the polygons themselves $\{\boldsymbol{\gamma}_i \mid 1 \leq i \leq F\}$, and $\{\beta_{ij}\}$ the interior angles of $\boldsymbol{\gamma}_i$. An application of the Gauss-Bonnet Formula shows that

$$\iint_M K\,dA = \sum_{i=1}^{F} \iint_{\mathcal{R}_i} K\,dA = \sum_{i=1}^{F} \Big(2\pi - \sum_j (\pi - \beta_{ij})\Big)$$
$$= 2\pi F - \sum_{i=1}^{F} \sum_j \pi + \sum_{i=1}^{F} \sum_j \beta_{ij}.$$

Now

$$\sum_j \pi = \pi \text{ (number of vertices of } \boldsymbol{\gamma}_i) = \pi \text{ (number of edges of } \boldsymbol{\gamma}_i).$$

Since each edge belongs to two polygons, $\sum_{i=1}^{F} \sum_j \pi = 2\pi E$. $\sum_{i=1}^{F} \sum_j \beta_{ij}$ is the sum of all interior angles. At each vertex the sum is 2π. Hence $\sum_{i=1}^{F} \sum_j \beta_{ij} = 2\pi V$. Thus $\iint_M K\,dA = 2\pi F - 2\pi E + 2\pi V = 2\pi\chi$. ∎

The quantity $\iint_M K\,dA$ is called the *total curvature* of M or the *curvatura integra*. Since it is a well defined number, χ cannot depend on how the polygons are chosen, so long as each is small enough to be contained in a simply connected geodesic coordinate patch! This fact alone is surprising but we may also look at the Gauss-Bonnet Theorem the other way. From its definition χ must be an integer, hence $(1/2\pi) \iint_M K\,dA$ is an integer! There is no *a priori* reason for this. (It is similar to the situation for plane curves, where $(1/2\pi) \int k\,ds$ is an integer.) The Gauss-Bonnet Theorem shows us one of the beautiful things about global differential geometry (and its present popularity): the mixing of topological and geometric results in a very unusual and striking manner. To us, this theorem is even more egregious than the *Theorema Egregium.*

To define χ we do not really need geodesic polygons, just piecewise regular curves. They were chosen to be geodesic so that the terms $\int_\gamma \kappa_g\,ds$ in the Gauss-Bonnet Formula would not appear. They would cancel anyway since each edge is traversed twice and in opposite directions because it bounds part of two regions.

On the other hand, if M is broken into polygons that are not small enough to be in simply connected geodesic coordinate patches, we could still define $\bar{\chi} = \bar{F} - \bar{E} + \bar{V}$, where $\bar{F}$, $\bar{E}$, $\bar{V}$ are the number of faces, edges, and vertices. If a particular polygon is not small enough, let P be a point inside it. Join P to each of the vertices of that polygon by a regular curve segment. (Since M is compact, these curves can even be geodesics.) If the polygon has r vertices, we have added r edges, 1 vertex, and $r - 1$ faces (where there was 1 face there are now r). $F = \bar{F} + r - 1$, $E = \bar{E} + r$, $V = \bar{V} + 1$.

Then $F - E + V = \bar{F} - \bar{E} + \bar{V} + r - 1 - r + 1 = \bar{\chi}$. Hence we have not changed the value of $\bar{\chi}$, but we have eliminated one large polygon. This can be continued until all polygons are small enough.

DEFINITION. χ is called the *Euler characteristic* of M.

EXAMPLE 5.2. $\chi(S^2) = \frac{1}{2\pi} \iint_{S^2} K\,dA = \frac{4\pi}{2\pi} = 2.$

χ has been defined combinatorially. Our theorem shows that it is an intrinsic geometric invariant. χ can also be defined topologically.

The proof of the following theorem (which gives the classification of surfaces) and the formal definition of "looks like" can be found in Massey [1967].

THEOREM 5.3. Every compact surface in $\mathbf{R}^3$ "looks like" a sphere with handles. The number of handles is called the *geometric genus* of M and is denoted g. (See Figures 6.10 and 6.11.)

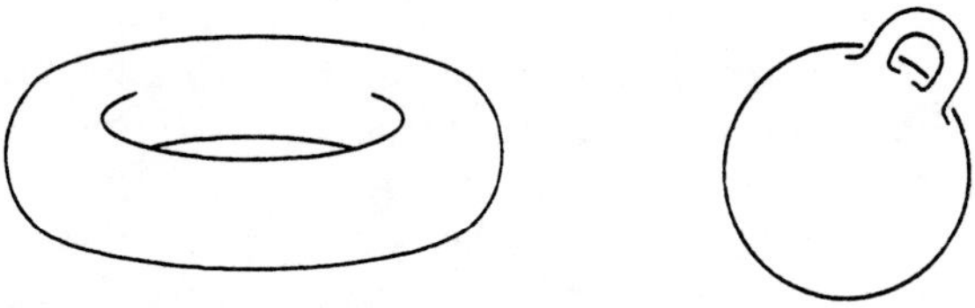

FIGURE 6.10

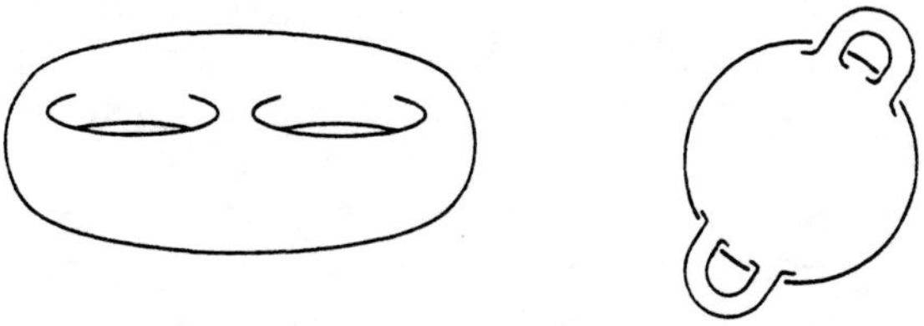

FIGURE 6.11

THEOREM 5.4. If M is a compact surface in $\mathbf{R}^3$, then $\chi = 2(1 - g)$.

Proof (by induction on g): If $g = 0$, then M looks like S^2 and $\chi(S^2) = 2 = 2(1 - 0)$. Assume the theorem is true for $g = n$ and let M have genus $n + 1$. Break M into polygons with F, E, V faces, edges, and vertices and so that one handle looks like that in Figure 6.12.

If three faces are removed as in Figure 6.13 and the resulting holes filled in, there is a net loss of one face and three edges. The new surface has Euler characteristic $\bar{\chi}$ which is two more than the original. It has genus n. By induction $\bar{\chi} = 2(1 - n)$. Hence the Euler characteristic of the original surface is $\chi = \bar{\chi} - 2 = 2(1 - n) - 2 = 2(1 - (n + 1)) = 2(1 - g)$. ∎

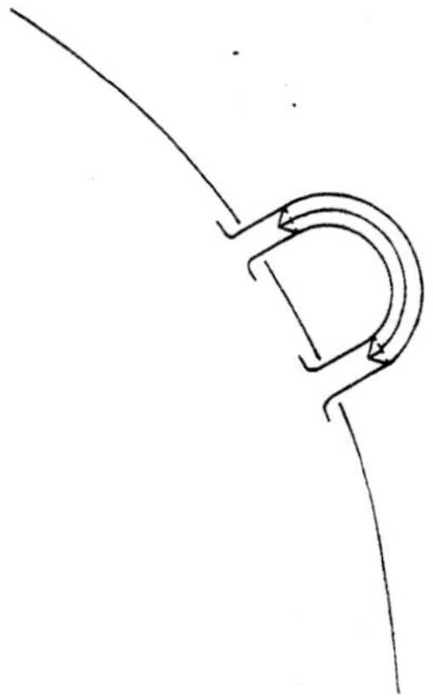

FIGURE 6.12

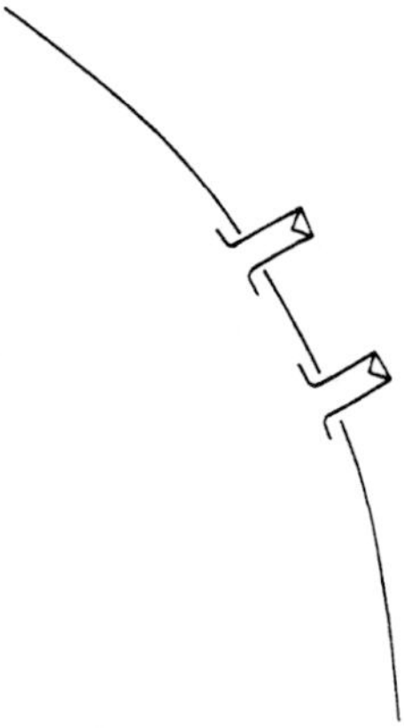

FIGURE 6.13

Note that the Euler characteristic of a compact surface in $\mathbf{R}^3$ is even and less than or equal to 2.

There is a more general version of the Gauss-Bonnet Theorem in our context.

THEOREM 5.5. Let γ be a piecewise regular curve in an oriented surface M. Suppose γ bounds a region $\mathcal{R}$. Then

$$\iint_{\mathcal{R}} K\,dA + \int_{\gamma} \kappa_g\,ds + \sum (\pi - \alpha_i) = 2\pi\chi(\mathcal{R}),$$

where the α_i are the interior angles of γ and $\chi(\mathcal{R})$ is the Euler characteristic of $\mathcal{R}$ found by breaking $\mathcal{R}$ into polygons and counting those edges and vertices lying on γ in addition to those in $\mathcal{R}$.

Proof: Problem 5.1. ▮

There are many generalizations of the Gauss-Bonnet Theorem. Note that in its simplest form it states that positive curvature everywhere implies positive Euler characteristic. The conjecture that this is true in higher even dimensions (in the context of Chapter 7) has long been one of the outstanding problems in differential geometry. It is called the Chern-Hopf Conjecture (Conjecture 8.16 of Chapter 7).

PROBLEMS

***5.1.** Prove Theorem 5.5.

5.2. Call a polygon an n-gon if it has n vertices (so a 0-gon is a closed geodesic). Let M be a surface with $K < 0$. Prove that there are no n-gons for $n = 0, 1, 2$.

5.3. Find an example of a surface with $K < 0$ that has a closed geodesic $\boldsymbol{\gamma}$. ($\boldsymbol{\gamma}$ cannot bound a simply connected region by Problem 5.2.)

5.4. Consider a "triangle" in the Poincaré disk whose vertices are on the boundary of the disk ("at infinity"). What do you expect the area to be? (See Figure 6.14.) What would be the area of an n-gon if all the vertices are at infinity? (Figure 6.15).

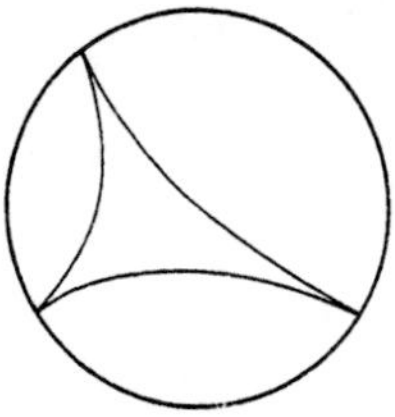

FIGURE 6.14

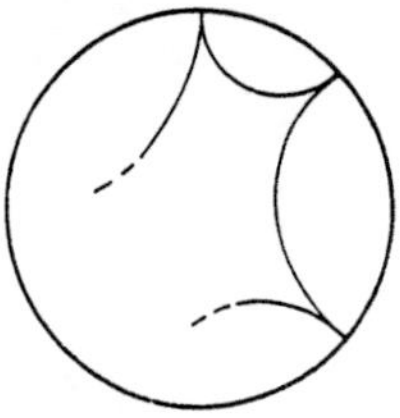

FIGURE 6.15

5.5. Prove that a compact surface in $\mathbf{R}^3$ which has $K \geq 0$ "looks like" a sphere. (*Hint:* You may use Theorem 5.3.)

6–6. THE THEOREMS OF JACOBI AND HADAMARD

In this section we give two more examples of theorems from global differential geometry. The first is due to C. G. J. Jacobi (1804–1851) and the second to J. Hadamard (1865–1963). Both are proven as applications of the Gauss-Bonnet Theorem.

Recall that if $\boldsymbol{\gamma}$ is a regular space curve with $\kappa > 0$, then its normal spherical image $\boldsymbol{\sigma}$ is the curve on S^2 given by $\boldsymbol{\sigma}(s) = \mathbf{N}(s)$, where $\mathbf{N}$ is the normal to $\boldsymbol{\gamma}$. Since $\mathbf{N}' = -\kappa\mathbf{T} + \tau\mathbf{B} \neq \mathbf{0}$, $\boldsymbol{\sigma}(s)$ is regular.

THEOREM 6.1 (Jacobi, 1842). Let $\boldsymbol{\gamma}$ be a regular closed unit speed space curve with $\kappa > 0$. Assume that $\boldsymbol{\sigma}$, the normal spherical image, is a simple curve. Then it divides the unit sphere into two regions of equal area.

Proof: Let t be arc length on $\boldsymbol{\sigma}$ and $\bar{\kappa}_g$ the geodesic curvature of $\boldsymbol{\sigma}$ viewed as a curve on S^2. Let $\mathcal{R}$ be the region bounded by $\boldsymbol{\sigma}$. We will show that the area of $\mathcal{R}$ is 2π by applying the Gauss-Bonnet Formula. (Since the area of S^2 is 4π, this will complete the proof.) To this end, we must compute $\bar{\kappa}_g$. We shall show that

$$\bar{\kappa}_g = \frac{d}{ds}\left(\arctan\frac{\tau}{\kappa}\right)\frac{ds}{dt}. \tag{6-1}$$

Since t is arc length on $\boldsymbol{\sigma}$, the Frenet-Serret equations of $\boldsymbol{\gamma}$ show that

$$\frac{d\boldsymbol{\sigma}}{dt} = \frac{d\mathbf{N}}{dt} = (-\kappa\mathbf{T} + \tau\mathbf{B})\frac{ds}{dt}, \tag{6-2}$$

which is a unit vector so that

$$\frac{ds}{dt} = \frac{1}{\sqrt{\kappa^2 + \tau^2}}. \tag{6-3}$$

Also

$$\frac{d^2\mathbf{N}}{dt^2} = \frac{d^2s}{dt^2}(-\kappa\mathbf{T} + \tau\mathbf{B}) + \left(\frac{ds}{dt}\right)^2(-\kappa'\mathbf{T} + \tau'\mathbf{B} - (\kappa^2 + \tau^2)\mathbf{N}),$$

or, after applying Equation (6-2),

$$\frac{d^2\mathbf{N}}{dt^2} = \frac{d^2s}{dt^2}\frac{dt}{ds}\frac{d\mathbf{N}}{dt} + \left(\frac{ds}{dt}\right)^2(-\kappa'\mathbf{T} + \tau'\mathbf{B} - (\kappa^2 + \tau^2)\mathbf{N}). \tag{6-4}$$

Now the surface normal to S^2 at $\mathbf{p}$ is $\mathbf{p}$. Hence the surface normal to S^2 at $\boldsymbol{\sigma}(t)$ is $\mathbf{N}(t)$ so that the intrinsic normal to $\boldsymbol{\sigma}$ is $\mathbf{N} \times d\mathbf{N}/dt$. By using (6-2), (6-4), and the fact that $\bar{\kappa}_g = \langle \mathbf{S}, d^2\mathbf{N}/dt^2 \rangle$, we obtain

$$\begin{aligned}
\bar{\kappa}_g &= \left\langle \mathbf{N} \times \frac{d\mathbf{N}}{dt}, \frac{d^2\mathbf{N}}{dt^2} \right\rangle = \left\langle \mathbf{N}, \frac{d\mathbf{N}}{dt} \times \frac{d^2\mathbf{N}}{dt^2} \right\rangle \\
&= \left\langle \mathbf{N}, \frac{d\mathbf{N}}{dt} \times \left(\frac{ds}{dt}\right)^2(-\kappa'\mathbf{T} + \tau'\mathbf{B} - (\kappa^2 + \tau^2)\mathbf{N}) \right\rangle \\
&= \left(\frac{ds}{dt}\right)^3 \langle \mathbf{N}, (-\kappa\mathbf{T} + \tau\mathbf{B}) \times (-\kappa'\mathbf{T} + \tau'\mathbf{B} - (\kappa^2 + \tau^2)\mathbf{N}) \rangle \\
&= \left(\frac{ds}{dt}\right)^3 (-\tau\kappa' + \kappa\tau').
\end{aligned}$$

Because of Equation (6-3), this last equation is

$$\bar{\kappa}_g = \frac{ds}{dt}\left(\frac{\tau'\kappa - \kappa'\tau}{\tau^2 + \kappa^2}\right) = \frac{ds}{dt}\frac{d}{ds}\left(\text{arc tan}\,\frac{\tau}{\kappa}\right)$$

as desired.

We integrate Equation (6-1) to obtain

$$\int_{\sigma} \bar{\kappa}_g \, dt = \int_{\sigma} \frac{d}{ds}\left(\text{arc tan}\,\frac{\tau}{\kappa}\right)\frac{ds}{dt}\, dt = \int_{\gamma} \frac{d}{ds}\left(\text{arc tan}\,\frac{\tau}{\kappa}\right) ds = 0$$

since $\boldsymbol{\gamma}$ is closed. Thus the Gauss-Bonnet Formula gives us

$$\iint_{\mathcal{R}} dA = \iint_{\mathcal{R}} K\, dA = 2\pi - \int_{\sigma} \bar{\kappa}_g \, dt = 2\pi.$$

Thus the area of $\mathcal{R}$ is 2π and $\boldsymbol{\sigma}$ breaks S^2 into two regions of equal area. ∎

The next result generalizes Theorem 3.2 of Chapter 3 relating convexity to the sign of the curvature of a plane curve.

THEOREM 6.2 (Hadamard, 1897). Let M be a compact surface in $\mathbf{R}^3$. If $K > 0$ everywhere, then the surface is convex (lies on one side of each tangent plane).

Proof: $K > 0$ implies $0 < \iint_M K\,dA = 2\pi\chi = 2\pi 2(1-g)$ and $g < 1$. Hence $g = 0$ and $\iint_M K\,dA = 4\pi$.

Let $\nu\colon M \longrightarrow S^2$ be the Gauss normal map defined by the outward pointing normal. We show that ν is onto. Let $\mathbf{n} \in S^2$. Let Π_r be the plane in $\mathbf{R}^3$ given by $\langle \mathbf{x}, \mathbf{n}\rangle = r$. Since M is bounded, there is an $r > 0$ with $-r \leq \langle \mathbf{p}, \mathbf{n}\rangle \leq r$ for all $\mathbf{p} \in M$. Let R be the smallest real number such that $\langle \mathbf{p}, \mathbf{n}\rangle \leq R$ for all $\mathbf{p} \in M$. R may be negative. Then $M \cap \Pi_R \neq 0$. If $P \in M \cap \Pi_R$, then Π_R is the tangent plane at P, $\nu(P) = \mathbf{n}$, and $\nu(M) = S^2$. Note $\iint_{\mathcal{R}} K\,dA$ is the area of $\nu(\mathcal{R})$ counted with multiplicity.

Next we show that ν is one-to-one. Suppose $\nu(P) = \nu(Q)$ with $P \neq Q$. Then there is a small open set $\mathcal{U}$ about Q such that $\nu(M - \mathcal{U}) = S^2$ also since ν is continuous. Hence $\iint_{M-\mathcal{U}} K\,dA \geq 4\pi$. $\iint_{\mathcal{U}} K\,dA > 0$. Hence $\iint_M K\,dA = \iint_{M-\mathcal{U}} K\,dA + \iint_{\mathcal{U}} K\,dA > 4\pi$, a contradiction. Hence ν is one-to-one.

Now suppose there is a point P on M such that there are points of M on both sides of the tangent plane at P. Let Q, R be points of M on opposite sides of this plane and farthest from the plane. The normals at P, Q, R must be parallel. Hence two are equal and ν is not one-to-one, a contradiction. Thus M is convex. ∎

PROBLEM

The following problem generalizes the notion of curves of constant width to surfaces. (See Section 3-5.)

†6.1. Let M be compact with $K > 0$ (hence M is convex).

(a) If $P \in M$, prove there is a unique point $\bar{P} \neq P$ with the tangent planes at P and $\bar{P}$ parallel.

(b) From now on assume that the distance between these planes is a constant independent of the choice of P. Prove that the line joining P to $\bar{P}$ is normal to M at both P and $\bar{P}$.

(c) Prove that the principal directions at P and $\bar{P}$ are parallel (in the Euclidean sense).

(d) Let $\mathbf{X}$ be a principal direction at P with principal curvature κ_1 and $\bar{\mathbf{X}}$ the corresponding principal direction at $\bar{P}$ with principal curvature $\bar{\kappa}_1$. Prove that $1/\kappa_1 + 1/\bar{\kappa}_1$ is a constant independent of P. Does $H(P) = H(\bar{P})$, where H is the mean curvature? (See Struik [1961, p. 202].)

(e) Prove $\iint_M H\,dA = 2\pi c$, where $c =$ width (constant distance from P to $\bar{P}$). (You will need the Gauss-Bonnet Theorem for this last result.)

6–7. THE INDEX OF A VECTOR FIELD

Let M be a compact surface with a vector field $\mathbf{V}$ on M. We say $P \in M$ is an *isolated zero* of $\mathbf{V}$ if there is an open set $\mathcal{U}$ about P with P the only point in $\mathcal{U}$ where $\mathbf{V}$ is zero. Suppose P is an isolated zero of $\mathbf{V}$, and let $\boldsymbol{\gamma}$ be a simple closed piecewise regular curve which bounds a simply connected region $\mathcal{R}$ with P the only zero of $\mathbf{V}$ in $\mathcal{R}$.

DEFINITION. The *index* of $\mathbf{V}$ at P is $i_P(\mathbf{V}) = (1/2\pi)\, \delta\sphericalangle(\mathbf{U}, \mathbf{V})$, where $\mathbf{U}$ is any field of unit vectors in $\mathcal{R}$.

We know that $i_P(\mathbf{V})$ does not depend on $\mathbf{U}$. It is an integer which does not really depend on $\boldsymbol{\gamma}$ either—a continuous deformation of $\boldsymbol{\gamma}$ cannot change the integer.

EXAMPLE 7.1. The vector fields in Figure 6.16 have indices as indicated.

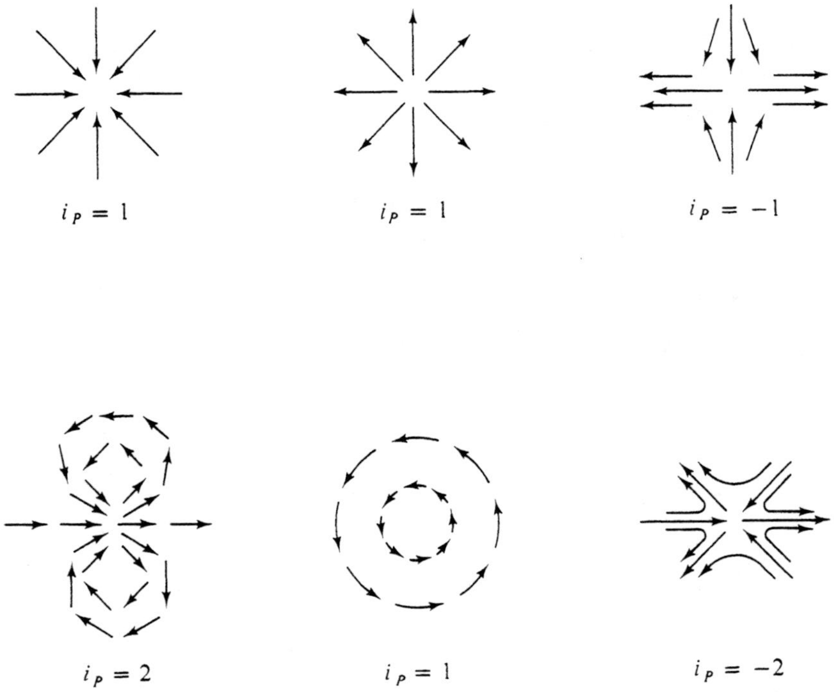

FIGURE 6.16

Let M be a compact surface in $\mathbf{R}^3$ with $\mathbf{V}$ a vector field on M with only a finite number of zeros. The *total index of* $\mathbf{V}$ is $I(\mathbf{V}) = \sum i_p(\mathbf{V})$.

PROPOSITION 7.2. If $\mathbf{V}$ and $\mathbf{W}$ are two vector fields on a compact surface M with only a finite number of zeros, then $I(\mathbf{V}) = I(\mathbf{W})$.

Proof: Break M up into triangles $\boldsymbol{\gamma}_i$ so that each triangle is in a simply connected coordinate patch, each triangle has at most one zero of $\mathbf{V}$ and at most one zero of $\mathbf{W}$ in its interior, and all zeros are inside triangles. Let $\mathbf{U}_i$ be a field of unit vectors in the patch about $\boldsymbol{\gamma}_i$.

$$I(\mathbf{V}) = \frac{1}{2\pi} \sum \delta\sphericalangle(\mathbf{U}_i, \mathbf{V}).$$

$$I(\mathbf{V}) - I(\mathbf{W}) = \frac{1}{2\pi} \sum (\delta\sphericalangle(\mathbf{U}_i, \mathbf{V}) - \delta\sphericalangle(\mathbf{U}_i, \mathbf{W})) = \frac{1}{2\pi} \sum \delta\sphericalangle(\mathbf{W}, \mathbf{V})$$

$$= \frac{1}{2\pi} \sum_{\text{triangles}} \sum_{\text{edges}} \delta\sphericalangle(\mathbf{W}, \mathbf{V}) = 0,$$

since each edge is traversed twice, in opposite directions. Thus $I(\mathbf{V}) = I(\mathbf{W})$. ▮

THEOREM 7.3 (Poincaré-Brouwer). If M is a compact surface and $\mathbf{V}$ is a vector field on M with only a finite number of zeros, then $I(\mathbf{V}) = \chi(M)$, the Euler characteristic of M.

Proof: We break M into triangles. We can define a vector field $\mathbf{W}$ on M which has a zero at each vertex, at the midpoint of each edge, and at the center of each triangle as in Figure 6.17. The zeros at the vertices and at the centers have index $+1$ while those on the edges have index -1. Thus

$$I(\mathbf{W}) = \sum i_p = (+1)V + (-1)E + (+1)F = V - E + F = \chi.$$

But $I(\mathbf{V}) = I(\mathbf{W})$ by Proposition 7.2. ▮

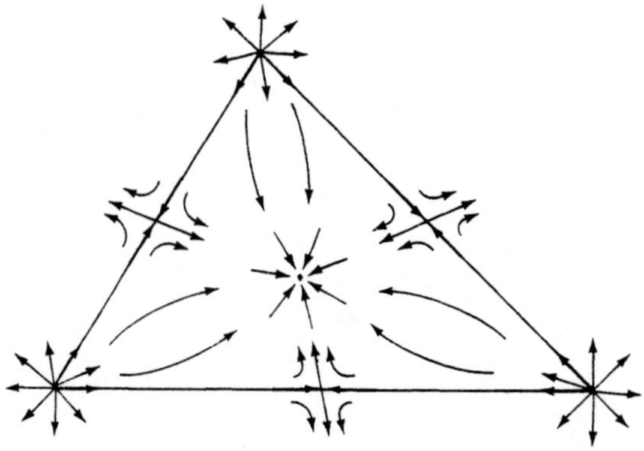

FIGURE 6.17

This theorem implies that any vector field on S^2 must have a zero. This fact is sometimes referred to by saying "you can't grow hair on a coconut," meaning somewhere there will be a swirl or a cowlick.

PROBLEMS

7.1. Let M be a compact surface. Prove that M has a vector field with no zeros if and only if M "looks like" a torus.

7.2. Prove that on $M = \mathbf{R}^2$ there are vector fields with infinitely many zeros P_n such that the sequence $\{P_n\}$ does not have a convergent subsequence.

7.3. Prove that on $M = \mathbf{R}^2$ there are vector fields with no zeros.

7.4. Find an example to show that Proposition 7.2 is false if the hypothesis "compact" is omitted.

7

Introduction to Manifolds

In this chapter we return to the theme of the introduction to Chapter 4—that of geometry being built with the concepts of "space" and "line." In Chapter 4 we used surfaces for the concept of space and geodesics for the concept of line. In this chapter we will take a much more abstract view of space by defining "space" to be anything that looks locally like Euclidean n-space (a manifold). Note that this is similar to what we did for surfaces; they were locally like $\mathbf{R}^2$. One important difference here is that we are not assuming that our manifolds lie in a Euclidean space of one dimension higher (as we did for surfaces) or even that they lie in Euclidean space at all!

Having settled on a definition for space (which occupies the first two sections), we must find a suitable geometric structure to add to space to give the concept of "lines" (which we again call geodesics). A careful reading of Chapter 4 shows that there are two geometric structures which led inexorably to the notion of lines. One was the notion of differentiating a vector field with respect to a vector field and the other was the notion of measuring lengths and angles (the metric tensor). We shall take the view here that the first idea is primitive and call such a structure a "covariant differentiation" in Section 7-6. We will also introduce the notion of a Riemannian metric on a manifold and show in Section 7-8 that every Riemannian metric gives rise to a covariant differentiation. We conclude this last section with a discussion of the various notions of curvature which can be defined on a manifold with Riemannian metric and their relationship with curvature as defined in Chapter 4.

There are a couple of things to notice. The first is that we must be very careful to define everything intrinsically because there is no such thing as extrinsic! This is one reason why it was so important in Chapter 4 to get an intrinsic description of something we had already defined extrinsically. In this chapter we will use the intrinsic description for the very definition of the concept. (This set of ideas is best exemplified by the notion of curvature. See Gauss's *Theorema Egregium* of Section 4-9. Compare that section with Section 7-8.) A more philosophical question is: Since this material has become so very abstract, has it lost all possible "relevance" to anybody but a mathematician? The answer is an emphatic no! There are many applications of this material to high energy physics (the theory of Lie groups is useful in quantum mechanics) and relativity theory. In fact Albert Einstein gave much impetus to differential geometry at the turn of the twentieth century by introducing relativity theory.

On the question of what is the proper level of abstraction, we would certainly want our approach to be abstract enough to include all of the classical geometries. In fact, we do have as special cases Euclidean, spherical, hyperbolic, and projective geometries. There are, of course, other ways of obtaining these classical geometries. For example, a group theoretic approach is taken in Millman [1977].

Another point to make about this chapter is that we have written it in a different manner than the previous six. We have written it in an open-ended style; that is, we have not tried to be encyclopedic but rather have given references for the results which we do not prove here. What we are doing here is motivating the abstract definitions of manifold theory so that if the reader wishes to study some of the more advanced works in the field he or she may do so without being attacked by a formalism which seems to be totally unmotivated. We have also adopted the notation which is most common in manifold theory. In this we follow Kobayashi and Nomizu [1963].

We make one further comment on this chapter and that is about the level of sophistication required to read it. We outline in Section 7-1 some of the analytic preliminaries necessary. The reader should have some familiarity with metric spaces and with advanced calculus (at least to the point of the Inverse and Implicit Function Theorems).

7–1. SOME ANALYTIC PRELIMINARIES

In this section we review some basic results from analysis, especially metric spaces and the Inverse and Implicit Function Theorems.

Definition. Let X be a set. A *metric* d on X is an assignment of a nonnegative number $d(x, y)$ to each pair of points $x, y \in X$ such that:

(a) $d(x, y) = 0$ if and only if $x = y$;

(b) $d(x, y) = d(y, x)$; and

(c) if $z \in X$ then $d(x, z) \leq d(x, y) + d(y, z)$.

A *metric space* is a set X together with a metric d.

Example 1.1. Let $X = \mathbf{R}^N$. If $x = (x^1, \ldots, x^N)$ and $y = (y^1, \ldots, y^N)$, we define $d(x, y) = \sqrt{\sum_{i=1}^{N} (x^i - y^i)^2} = |x - y|$. This is the usual Euclidean distance on $\mathbf{R}^N$.

The next example is inserted to show that "strange looking" spaces can be higher dimensional surfaces. This example may be skipped on a first reading.

Example 1.2. Let P^n be the set of all straight lines through the origin in $\mathbf{R}^{n+1}$. P^n is called *real projective n-space*. P^2 is called the *projective plane*. If $l \in P^n$, then l may be represented by any nonzero point in $\mathbf{R}^{n+1}$ (say $a = (a^1, \ldots, a^{n+1})$) through which l passes. Note that a and $b \in \mathbf{R}^{n+1}$ represent the same line if and only if there is a $0 \neq \lambda \in \mathbf{R}$ such that $a = \lambda b$. Motivated by this fact, we define an equivalence relation on $\mathbf{R}^{n+1} - \{0\}$ by $a \sim b$ if there is a $0 \neq \lambda \in \mathbf{R}$ such that $a = \lambda b$. We write $[a]$ for the equivalence class of $a \in \mathbf{R}^{n+1}$. Note that $P^n = (\mathbf{R}^{n+1} - \{0\})/\sim$. The coordinates of any $a \in \mathbf{R}^{n+1}$ such that $l = [a]$ are called *homogeneous coordinates* for l. (They are only determined up to a real multiple, of course.) Let $l_1, l_2 \in P^n$ and write $l_i = [a_i]$ for $a_i \in \mathbf{R}^{n+1}$ where $i = 1, 2$. Define the distance from l_1 to l_2 to be

$$d(l_1, l_2) = \text{minimum of} \left| \frac{a_1}{|a_1|} - \frac{a_2}{|a_2|} \right| \quad \text{and} \quad \left| \frac{a_1}{|a_1|} + \frac{a_2}{|a_2|} \right|.$$

The motivation for this is as follows: For each $i = 1$ or 2, $a_i/|a_i|$ is one of the two points on S^n which represent l_i. (They are antipodal points.) We then measure distance by computing the Euclidean distance between the nearest pair of representatives of l_1 and l_2. See Figure 7.1.

Definition. If $x \in X$ and $\epsilon > 0$ then the (open) *ball of radius* ϵ *about* x is $B_\epsilon(x) = \{y \in X \mid d(x, y) < \epsilon\}$. A subset S of X is *open* in X if for each $x \in S$ there is an $\epsilon > 0$ such that $B_\epsilon(x) \subset S$. If S is open and $x \in S$, then S is a *neighborhood* of x.

The following is immediate.

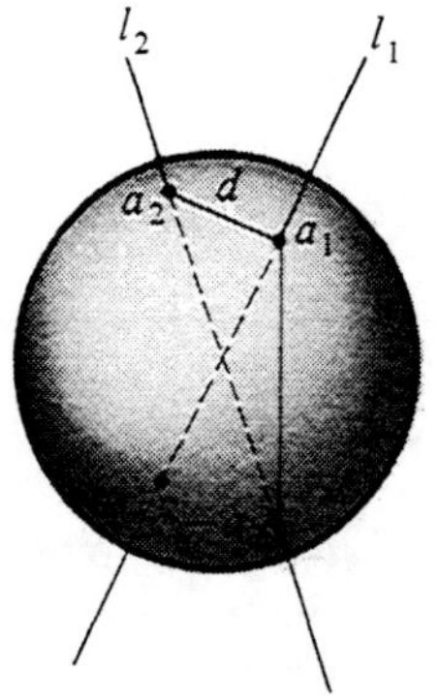

FIGURE 7.1

Proposition 1.3. Let X be a set with a metric d.

(a) If U and V are open in X, then $U \cap V$ is open.

(b) If U_α is open in X for each $\alpha \in I$, then $\bigcup_{\alpha \in I} U_\alpha$ is open.

(c) If $p, q \in X$ and $p \neq q$, then there are open sets U and V in X such that $p \in U, q \in V$, and $U \cap V$ is empty.

For students who have seen the definition, the following proposition is immediate from the definition of continuity in a metric space. For students who have not, the following may be taken as the definition of continuity.

Proposition 1.4. Let X and Y be metric spaces and $f: X \longrightarrow Y$. f is continuous if and only if for every open set V in Y, $f^{-1}(V)$ is open in X.

We now recall some facts about calculus. If $f: \mathbf{R}^n \longrightarrow \mathbf{R}^m$, then $f(x)$ may be written as $(f^1(x), \ldots, f^m(x))$ for some real-valued functions $f^1, \ldots, f^m$ of n variables. f is C^k if all partial derivatives of each f^i exist and are continuous up to order k. We will usually omit reference to the k by saying that f is differentiable. As in the case $m = n = 2$, we may form the Jacobian matrix $J(f)$ at each point $p \in \mathbf{R}^n$:

$$J(f)(p) = \begin{pmatrix} \dfrac{\partial f^1}{\partial x^1}(p) & \cdots & \dfrac{\partial f^1}{\partial x^n}(p) \\ \dfrac{\partial f^2}{\partial x^1}(p) & \cdots & \dfrac{\partial f^2}{\partial x^n}(p) \\ \vdots & & \vdots \\ \dfrac{\partial f^m}{\partial x^1}(p) & \cdots & \dfrac{\partial f^m}{\partial x^n}(p) \end{pmatrix}.$$

EXAMPLE 1.5. Let $f\colon \mathbf{R}^2 \longrightarrow \mathbf{R}^3$ be given by $f(x^1, x^2) = (\sin x^1, x^2x^1, x^2 \cos x^1)$. At $p = (x^1, x^2)$ we have

$$J(f)(p) = \begin{pmatrix} \cos x^1 & 0 \\ x^2 & x^1 \\ -x^2 \sin x^1 & \cos x^1 \end{pmatrix}.$$

DEFINITION. Suppose U is open in $\mathbf{R}^n$, V is open in $\mathbf{R}^n$, and $f\colon U \longrightarrow V$. f is a C^k *diffeomorphism* if f is C^k and there is a C^k function $g\colon V \longrightarrow U$ such that $f \circ g$ (resp. $g \circ f$) is the identity map on V (resp. on U). g is called the *inverse* of f. (These were called coordinate transformations in Chapter 4.)

EXAMPLE 1.6. Let $U = V =$ the open interval $(-1, 1)$ in $\mathbf{R}$ and $f(x) = x^3$. The inverse of f is $g(y) = \sqrt[3]{y}$ so that f is a C^k diffeomorphism if and only if $k = 0$.

For a proof of the following theorems see Fulks [1969].

THEOREM 1.7 (Inverse Function Theorem). Let U be an open set in $\mathbf{R}^n$, $p \in U$ and $f\colon U \longrightarrow \mathbf{R}^n$. If $\det J(f)(p) \neq 0$, then there are neighborhoods N_p of p and $N_{\phi(p)}$ of $\phi(p)$ such that $f|_{N_p}\colon N_p \longrightarrow N_{\phi(p)}$ is a diffeomorphism.

EXAMPLE 1.8. Let $U = \{(x^1, x^2) \mid -\frac{1}{2} < x^1 < \frac{1}{2}\}$ and define

$$f\colon U \longrightarrow \mathbf{R}^2 \text{ by } f(x^1, x^2) = (e^{-x^1} \cos x^2, e^{-x^1} \sin x^2).$$

An easy computation shows that $\det(J(f)(x^1, x^2)) = -e^{-2x^1} \neq 0$ so the Inverse Function Theorem applies for any point $p = (x^1, x^2)$. Note, however, that although f is one-to-one when restricted to small enough neighborhoods of any point, $f\colon U \longrightarrow \mathbf{R}^2$ is not one-to-one! However, f is one-to-one on the open set

$$N_{(0,0)} = \{(x^1, x^2) \mid -\tfrac{1}{2} < x^1 < \tfrac{1}{2}, -\pi < x^2 < \pi\}.$$

Let i be a fixed integer between 1 and $n + 1$. If

$$w = (w^1, \ldots, w^{n+1}) \in \mathbf{R}^{n+1},$$

then we shall write $\hat{w} = (w^1, \ldots, w^{i-1}, w^{i+1}, \ldots, w^{n+1}) \in \mathbf{R}^n$. If $W \subset \mathbf{R}^{n+1}$, then $\hat{W} = \{\hat{w} \mid w \in W\} \subset \mathbf{R}^n$. Note that if W is open in $\mathbf{R}^{n+1}$, then $\hat{W}$ is open in $\mathbf{R}^n$ (Problem 1.3). We shall now state the Implicit Function Theorem, which roughly says that if $\partial f/\partial u^i \neq 0$, then we can solve for the ith variable in terms of the others.

THEOREM 1.9 (Implicit Function Theorem). Let $f\colon \mathbf{R}^{n+1} \longrightarrow \mathbf{R}$ be a C^k function ($k \geq 1$), $a \in \mathbf{R}^{n+1}$ and assume that $(\partial f/\partial u^i)(a) \neq 0$ for some fixed

i. Then there is a neighborhood W of a in $\mathbf{R}^{n+1}$ and a C^k function $g\colon \hat{W} \to R$ such that, for $w = (w^1, \ldots, w^{n+1}) \in \mathbf{R}^{n+1}$, $f(w^1, \ldots, w^{n+1}) = 0$ if and only if $w^i = g(\hat{w})$.

PROBLEMS

1.1. Prove that P^n is a metric space.

***1.2.** Let $U_i = \{l \in P^n \mid l = [(a^1, \ldots, a^{n+1})] \text{ and } a^i \neq 0\}$. Prove that U_i is well defined and open in P^n.

***1.3.** If $W \subset \mathbf{R}^{n+1}$ is open, show that $\hat{W}$ is open in $\mathbf{R}^n$. Show that the converse is false.

1.4. Prove Theorem 1.9 for $f(x^1, \ldots, x^{n+1}) = -1 + \sum (x^i)^2$.

7–2. MANIFOLDS—DEFINITION AND EXAMPLES

In this section we define the main object of study of this chapter, a differentiable n-manifold, and give a large collection of examples. It will be obvious from the definition of manifolds that surfaces (as defined in Chapter 4) are exactly those 2-manifolds which are subsets of $\mathbf{R}^3$. (See Section 7-5 for a more precise description.) We shall prove the important fact that if $f\colon \mathbf{R}^{n+1} \to \mathbf{R}$ is a differentiable function, then $f^{-1}(0)$ (the *hypersurface defined by f*) is an n-manifold if a certain condition on f holds.

Just as a surface looked locally like $\mathbf{R}^2$, an n-manifold will look locally like $\mathbf{R}^n$.

DEFINITION. Let M be a metric space and $p \in M$. A *coordinate chart about p of dimension n* is a neighborhood U of p and a one-to-one continuous function $\phi\colon U \to \mathbf{R}^n$ such that $\tilde{U} = \phi(U)$ is open in $\mathbf{R}^n$. (U, ϕ) is a *proper coordinate chart* if $\phi^{-1}\colon \phi(U) \to U \subset M$ is continuous.

EXAMPLE 2.1. Let $S^n = \left\{(x^1, \ldots, x^{n+1}) \in \mathbf{R}^{n+1} \,\middle|\, \sum_{i=1}^{n+1} (x^i)^2 = 1\right\}$ be the n-sphere and let $p = (0, \ldots, 1)$. A proper coordinate chart about p of dimension n is given by $U = \{(x^1, \ldots, x^{n+1}) \in S^n \mid x^{n+1} > 0\}$ and $\phi\colon U \to \mathbf{R}^n$ by $\phi(x^1, \ldots, x^{n+1}) = (x^1, \ldots, x^n)$. Note that

$$\tilde{U} = \phi(U) = B_1(0),$$

which is certainly open. Clearly

$$\phi^{-1}(u^1, \ldots, u^n) = \left(u^1, \ldots, u^n, \sqrt{1 - \sum_{i=1}^{n} (u^i)^2}\right)$$

which is continuous so that (U, ϕ) is a proper coordinate chart.

Definition. Let M be a metric space. M is an *n-dimensional C^∞ manifold* if there is a collection $\mathcal{A}$ of coordinate charts (U, ϕ), called the *atlas* of M, such that

(a) for each $p \in M$ there is a proper coordinate chart $(U, \phi) \in \mathcal{A}$ of dimension n with $p \in U$;

(b) if (U, ϕ), $(V, \psi) \in \mathcal{A}$ with $U \cap V \neq \varnothing$ then

$$\psi \circ \phi^{-1} : \phi(U \cap V) \to \psi(U \cap V)$$

is a C^∞ diffeomorphism (of open sets of $\mathbf{R}^n$); and

(c) $\mathcal{A}$ is maximal with respect to conditions (a) and (b), i.e., $\mathcal{A}$ contains all possible charts with these properties. (See Figure 7.2.)

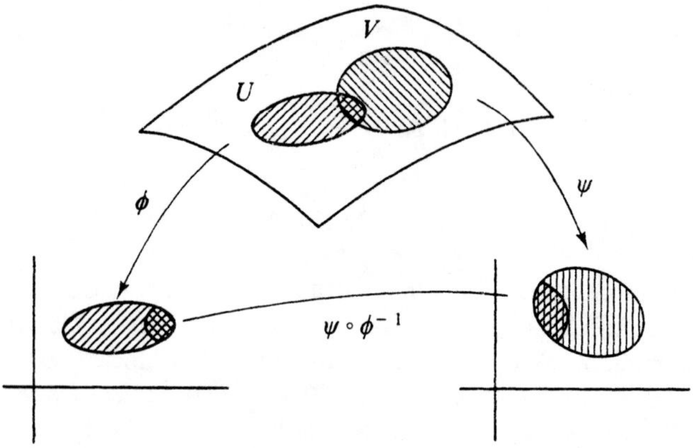

FIGURE 7.2

The content of condition (b) is that $\psi \circ \phi^{-1}$ is C^∞. Having proven this for all ψ, ϕ shows that $\psi \circ \phi^{-1}$ is a diffeomorphism since

$$(\psi \circ \phi^{-1})^{-1} = \phi \circ \psi^{-1}.$$

We will usually abbreviate "n-dimensional C^∞ manifold" by "differentiable n-manifold" or "n-manifold," or write M^n. It is possible to define a C^k manifold for any $1 \leq k < \infty$. However, in this case it is necessary to continually check the degree of differentiability. We prefer to avoid this.

In practice we usually give a collection of proper coordinate charts (U_i, ϕ_i) on M which *cover* M (i.e., $M = \bigcup_i U_i$). There is then a unique atlas determined which includes the given subcollection of (U_i, ϕ_i), much as a subbasis determines a topology. It is possible to have two different atlases on a metric space M, making M into a manifold in different ways. We shall not get involved with such problems here.

Note that this definition is essentially the same as that of a surface as given in Section 4-2, with two exceptions: the direction of the maps has been reversed and U is a subset of M instead of being a subset of $\mathbf{R}^2$. A coordinate patch was a map from $\mathbf{R}^2$ to the surface; a coordinate chart is a map from the manifold (surface) to $\mathbf{R}^n$ ($\mathbf{R}^2$). This change emphasizes the fact that the manifold is the primary object of study, not Euclidean space.

EXAMPLE 2.2. $\mathbf{R}^n$ itself is an n-manifold, for example, with the single coordinate chart ($\mathbf{R}^n$, identity). Similarly any open set in $\mathbf{R}^n$ is an n-manifold.

EXAMPLE 2.3. S^n is an n-manifold. This can be seen by using Example 2.1 and imitating the proof that S^2 is a surface. See also the proof of Theorem 2.6 below.

EXAMPLE 2.4. Let $G = GL(n, \mathbf{R})$ be the group of all nonsingular $n \times n$ matrices. We show that G is an n^2-dimensional manifold. G is a metric space with distance function $d(A, B) = \sqrt{\sum_{i,j=1}^{n} (a_{ij} - b_{ij})^2}$, where $A = (a_{ij})$ and $B = (b_{ij})$. If $A = (a_{ij}) \in G$, let

$$\phi(A) = (a_{11}, a_{12}, \ldots, a_{1n}, a_{21}, \ldots, a_{nn}) \in \mathbf{R}^{n^2}.$$

To show that (G, ϕ) is a proper coordinate chart we need only show that $\phi(G)$ is open in $\mathbf{R}^{n^2}$. We define a function $\Delta\colon \mathbf{R}^{n^2} \longrightarrow \mathbf{R}$ by

$$\Delta(x_{11}, \ldots, x_{nn}) = \sum_{\sigma \in S_n} (-1)^\sigma x_{1,\sigma(1)}\, x_{2,\sigma(2)} \cdots x_{n,\sigma(n)},$$

where S_n is the group of permutations of n letters. Note that Δ is continuous and $\Delta \circ \phi(A) = \det A$. Therefore $\phi(G) = \Delta^{-1}(\mathbf{R} - \{0\})$ is open by Proposition 1.4.

If we write $M(n)$ for the set of all $n \times n$ matrices, then we have identified $M(n)$ and $\mathbf{R}^{n^2}$ via the mapping ϕ.

EXAMPLE 2.5. We show that projective n-space P^n is an n-manifold. We define $(n + 1)$ sets $U_i = \{l \in P^n \mid l = [(a_1, \ldots, a_{n+1})]$ and $a_i \neq 0\}$. By Problem 1.2 these are open subsets of P^n. Note that $\{U_1, U_2, \ldots, U_{n+1}\}$ covers P^n. We define $\phi_i\colon U_i \longrightarrow \mathbf{R}^n$ by

$$\phi_i(l) = \left(\frac{a_1}{a_i}, \ldots, \frac{a_{i-1}}{a_i}, \frac{a_{i+1}}{a_i}, \ldots, \frac{a_n}{a_i}\right) \in \mathbf{R}^n \qquad \text{for } 1 \leq i \leq n + 1.$$

($\phi_i(l)$ is called the ith *nonhomogeneous coordinates* of l.) Note that ϕ_i is well defined and that $\phi_i(U_i) = \mathbf{R}^n$, which is certainly open. (U_i, ϕ_i) is proper because

(2-1) $$\phi_i^{-1}(u^1, \ldots, u^n) = [(u^1, u^2, \ldots, u^{i-1}, 1, u^i, \ldots, u^n)].$$

To save on notation we will only show that $\phi_2 \circ \phi_3^{-1}$ is C^∞. By (2-1)

$$\phi_2 \circ \phi_3^{-1}(u^1, \ldots, u^n) = \phi_2(u^1, u^2, 1, u^3, \ldots, u^n) = \left(\frac{u^1}{u^2}, \frac{1}{u^2}, \frac{u^3}{u^2}, \ldots, \frac{u^n}{u^2}\right).$$

Since $u^2 \neq 0$ on $\phi_3(U_2 \cap U_3)$, which is the domain of $\phi_2 \circ \phi_3{}^{-1}$, $\phi_2 \circ \phi_3{}^{-1}$ is C^∞.

Although P^2 is a 2-manifold, it is not a surface in $\mathbf{R}^3$ (see Section 7-5).

The next result is really another example, but because it includes so many familiar examples, we separate it out as a theorem. This theorem is the rigorous basis for the assertion of Section 4-2 about level surfaces of a function. Because most students have trouble with the constructions involved in this theorem, we will carry along (inside bold-face brackets **[ ]**) the special case which gives S^2.

DEFINITION. If $f\colon \mathbf{R}^{n+1} \longrightarrow \mathbf{R}$ is C^∞ (with $k \geq 1$), then the *gradient* of f is the function grad $f\colon \mathbf{R}^{n+1} \longrightarrow \mathbf{R}^{n+1}$ given by

$$(\text{grad } f)(p) = \left(\frac{\partial f}{\partial u^1}(p), \ldots, \frac{\partial f}{\partial u^{n+1}}(p)\right)\cdot$$

THEOREM 2.6. Let $f\colon \mathbf{R}^{n+1} \longrightarrow \mathbf{R}$ be a C^∞ function and

$$M_f = \{x \in \mathbf{R}^{n+1} \,|\, f(x) = 0\}.$$

If $(\text{grad } f)(p) \neq 0$ for all $p \in M_f$, then M_f is a C^∞ n-manifold (called the *hypersurface defined by* f).

Proof: We write M for M_f. **[**$f(x, y, z) = x^2 + y^2 + z^2 - 1$ for S^2**]**. Let $m_0 \in M$ be fixed. Since $(\text{grad} f)(m_0) \neq 0$, there is an $1 \leq i \leq n + 1$ such that $\partial f/\partial u^i (m_0) \neq 0$. For notational convenience we assume that $i = n + 1$. **[**Take $m_0 = (0, 0, 1)$ for S^2.**]** This can be accomplished by renumbering the variables. As in Section 7-1 if $w = (w^1, \ldots, w^{n+1}) \in \mathbf{R}^{n+1}$ and $W \subset \mathbf{R}^{n+1}$, we will write $\hat{w} = (w^1, \ldots, w^n) \in \mathbf{R}^n$ and $\hat{W} = \{\hat{w} \,|\, w \in W\}$. By Theorem 1.9 there is a neighborhood W of m_0 in $\mathbf{R}^{n+1}$ and a C^∞ function $g\colon \hat{W} \longrightarrow \mathbf{R}$ such that $w = (w^1, \ldots, w^{n+1}) \in W \cap M$ if and only if $w^{n+1} = g(w^1, \ldots, w^n)$. **[**For S^2: $W = \{(x, y, z) \,|\, x > 0\}$ and $g\colon \hat{W} \longrightarrow \mathbf{R}$ is given by

$$g(u^1, u^2) = \sqrt{1 - (u^1)^2 - (u^2)^2}.\mathbf{]}$$

Since W is open in $\mathbf{R}^{n+1}$, $U = W \cap M$ is open in M **[**U is the upper hemisphere for S^2**]**. Let $\phi\colon U \longrightarrow \mathbf{R}^n$ be defined by

$$\phi(w^1, \ldots, w^{n+1}) = (w^1, \ldots, w^n).$$

The point is that (U, ϕ) is a chart about m_0 for M. First note that $\phi(U) = \tilde{U}$, which is open by Problem 1.3. To show that ϕ is one-to-one and proper we need only write down $\phi^{-1}\colon \phi^{-1}(u^1, \ldots, u^n) = (u^1, \ldots, u^n, g(u^1, \ldots, u^n))$. **[**For S^2: $\phi^{-1}(u^1, u^2) = (u^1, u^2, \sqrt{1 - (u^1)^2 - (u^2)^2})$.**]**

We now check the overlap condition: Suppose that (V, ψ) is another patch about m_0. We may assume that this corresponds to $\partial f/\partial x^1(m_0) \neq 0$ and so by the construction above $\psi: V \longrightarrow \mathbf{R}^n$ is given by

$$\psi(x^1, \ldots, x^{n+1}) = (x^2, \ldots, x^{n+1}).$$

Therefore.

$$\begin{aligned}\psi \circ \phi^{-1}(u^1, \ldots, u^n) &= \psi(u^1, \ldots, u^n g(u^1, \ldots, u^n)) \\ &= (u^2, \ldots, u^n, g(u^1, \ldots, u^n)),\end{aligned}$$

which is C^∞ on $\phi^{-1}(V \cap U)$. [For S^2:

$$\psi \circ \phi^{-1}(u^1, u^2) = (u^2, \sqrt{1 - (u^1)^2 - (u^2)^2}).] \quad \blacksquare$$

The next example shows that Theorem 2.6 can also be used to produce fairly abstract examples.

EXAMPLE 2.7. Let $SL(n)$ be the set of all matrices of determinant one. We may view $SL(n)$ as a subset of $\mathbf{R}^{n^2}$ by stringing out coordinates as in Example 2.4. Showing that $SL(n)$ is a manifold of dimension $n^2 - 1$ amounts to investigating the function $f: \mathbf{R}^{n^2} \longrightarrow \mathbf{R}$ given by $f(A) = \det A$. Since $f^{-1}(1) = SL(n)$ we can use Theorem 2.6 once we know that $(\text{grad } f)(A) \neq 0$ for all $A \in SL(n)$. This computation is done in Guillemin and Pollack [1974].

EXAMPLE 2.8. Let $C = \{(x, y, z) \in \mathbf{R}^3 \mid x^2 + y^2 - z^2 = 0\}$ be a double cone. Then C is not a 2-manifold. (See Figure 7.3.) Suppose C were a 2-manifold. Intuitively, we can disconnect C by removing the origin which could not happen to a 2-manifold. More formally (for those who have had some topology) let (U, ϕ) be a coordinate chart about $p = (0, 0, 0)$ with U connected. Then $U - \{p\}$ must be connected since $\phi(U) - \{\phi(p)\}$ is connected. On the other hand, $U - \{p\}$ is obviously not connected (from the picture). Note that if $f(x, y, z) = x^2 + y^2 - z^2$ then $(\text{grad } f)(0, 0, 0) = \mathbf{0}$.

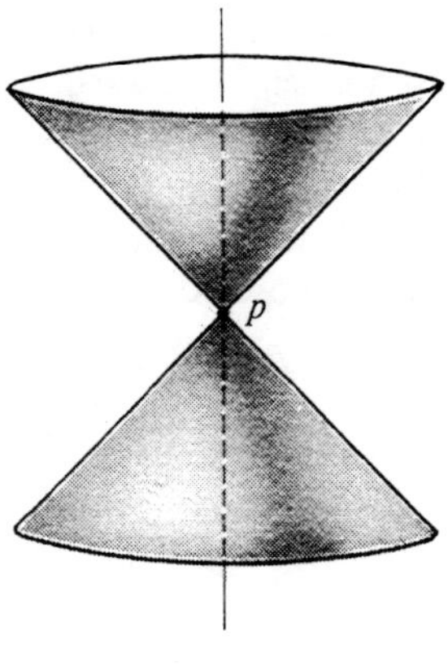

FIGURE 7.3

PROBLEMS

2.1. Prove that S^n is an n-manifold directly from the definition (i.e., without Theorem 2.6).

2.2. If M_1 (resp. M_2) is a C^∞ n_1-manifold (resp. n_2-manifold), prove that $M_1 \times M_2$ is a C^∞ n-manifold for $n = n_1 + n_2$. (*Hint:* If (U_i, ϕ_i) is a chart about $m_i \in M_i$, then make $(U_1 \times U_2, \phi_1 \times \phi_2)$ a chart about (m_1, m_2).)

2.3. Let $T^n =$

$$\{(x^1, \ldots, x^{2n}) \in \mathbf{R}^{2n} \mid (x^1)^2 + (x^2)^2 = 1, \ldots, (x^{2n-1})^2 + (x^{2n})^2 = 1\}.$$

Prove that T^n is an n-manifold. T^n is called the n-torus. (For those who have had topology: T^2 is homeomorphic to the inner tube of Chapter 4.)

2.4. Let $M = \{A \in M(n) \mid \text{trace}\,(A) = 0\}$. Prove that M is a manifold (of what dimension?). Can you picture M?

†2.5. Let $O(n) = \{A \in M(n) \mid AA^t = I\}$ be the set of orthogonal matrices. Since $O(n) \subset M(n) = \mathbf{R}^{n^2}$, we may view $O(n) \subset \mathbf{R}^{n^2}$.
(a) Show that $O(n)$ is compact (i.e., closed and bounded).
(b) Show that $O(n)$ is an $(n(n-1)/2)$-manifold.

7–3. TANGENT VECTORS AND THE TANGENT SPACE

Because the concept of a tangent vector was so useful in surface theory, we would like to define an analogous concept on an arbitrary manifold. An immediate generalization of the definition is not possible. On a surface we defined a tangent vector to be the velocity vector of some curve on the surface. This meant we viewed the curve as lying in $\mathbf{R}^3$, *where we knew how to differentiate*. In our current setting a curve on M is only a curve on M (there is no ambient Euclidean space in which it is included) and it does not make sense to differentiate in a metric space. The reader may argue that locally M is an open set in $\mathbf{R}^n$, where we do know how to differentiate. This is true, but there is no canonical way to make M look locally like Euclidean space. Different people could choose different coordinate charts and obtain different notions of differentiation.

In elementary calculus, a vector $\mathbf{v}$ at a point $p \in R^n$ may be viewed as a directional derivative. If $\mathbf{v} = (a^1, a^2, \ldots, a^n)$ and $f\colon \mathbf{R}^n \longrightarrow \mathbf{R}$ is differentiable, then the directional derivative of f at p in the direction $\mathbf{v}$ is

$$\mathbf{v}(f) = \sum a^i \frac{\partial f}{\partial u^i}(p).$$

We will imitate this idea by defining a tangent vector as a real-valued operator on the set of differentiable functions on M which obeys the properties of a derivative. First we will need the concept of a differentiable function defined on a manifold.

Let M be a C^∞ n-manifold. We shall use the word "chart" for "proper coordinate chart."

DEFINITION. Let $p \in M$ and let $f\colon M \longrightarrow \mathbf{R}$. f is *differentiable of class* C^l $(l \leq \infty)$ at p if there is a chart (U, ϕ) about p such that

$$f \circ \phi^{-1}\colon \phi(U) \longrightarrow \mathbf{R}$$

is of class C^l at $\phi(p) \in \mathbf{R}^n$.

There are two things to note about this definition. First of all, it is not circular because $\phi(U)$ is an open set in $\mathbf{R}^n$, where we know already what it means to be differentiable. Also, the definition does not really depend upon the choice of chart because if (V, ψ) is another chart, then

$$f \circ \psi^{-1} = (f \circ \phi^{-1}) \circ (\phi \circ \psi^{-1})$$

is also of class C^l.

EXAMPLE 3.1. Let $f\colon S^2 \longrightarrow \mathbf{R}$ be given by $f(x, y, z) = x^2 + z$. We show that f is differentiable at $p = (0, 0, 1)$ by using the chart (U, ϕ) in which U is the upper hemisphere and $\phi(x, y, z) = (x, y)$. Since

$$\phi^{-1}(u^1, u^2) = (u^1, u^2, \sqrt{1 - (u^1)^2 - (u^2)^2}),$$

we have $f \circ \phi^{-1}(u^1, u^2) = (u^1)^2 + \sqrt{1 - (u^1)^2 - (u^2)^2}$, which is of class C^∞ at $\phi(0, 0, 1) = (0, 0)$.

DEFINITION. If $f\colon M^m \longrightarrow N^n$ is a function between C^∞ manifolds, f is *differentiable of class C^l $(l \leq \infty)$ at $p \in M$* if there is a chart (U, ϕ) about p on M and a chart (V, ψ) about $f(p)$ on N such that $f(U) \subset V$ and $\psi \circ f \circ \phi^{-1}\colon \phi(U) \longrightarrow \psi(V)$ is *of class* C^l at $\phi(p)$. f is of class C^l if it is of class C^l at each point. If the precise value of l is unimportant, then we simply say that f is *differentiable*.

DEFINITION. $\mathfrak{F}(M) = \{f : M \longrightarrow \mathbf{R} \mid f \text{ is of class } C^\infty\}$.

EXAMPLE 3.2. The function of Example 3.1 is an element of $\mathfrak{F}(S^2)$.

DEFINITION. If G is a manifold which is also a group such that both $\mu\colon G \times G \longrightarrow G$ by $\mu(x, y) = xy$ and $\iota\colon G \longrightarrow G$ by $\iota(x) = x^{-1}$ are of class C^∞, then G is called a *Lie group*.

Under addition, Example 2.2 is a Lie group. Under matrix multiplication, Examples 2.4 and 2.7 and Problem 2.5 are Lie groups. We have defined

Lie groups because of their importance in both mathematics and physics. We shall not pursue the study of Lie groups here. The reader is urged to consult Warner [1971] for more information.

Before we continue we need certain technical results on the existence of differentiable functions with certain properties on a manifold M. We shall only sketch the proofs and leave the details to the reader.

LEMMA 3.3. Let W be an open set in M with $p \in W$. Then there is an open set V with $p \in V \subset W$ and a C^∞ function $f: M \longrightarrow \mathbf{R}$ such that
(a) $f(x) = 1$ if $x \in V$;
(b) $f(x) = 0$ if $x \notin W$; and
(c) $0 \leq f(x) \leq 1$ for all $x \in M$.

Proof: Let (U, ϕ) be a chart about p with $U \subset W$ and $\phi(p) = 0 \in \mathbf{R}^n$. Without loss of generality we may assume that $B_2(0) \subset \phi(U)$. Let h be the function $h: \mathbf{R} \longrightarrow \mathbf{R}$ of Problem 3.6. Define $g: \mathbf{R}^n \longrightarrow \mathbf{R}$ by $g(y) = h(|y|^2)$ and define $f: M \longrightarrow \mathbf{R}$ by

$$f(x) = \begin{cases} (g \circ \phi)(x) & \text{if } x \in \phi^{-1}(B_2(0)) \\ 0 & \text{otherwise.} \end{cases}$$

Then this f satisfies the conditions of the lemma if $V = \phi^{-1}(B_{1/2}(0))$. ∎

LEMMA 3.4. Let W be a neighborhood of $p \in M$ and let $F: W \longrightarrow \mathbf{R}$ be differentiable. Then there exists a differentiable function $\tilde{F}: M \longrightarrow \mathbf{R}$ and an open set V with $p \in V \subset W$ such that $\tilde{F}(x) = F(x)$ for all $x \in V$.

Proof: Let f and V be as in Lemma 3.3 and define $\tilde{F}$ by

$$\tilde{F}(x) = \begin{cases} f(x)F(x) & \text{if } x \in W \\ 0 & \text{if } x \notin W. \end{cases}$$ ∎

LEMMA 3.5. Let $p \in M$ and let W be a neighborhood of p. Then
(a) there is a differentiable function $f: M \longrightarrow \mathbf{R}$ with $f(p) = 1, f(x) = 0$ if $x \notin W$; and
(b) there is a differentiable function $g: M \longrightarrow \mathbf{R}$ with $g(p) = 0, g(x) = 1$ if $x \notin W$.

Proof: Problem 3.7. ∎

We are now able to define the concept of a tangent vector and prove that the set of tangent vectors to M^n at a point forms a vector space of dimension n.

DEFINITION. A *tangent vector to M at p* is a function $X_p: \mathfrak{F}(M) \longrightarrow \mathbf{R}$, whose value at f is denoted $X_p f$ or $X_p(f)$, such that for all $f, g \in \mathfrak{F}(M)$ and $r \in \mathbf{R}$,

(a) $X_p(f+g) = X_p(f) + X_p(g)$;
(b) $X_p(rf) = rX_p(f)$; and
(c) $X_p(fg) = f(p)X_p(g) + g(p)X_p(f)$, where fg is the ordinary product of functions f and g and $f(p)X_p(g)$ is the product of real numbers $f(p)$ and $X_p(g)$.

We urge the reader to read "$X_p f$" as "the directional derivative of f in the direction X_p at p." Condition (c) is the product rule for derivatives.

LEMMA 3.6. Let X_p be a tangent vector at $p \in M$.
(a) If $f, g \in \mathfrak{F}(M)$ with $f(p) = g(p) = 0$, then $X_p(fg) = 0$.
(b) If $\mathbf{r}$ is the constant function $\mathbf{r}(m) \equiv r$, then $X_p(\mathbf{r}) = 0$.
(c) If $f, g \in \mathfrak{F}(M)$ and $f(x) = g(x)$ for all x in some neighborhood W of p, then $X_p(f) = X_p(g)$.

Proof: (a) This is immediate from the definition of a tangent vector.
(b) Let $\mathbf{1}$ be the constant function $\mathbf{1}(x) \equiv 1$. Then

$$X_p(\mathbf{1}) = X_p(\mathbf{1}\cdot\mathbf{1}) = X_p(\mathbf{1}) + X_p(\mathbf{1}) = 2X_p(\mathbf{1}).$$

Hence $X_p(\mathbf{1}) = 0$ and $X_p(\mathbf{r}) = X_p(r\mathbf{1}) = rX_p(\mathbf{1}) = 0$.
(c) It suffices to show that if $h \equiv 0$ on W, then $X_p(h) = 0$. (Let $h = f - g$.) By Lemma 3.5 there is a differentiable function $\mu: M \longrightarrow \mathbf{R}$ such that $\mu(p) = 0$ and $\mu(x) = 1$ if $x \notin W$.
We first show that $(h\mu)(x) = h(x)$ for all $x \in M$. If $x \in W$, then

$$(h\mu)(x) = h(x)\mu(x) = 0\cdot\mu(x) = 0 = h(x).$$

If $x \notin W$, then $(h\mu)(x) = h(x)\mu(x) = h(x)$. Thus $h\mu = h$. Hence

$$X_p(h) = X_p(h\mu) = h(p)X_p(\mu) + \mu(p)X_p(f) = 0. \quad \blacksquare$$

Because of part (c) of this lemma, it makes sense to differentiate a function defined only in a neighborhood U of p: if $f \in \mathfrak{F}(U)$, then $X_p(f)$ is defined to be $X_p(\tilde{f})$ where $\tilde{f}$ is any function in $\mathfrak{F}(M)$ that agrees with f on some neighborhood of p. Lemma 3.6(c) says this does not depend upon the choice of $\tilde{f}$ and Lemma 3.4 says that such an extension $\tilde{f}$ of f exists.

EXAMPLE 3.7. Let $\alpha: (-\epsilon, \epsilon) \longrightarrow M$ be a differentiable curve in M with $\alpha(0) = p$. Let X_p^α be defined by $X_p^\alpha(f) = (d(f \circ \alpha)/dt)(0)$, where the derivative on the right-hand side is the usual derivative of the real-valued function of a real variable, $f \circ \alpha$. That X_p^α is a tangent vector at p is the content of Problem 3.5.

This example shows that our new concept of tangent vector agrees with that on surfaces. In Proposition 7.4 of Chapter 4 we showed that a tangent vector X_p to a surface may be viewed as a directional derivative by choosing some curve whose velocity vector is X_p and then differentiating along the curve. This is what Example 3.7 is doing.

We now define some important tangent vectors $(\partial/\partial x^i)_p$ to M at p. They will play the same role for manifolds as the $\mathbf{x}_i$ did in surface theory. In particular, they will serve as a basis of the tangent space of M at p.

DEFINITION. Let (U, ϕ) be a chart about $p \in M$ and let $u^1, u^2, \ldots, u^n$ be Cartesian coordinates in $\mathbf{R}^n$. Then $(\partial/\partial x^i)_p$ is the tangent vector given by

$$\left(\frac{\partial}{\partial x^i}\right)_p (f) = \frac{\partial (f \circ \phi^{-1})}{\partial u^i}(\phi(p)). \tag{3-1}$$

(That $(\partial/\partial x^i)_p$ really is a tangent vector is Problem 3.4.)

EXAMPLE 3.8. Let $f\colon S^2 \longrightarrow \mathbf{R}$ be given by $f(x, y, z) = x^2 + z$ and let $p = (1/\sqrt{2}, 0, 1/\sqrt{2})$. If (U, ϕ) is the chart given by the upper hemisphere U and the function $\phi(x, y, z) = (x, y)$, then

$$(f \circ \phi^{-1})(u^1, u^2) = (u^1)^2 + \sqrt{1 - (u^1)^2 - (u^2)^2}$$

and $\phi(p) = (1/\sqrt{2}, 0)$. Hence

$$\begin{aligned}\left(\frac{\partial}{\partial x^1}\right)_p (f) &= \frac{\partial (f \circ \phi^{-1})}{\partial u^1}\left(\frac{1}{\sqrt{2}}, 0\right) \\ &= \left(2u^1 - \frac{u^1}{\sqrt{1 - (u^1)^2 - (u^2)^2}}\right)\left(\frac{1}{\sqrt{2}}, 0\right) \\ &= \sqrt{2} - 1.\end{aligned}$$

Similarly

$$\left(\frac{\partial}{\partial x^2}\right)_p (f) = \left(\frac{-u^2}{\sqrt{1 - (u^1)^2 - (u^2)^2}}\right)\left(\frac{1}{\sqrt{2}}, 0\right) = 0.$$

We cannot stress enough that the value of $(\partial/\partial x^i)_p f$ depends critically on the chart used. See Problem 3.1.

DEFINITION. The *i*th (*local*) *coordinate function* on M with respect to a chart (U, ϕ) is the function $x^i\colon U \longrightarrow \mathbf{R}$ given by $x^i(m) = u^i(\phi(m))$.

Note that since $x^i\colon U \longrightarrow \mathbf{R}$, it makes sense to compute $X_p x^i$ if $p \in U$. In particular, we have as a result of Equation (3-1)

$$\left(\frac{\partial}{\partial x^i}\right)_p (x^j) = \delta_i^{\,j} \qquad \text{for all } 1 \leq i, j \leq n. \tag{3-2}$$

As on a surface, we have parametric curves given by all but one parameter u^i held constant. It should be true (by the general "poetry" of mathematics) that if α_i is a parametric curve then $X_p^{\alpha_i}$ (which on a surface was

$\mathbf{x}_i(p)$) corresponds to $(\partial/\partial x^i)_p$. The following computation shows that this is indeed the case.

Let $p = \phi^{-1}(a)$ and write $a = (a^1, \ldots, a^n) \in \mathbf{R}^n$. Let

$$\alpha_i(t) = \phi^{-1}(a^1, \ldots, a^i + t, \ldots, a^n).$$

This is exactly the analogue of a parametric curve. (Compare with the definition of parametric curve in Section 4-1, remembering that for a surface ϕ^{-1} is $\mathbf{x}$.) We show that $X_p^{\alpha_i} = (\partial/\partial x^i)_p$ by showing that $X_p^{\alpha_i} f = (\partial f/\partial x^i)(p)$ for all $f \in \mathfrak{F}(M)$.

$$X_p^{\alpha_i} f = \frac{d(f \circ \alpha_i)}{dt}(0) = \frac{\partial(f \circ \phi^{-1})}{\partial u^i}(a^1, \ldots, a^n) = \left(\frac{\partial}{\partial x^i}\right)_p (f)$$

where the next to the last equality uses the chain rule and the last equality is Equation (3-1).

Definition. The *tangent space* to M at p, T_pM, is the set of all tangent vectors to M at p.

Proposition 3.9. T_pM is a vector space.

Proof: If $X_p, Y_p \in T_pM$ and $r \in \mathbf{R}$, we define $X_p + Y_p$ and rX_p by

$$(X_p + Y_p)(f) = X_pf + Y_pf \quad \text{and} \quad (rX_p)f = r \cdot X_p(f).$$

By Problem 3.3, $X_p + Y_p$ and rX_p belong to T_pM. It is then trivial to check the rest of the axioms of a vector space. Note that the 0 of this vector space is the operator which is zero on all functions. ∎

We next wish to show that the set $\{(\partial/\partial x^i)_p \mid 1 \leq i \leq n\}$ is a basis of T_pM. This requires the following technical result which is essentially Taylor's Theorem with remainder.

Lemma 3.10. If $F: \mathbf{R}^n \to \mathbf{R}$ is a function of class C^k and $(a^1, \ldots, a^n) \in \mathbf{R}^n$, then there exists functions h_{ij} of class C^{k-2} such that

$$F(u^1, \ldots, u^n) = F(a^1, \ldots, a^n) + \sum \frac{\partial F}{\partial u^i}(a^1, \ldots a^n)(u^i - a^i)$$
$$+ \sum h_{ij}(u^1, \ldots, u^n)(u^i - a^i)(u^j - a^j)$$

Proof: Let $u = (u^1, \ldots, u^n)$ and $a = (a^1, \ldots, a^n)$. Then

$$\begin{aligned}F(u) - F(a) &= \int_0^1 \frac{d}{dt}(F(t(u - a) + a))\, dt \\ &= \int_0^1 \left(\sum \frac{\partial F}{\partial u^i}(t(u - a) + a)(u^i - a^i)\right) dt \\ &= \sum (u^i - a^i) \int_0^1 \frac{\partial F}{\partial u^i}(t(u - a) + a)\, dt = \sum g_i(u)(u^i - a^i).\end{aligned}$$

Note that $g_i(a) = \int_0^1 (\partial F/\partial u^i)(t(a-a)+a)\,dt = (\partial F/\partial u^i)(a)$. Similarly $g_i(u) - g_i(a) = \sum h_{ij}(u)(u^j - a^j)$. Hence

$$\begin{aligned} F(u) &= F(a) + \sum g_i(u)(u^i - a^i) \\ &= F(a) + \sum (g_i(a) + \sum h_{ij}(u)(u^j - a^j))(u^i - a^i) \\ &= F(a) + \sum g_i(a)(u^i - a^i) + \sum h_{ij}(u)(u^i - a^i)(u^j - a^j) \\ &= F(a) + \sum \frac{\partial F}{\partial u_i}(a)(u^i - a) + \sum h_{ij}(u)(u^i - a^i)(u^j - a^j). \quad \blacksquare \end{aligned}$$

PROPOSITION 3.11. Let (U, ϕ) be a chart about $p \in M$. Then

$$\mathcal{B} = \{(\partial/\partial x^i)_p \,|\, 1 \leq i \leq n\}$$

is a basis of T_pM. In particular, T_pM has dimension n.

Proof: First we show $\mathcal{B}$ is linearly independent. Suppose $\sum c^i(\partial/\partial x^i)_p = 0$. Then we must have $0 = \sum c^i(\partial/\partial x^i)_p(x^j) = \sum c^i \delta_i^j = c^j$. Hence $\mathcal{B}$ is linearly independent.

Now we must show that $\mathcal{B}$ spans T_pM. Let $f \in \mathfrak{F}(M)$, $a = \phi(p)$, and set $F = f \circ \phi^{-1}$. By Lemma 3.10 we have for $m \in U$

$$f(m) = F \circ \phi(m) = F \circ \phi(p) + \sum \frac{\partial F}{\partial u^i}(\phi(p))(x^i(m) - a^i)$$
$$+ \sum h_{ij}(\phi(m))(x^i(m) - a^i)(x^j(m) - a^j)$$

or

$$f = f(p) + \sum \left[\left(\frac{\partial}{\partial x^i}\right)_p (f)\right](x^i - a^i) + \sum (h_{ij} \circ \phi)(x^i - a^i)(x^j - a^j).$$

Hence, since $f(p)$ is a constant, $X_p(f(p)) = 0$ by Lemma 3.6(b) and

$$\begin{aligned} X_p(f) &= X_p(f(p)) + \sum \left[\left(\frac{\partial}{\partial x^i}\right)_p (f)\right] X_p(x^i - a^i) \\ &\quad + \sum X_p[(h_{ij} \circ \phi)(x^i - a^i)(x^j - a^j)] \\ &= 0 + \sum X_p(x_i)\left(\frac{\partial}{\partial x^i}\right)_p (f) + 0, \end{aligned}$$

where the last term is zero by Lemma 3.6(a) because it is X_p on the product of the two functions $(h_{ij} \circ \phi)(x^i - a^i)$ and $(x^j - a^j)$ both of which vanish at $p = \phi^{-1}(a^1, \ldots, a^n)$. Thus

$$X_p = \sum X_p(x^i)\left(\frac{\partial}{\partial x^i}\right)_p \tag{3-3}$$

and $\mathcal{B}$ spans T_pM. Hence $\mathcal{B}$ is a basis of T_pM. $\blacksquare$

EXAMPLE 3.12. Let $M = \mathbf{R}^n$ and $p \in \mathbf{R}^n$. Since $(\mathbf{R}^n, id)$ is a chart for all of R^n, we have $\{(\partial/\partial x^i)_p \,|\, 1 \leq i \leq n\}$ is a basis for T_pM for each $p \in \mathbf{R}^n$. If $X_p \in T_p\mathbf{R}^n$, then $X_p = \sum a^i(\partial/\partial x^i)_p$ for some numbers $a^1, a^2, \ldots, a^n$.

Hence we may associate with X_p the $2n$-tuple $(p; a^1, \ldots, a^n)$. Under this identification, the effect of $(p; a^1, \ldots, a^n)$ on a function f is $\sum a^i(\partial f/\partial u^i)_p$, which is exactly the Euclidean idea of directional derivative. We frequently write $X = (p; a^1, \ldots, a^n)$ (if $X \in T_p\mathbf{R}^n$) and $T_p\mathbf{R}^n = \mathbf{R}^n$.

If M_f is the hypersurface defined by f, we will show in Section 7-5 a way to visualize T_pM_f as a subspace of $T_p\mathbf{R}^{n+1}$ for each $p \in M_f$.

PROBLEMS

3.1. Let $f\colon S^2 \longrightarrow \mathbf{R}$ and $p \in S^2$ as in Example 3.8. Let (V, ψ) be the chart defined by $V =$ right hemisphere and $\psi(x, y, z) = (x, z)$. Show that with respect to this chart $(\partial f/\partial x^1)_p = \sqrt{2}$ and $(\partial f/\partial x^2)_p = 1$. Note that this is a different result than was obtained by using the chart of Example 3.8.

3.2. Let $\Phi\colon S^n \longrightarrow P^n$ be defined by $\Phi(a) =$ the line through the origin and a. Show that Φ is C^k for all k and that Φ is two-to-one. (If you have had some algebraic topology, Φ is actually a (double) covering map.)

***3.3.** If $X_p, Y_p \in T_pM$ and $r \in \mathbf{R}$, prove that both $X_p + Y_p$ and rX_p are in T_pM.

***3.4.** If (U, ϕ) is a coordinate chart about p, prove that $(\partial/\partial x^i)_p$ as defined by Equation (3-1) is a tangent vector at p for each i.

***3.5.** If $\alpha\colon (-\epsilon, \epsilon) \longrightarrow M$ is a curve with $\alpha(0) = p$, show that $X_p^\alpha \in T_pM$.

***3.6.** The point of this problem is to show that there is a C^∞ function $h\colon \mathbf{R} \longrightarrow \mathbf{R}$ such that $0 \leq h(x) \leq 1$ for all $x \in \mathbf{R}$, $h(x) = 1$ if $|x| \leq \frac{1}{4}$ and $h(x) = 0$ if $|x| \geq 1$. The graph of h will look like that of Figure 7.4. h is called a *bump function.*

(a) Let

$$h_1(x) = \begin{cases} e^{-1/x^2} & \text{if } x > 0 \\ 0 & \text{if } x \leq 0. \end{cases}$$

Show that $h_1(x)$ is C^k for all k.

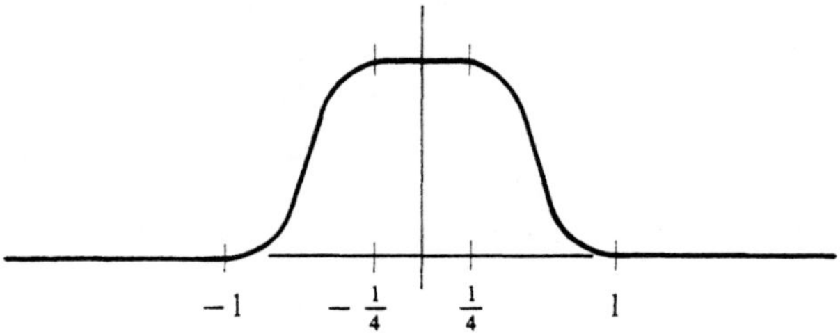

FIGURE 7.4

(b) Let $h_2(x) = h_1(x+1)h_1(-\frac{1}{4} - x)$. Show that $h_2(x) > 0$ if $-1 < x < -\frac{1}{4}$ but $h_2(x) = 0$ otherwise.

(c) Show that $h_3(x) = 0$ if $x \leq -1$, $h_3(x) = 1$ if $x \geq -\frac{1}{4}$, and $0 \leq h_3(x) \leq 1$ for all $x \in \mathbf{R}$, where

$$h_3(x) = \frac{\int_{-1}^{x} h_2(t)\,dt}{\int_{-1}^{-1/4} h_2(t)\,dt}.$$

(d) Show the existence of the function $h\colon \mathbf{R} \to \mathbf{R}$ described above. (*Hint:* Let $h(x) = h_3(x)h_3(-x)$.)

***3.7.** Prove Lemma 3.5.

7–4. VECTOR FIELDS AND LIE BRACKETS

In this section we introduce two concepts (vector fields and Lie brackets) which we will need later. Let M be a fixed C^∞ n-manifold.

DEFINITION. A *field of vectors* X is an assignment of a tangent vector $X_p \in T_pM$ to each $p \in M$. If X is a field of vectors and $f \in \mathfrak{F}(M)$, then we may define a real-valued function Xf on M by $(Xf)(p) = X_pf$.
If $Xf \in \mathfrak{F}(M)$ for each $f \in \mathfrak{F}(M)$, then X is called a *vector field*.

Note that if X is a vector field and (U, ϕ) is a patch, then (since the x^i are differentiable functions) Equation (3-3) shows that X can be written as $X_p = \sum X^i(p)(\partial/\partial x^i)_p$ for $p \in U$.

This is usually written $X = \sum X^i(\partial/\partial x^i)$. This definition makes sense if we only want X to be defined on an open set U of M, in which case we have a vector field on U. We will write $\mathfrak{X}(M)$ for the set of all vector fields on M. The zero vector field, X, is defined by $X_p = 0 \in T_pM$. There are certainly nonzero vector fields in $\mathfrak{X}(M)$, as Problem 4.2 shows. The question of whether, given M, there is a vector field which is never zero on M is much more difficult. (*Remember:* X is nonzero if $X_p \neq 0$ for some p. We are looking for a vector field X with $X_p \neq 0$ for all p.) In Theorem 7.3 of Chapter 6 we proved that for a compact surface there was a vector field on M which is never zero only if the Euler characteristic of M is zero. This theorem generalizes to compact, orientable, even-dimensional manifolds for a suitable definition of Euler characteristic. See Milnor [1965].

Since T_pM is a real vector space, it comes as no surprise that $\mathfrak{X}(M)$ is also a vector space (of infinite dimension). We can also define fX for $f \in \mathfrak{F}(M)$ and $X \in \mathfrak{X}(M)$. More precisely, if $X, Y \in \mathfrak{X}(M)$, $r \in \mathbf{R}$, and $f \in \mathfrak{F}(M)$, we may define $X + Y$, rX, and fX by

$$(X+Y)_p = X_p + Y_p$$

$$(rX)_p = r \cdot X_p$$

and

$$(fX)_p = f(p)X_p.$$

The reader is warned: fX is a vector field while Xf is a function.

Note that Xf is a C^∞ function so that $Y_p(Xf)$ makes sense.

DEFINITION. If $X, Y \in \mathfrak{X}(M)$, then the *Lie bracket* of X and Y, $[X, Y]$, is the field of vectors defined by

(4-1) $\quad [X, Y]_p f = X_p(Yf) - Y_p(Xf) \qquad \text{for } f \in \mathfrak{F}(M) \text{ and } p \in M.$

LEMMA 4.1. $[X, Y]$ is a vector field on M.

Proof: It is immediate from (4-1) that $[X, Y]_p(f+g) = [X, Y]_p f + [X, Y]_p g$ and $[X, Y]_p(rf) = r[X, Y]_p f$ for $r \in R$ and $f, g \in \mathfrak{F}(M)$. To show that $[X, Y]_p \in T_pM$ we compute $[X, Y]_p(fg)$. Note that since

$$Y_p(fg) = f(p)Y_p g + g(p)Y_p f,$$

we have $Y(fg) = fYg + gYf$, where we are multiplying the two differentiable functions in each term of the right-hand side together (i.e., f and Yg are multiplied as functions). Now

$$\begin{aligned}
[X, Y]_p(fg) &= X_p(Y(fg)) - Y_p(X(fg)) \\
&= X_p(fYg + gYf) - Y_p(fXg + gXf) \\
&= f(p)X_p Yg + (Yg)(p)X_p f + g(p)X_p Yf + (Yf)(p)X_p g \\
&\quad - f(p)Y_p Xg - (Xg)(p)Y_p f - g(p)Y_p Xf - (Xf)(p)Y_p g \\
&= f(p)[X, Y]_p g + g(p)[X, Y]_p f.
\end{aligned}$$

Therefore $[X, Y]_p \in T_pM$.

To show that $[X, Y] \in \mathfrak{X}(M)$ we note that Yf and Xf are in $\mathfrak{F}(M)$; then so are $X(Yf)$, $Y(Xf)$, and their difference $[X, Y]f$. ∎

EXAMPLE 4.2. Let $M = \mathbf{R}^2$, $X = x^1x^2(\partial/\partial x^1)$ and $Y = x^2(\partial/\partial x^2)$. If $g(x^1, x^2) = (x^1)^2x^2$, then $gX = (x^1)^3(x^2)^2(\partial/\partial x^1)$ but

$$Xg = x^1x^2(2x^1x^2) = 2(x^1)^2(x^2)^2.$$

If $f \in \mathfrak{F}(R^2)$, then

$$\begin{aligned}
[X, Y]f &= x^1x^2 \frac{\partial}{\partial x^1}\left(x^2 \frac{\partial f}{\partial x^2}\right) - x^2 \frac{\partial}{\partial x^2}\left(x^1x^2 \frac{\partial f}{\partial x^1}\right) \\
&= x^1(x^2)^2 \frac{\partial^2 f}{\partial x^1\,\partial x^2} - x^2x^1 \frac{\partial f}{\partial x^1} - x^1(x^2)^2 \frac{\partial^2 f}{\partial x^2\,\partial x^1} \\
&= -x^2x^1 \frac{\partial f}{\partial x^1}.
\end{aligned}$$

Thus $[X, Y] = -x^1x^2(\partial/\partial x^1)$.

The next to last line in the above computation is quite revealing because in it we see that second order terms (the mixed partials) cancel out. This is the reason that, although $f \longrightarrow X_p(Yf)$ is not a tangent vector, $f \longrightarrow X_p(Yf) - Y_p(Xf)$ is a tangent vector.

The following lemma is left as an exercise.

LEMMA 4.3. If $X, Y, Z \in \mathfrak{X}(M)$ and $r \in \mathbf{R}$, then

(a) $[X, Y] = -[Y, X]$ and $[rX, Y] = r[X, Y]$;

(b) $[X + Y, Z] = [X, Z] + [Y, Z]$, $[Z, X + Y] = [Z, X] + [Z, Y]$;

(c) (Jacobi's Identity.) $[[X, Y], Z] + [[Y, Z], X] + [[Z, X], Y] = 0$.

Note in particular that the Lie bracket "multiplication" of vector fields is not associative but instead satisfies the rather strange substitute of Lemma 4.3(c). Conditions (a), (b), and (c) should remind you of the cross product of vectors in $\mathbf{R}^3$.

PROBLEMS

***4.1.** Let $\tilde{X}_p \in T_pM$ for some point $p \in M$. Show that there is a neighborhood U of p and a vector field X defined on U such that $X_p = \tilde{X}_p$. (*Hint:* Use local coordinates.)

***4.2.** Let $\tilde{X}_p \in T_pM$ for some point $p \in M$. Show that there is a vector field, X, on M such that $X_p = \tilde{X}_p$. (*Hint:* Use Problem 4.1 and Lemma 3.5.) In particular, there is a nonzero vector field on M.

4.3. Let (U, ϕ) be a chart with $\{(\partial/\partial x^1)_p, \ldots, (\partial/\partial x^n)_p\}$ as basis for T_pM for $p \in U$.

(a) Show that $\left[\frac{\partial}{\partial x^i}, \frac{\partial}{\partial x^j}\right] = 0$.

(b) Show that if $X = \sum X^i(\partial/\partial x^i)$ and $Y = \sum Y^j(\partial/\partial x^j)$, then

$$[X, Y] = \sum \left(X^i \frac{\partial Y^j}{\partial x_i} - Y^i \frac{\partial X^j}{\partial x^i}\right) \frac{\partial}{\partial x^j}.$$

Compare this with Problems 9.5 and 9.8 of Chapter 4.

4.4. If $f, g \in \mathfrak{F}(M)$ and $X, Y \in \mathfrak{X}(M)$, show that

$$[fX, gY] = fg[X, Y] + f(Xg)Y - g(Yf)X.$$

***4.5.** Prove Lemma 4.3.

4.6. M^n is called *parallelizable* if there are n vector fields, $X_1, \ldots, X_n$, such that $\{(X_1)_p, \ldots, (X_n)_p\}$ is a basis for T_pM for each $p \in M$. Prove that $\mathbf{R}^n$ and T^n are parallelizable. It is often quite difficult to decide if a manifold is parallelizable. S^n is parallelizable if and only if $n = 1, 3$, or 7. $S^p \times S^q$ is parallelizable if p and q are both odd. (Do not try to prove these last assertions!)

7–5. THE DIFFERENTIAL OF A MAP AND SUBMANIFOLDS

In this section we introduce the concept of the differential of a mapping between manifolds and then apply this notion to the idea of submanifolds of a given manifold. It is this last idea that will provide many examples of the geometric concepts which have been treated already and those which follow. In particular, we will have a concrete way of visualizing the tangent space of the hypersurface defined by a function f. We shall assume throughout that M and N are manifolds and that $\Phi : M \longrightarrow N$ is differentiable. We want to define for each $X_p \in T_pM$ an element $(\Phi_*)_p(X_p)$ of $T_{\Phi(p)}N$ in such a way that the assignment $X_p \longrightarrow (\Phi_*)_p(X_p)$ is a linear transformation for each $p \in M$. One way of doing this would be to define $(\Phi_*)_p$ on a basis of T_pM, i.e., view $(\Phi_*)_p$ as a matrix instead of a linear transformation. There is a natural candidate for the matrix definition of $(\Phi_*)_p$—just take the Jacobian of $\psi \circ \Phi \circ \phi^{-1}$ (evaluated at $\phi(p)$) where (U, ϕ) is a chart about p and (V, ψ) is a chart about $\Phi(p)$. The trouble with this definition is that it is not an invariant definition; that is, the definition itself depends on a coordinate chart. This is contrary to our desire for invariance. It also causes many technical problems (in the proof of other results) involving picking the "right" coordinate chart to do the computations. The definition which we will use avoids these technical and aesthetic problems but is, unfortunately, quite abstract.

DEFINITION. If $\Phi: M \longrightarrow N$ is differentiable, the *differential of* Φ *at* p is the function $(\Phi_*)_p: T_pM \longrightarrow T_{\Phi(p)}N$ defined by

$$(\Phi_*)_p(X_p)(f) = X_p(f \circ \Phi) \tag{5-1}$$

where $X_p \in T_pM$ and $f \in \mathfrak{F}(N)$.

LEMMA 5.1. $(\Phi_*)_p$ is well defined; that is, if $\Phi: M \longrightarrow N$ and $X_p \in T_pM$, then $(\Phi_p)_*(X_p) \in T_{\Phi(p)}N$.

Proof: For notational convenience we drop the subscript p in this proof. We must show that, for all $f, g \in \mathfrak{F}(N)$ and $r \in \mathbf{R}$,

(a) $\Phi_*(X)(f+g) = \Phi_*(X)(f) + \Phi_*(X)(g)$,

(b) $\Phi_*(X)(rf) = r\Phi_*(X)(f)$, and

(c) $\Phi_*(X)(fg) = f(\Phi(p))\Phi_*(X)(g) + g(\Phi(p))\Phi_*(X)(f)$.

First we prove (a):

$$\Phi_*(X)(f+g) = X((f+g)\circ\Phi) = X(f\circ\Phi) + X(g\circ\Phi)$$
$$= \Phi_*(X)(f) + \Phi_*(X)(g),$$

where we have used Equation (5-1) repeatedly.

Next we prove (b):

$$\Phi_*(X)(rf) = X(rf\circ\Phi) = rX(f\circ\Phi) = r\Phi_*(X)(f).$$

For (c) we note that $((fg)\circ\Phi)(q) = (fg)(\Phi(q)) = f(\Phi(q))g(\Phi(q))$ for any $q \in M$ so that

$$(fg)\circ\Phi = (f\circ\Phi)(g\circ\Phi).$$

Hence,

$$\Phi_*(X)(fg) = X((fg)\circ\Phi) = X((f\circ\Phi)(g\circ\Phi))$$
$$= (f\circ\Phi)(p)X(g\circ\Phi) + (g\circ\Phi)(p)X(f\circ\Phi)$$
$$= f(\Phi(p))\Phi_*(X)g + g(\Phi(p))\Phi_*(X)f. \quad \blacksquare$$

PROPOSITION 5.2. Let $\Phi: M \longrightarrow N$ and $p \in M$. Then $(\Phi_*)_p: T_pM \longrightarrow T_{\Phi(p)}N$ is a linear transformation.

Proof: Let $r \in \mathbf{R}$, $X_p, Y_p \in T_pM$. We shall prove that

$$(\Phi_*)_p(rX_p + Y_p) = r(\Phi_*)_pX_p + (\Phi_*)_pY_p.$$

For any $f \in \mathfrak{F}(N)$:

$$((\Phi_*)_p(rX_p + Y_p))(f) = (rX_p + Y_p)(f\circ\Phi)$$
$$= rX_p(f\circ\Phi) + Y_p(f\circ\Phi)$$
$$= r(\Phi_*)_p(X_p)f + (\Phi_*)_p(Y_p)f. \quad \blacksquare$$

PROPOSITION 5.3. If $\Phi: M \longrightarrow N$ and $\Psi: N \longrightarrow P$ are differentiable maps of the manifolds M, N, and P, if $p \in M$ and $q = \Phi(p)$, then

$$((\Psi\circ\Phi)_*)_p = (\Psi_*)_q \circ (\Phi_*)_p.$$

(More succinctly: $(\Psi\circ\Phi)_* = \Psi_*\circ\Phi_*$.)

Proof: Problem 5.1. $\blacksquare$

PROPOSITION 5.4. Let $\Phi: M \longrightarrow N$, $p \in M$. Let (U, ϕ) be a chart about p and (V, ψ) be a chart about $q = \Phi(p)$ such that $\Phi(U) \subset V$. If we write $\{(\partial/\partial x^i)_p \mid i = 1, \ldots, m\}$ for the basis of T_pM derived from (U, ϕ) and $\{(\partial/\partial y^i)_q \mid i = 1, \ldots, n\}$ for the basis of T_qN derived from (V, ψ), then the matrix of $(\Phi_*)_p$ with respect to these bases is the Jacobian of $\psi\circ\Phi\circ\phi^{-1}$:

$$(\Phi_*)_p\left(\left(\frac{\partial}{\partial x^i}\right)_p\right) = \sum_j \frac{\partial(y^j\circ\Phi\circ\phi^{-1})}{\partial u^i}(s)\left(\frac{\partial}{\partial y^j}\right)_q, \tag{5-2}$$

where $s = \phi(p)$.

Proof: By Equation (3-3) we know that

$$(\Phi_*)_p\left(\frac{\partial}{\partial x^i}\right)_p = \sum_j \left[(\Phi_*)_p\left(\frac{\partial}{\partial x^i}\right)_p(y^j)\right]\left(\frac{\partial}{\partial y^j}\right)_{\Phi(p)}$$

$$= \sum_j \left[\left(\frac{\partial}{\partial x^i}\right)_p(y^j \circ \Phi)\right]\left(\frac{\partial}{\partial y^j}\right)_{\Phi(p)}$$

An application of (3-1) gives (5-2). ▮

It is also convenient to write $\Phi^j = y^j \circ \Phi$ so that

$$\psi \circ \Phi(p) = (\Phi^1(p), \ldots, \Phi^n(p)).$$

With this notation (5-2) becomes

$$(5\text{-}3) \qquad (\Phi_*)_p\left(\frac{\partial}{\partial x^i}\right)_p = \sum_{j=1}^{n} \frac{\partial \Phi^j}{\partial x^i}(p)\left(\frac{\partial}{\partial y^j}\right)_q$$

Example 5.5. Let $M = M_f$ be the hypersurface induced by $f\colon \mathbf{R}^{n+1} \longrightarrow \mathbf{R}$. (See Theorem 2.6.) Let $\Phi\colon M \longrightarrow \mathbf{R}^{n+1}$ be inclusion. Let $p \in M$ such that $\partial f/\partial u^{n+1}(p) \neq 0$. In the natural coordinates for $\mathbf{R}^{n+1}$ and the coordinates about p given by Theorem 2.6 we see that

$$(5\text{-}4) \qquad (\Phi_p)_*\left(\frac{\partial}{\partial x^i}\right)_p = e_i(p) + \frac{\partial g}{\partial u^i}(p)e_{n+1}(p)$$

where $e_i(p) = (p; 0, \ldots, 1, 0, \ldots, 0)$ as discussed in Example 3.12.

Note that in the notation of the proposition, $V = \mathbf{R}^{n+1}$, ψ is the identity, and $\phi^{-1}(u^1, \ldots, u^n) = (u^1, \ldots, u^n, g(u^1, \ldots, u^n))$. Thus $\psi \circ \Phi \circ \phi^{-1}(u^1, \ldots, u^n) = (u^1, \ldots, u^n, g(u^1, \ldots, u^n))$ and Equation (5-4) follows by computing the Jacobian of $\psi \circ \Phi \circ \phi^{-1}$.

Definition. Let M^m and N^n be manifolds. M is a *submanifold* of N if there is a differentiable function $\Phi\colon M \longrightarrow N$ such that Φ is one-to-one and $(\Phi_*)_p$ is one-to-one for each $p \in M$. Such a map Φ is called an *embedding* of M in N.

We remark that: (1) if M is a submanifold of N, then $\dim M \leq \dim N$; (2) it is possible for $(\Phi_*)_p$ to be one-to-one for all p without Φ being an embedding (Example 1.8); and (3) it is possible for Φ to be one-to-one without Φ being an embedding (take $\Phi\colon \mathbf{R}^1 \longrightarrow \mathbf{R}^1$ with $\Phi(t) = t^3$).

Because $(\Phi_*)_p$ is one-to-one for an embedding we may identify T_pM and $(\Phi_*)_pT_pM \subset T_{\Phi(p)}N$. Using this identification we are simply viewing the tangent space to M as a subspace of the tangent space to N. This is very important intuitively in the case when $N = \mathbf{R}^{m+k}$ because it is exactly how we define the tangent plane to a surface (i.e., when $m = 2$ and $k = 1$). (Recall that the tangent plane to a surface at p was defined as a certain subspace of $T_p\mathbf{R}^3 = \mathbf{R}^3$.) One word of caution: we must be careful not to forget the embedding itself. (There are many ways to embed $\mathbf{R}^1$ in $\mathbf{R}^2$!)

The next proposition gives us even more faith in this abstract approach because it reaffirms our intuition.

PROPOSITION 5.6. If $M = M_f$ is the hypersurface defined by $f: \mathbf{R}^{n+1} \longrightarrow \mathbf{R}$, then T_pM is isomorphic (as a vector space) to

$$(\Phi_*)_p T_p M = \{X_p \in T_p\mathbf{R}^{n+1} \,|\, \langle(\text{grad } f)_p, X_p\rangle = 0\},$$

where Φ is the inclusion of M in $\mathbf{R}^{n+1}$.

Proof: We already know that $(\Phi_*)_p$ is an isomorphism, so we need only show the equality. Actually since both subspaces have the same dimension we need only show that $(\Phi_*)_p(T_pM) \subset \{X_p \in T_p\mathbf{R}^{n+1} \,|\, \langle(\text{grad } f)_p, X_p\rangle = 0\} = \mathcal{E}$. It is sufficient that we show that $(\Phi_*)_p(\partial/\partial x^i)_p \in \mathcal{E}$ for each $i = 1, 2, \ldots, n$, where we are now in some chart about p. We shall do this assuming that (at p) $\partial f/\partial u^{n+1} \neq 0$. (The computation for the other charts is similar.) Using Example 5.5 we see that

$$\left\langle (\Phi_*)_p\left(\frac{\partial}{\partial x^i}\right)_p, (\text{grad } f)_p \right\rangle = \left\langle e_i(p) + \frac{\partial g}{\partial u^i}(p)e_{n+1}(p), \sum_{j=1}^{n+1} \frac{\partial f}{\partial u^j}(p)e_j(p) \right\rangle$$

or

$$\left\langle (\Phi_*)_p\left(\frac{\partial}{\partial x^i}\right)_p, (\text{grad } f)_p \right\rangle = \frac{\partial f}{\partial u^i}(p) + \frac{\partial g}{\partial u^i}(p)\frac{\partial f}{\partial u^{n+1}}(p). \tag{5-5}$$

Recall, however, that $g = g(u^1, \ldots, u^n)$ was defined in such a way that $f(u^1, \ldots, u^n, g(u^1, \ldots, u^n)) = 0$. Differentiate this equation with respect to u^i to obtain

$$\frac{\partial f}{\partial u^i} + \frac{\partial f}{\partial u^{n+1}}\frac{\partial g}{\partial u^i} = 0,$$

so that the right-hand side of Equation (5-5) is zero. $(\Phi_*)_p(\partial/\partial x^i)_p$ is therefore in $\mathcal{E}$ for each $1 \leq i \leq n$. ▮

We have defined embeddings in this book so that we could present Proposition 5.6 which enables us to concretely visualize T_pM if M is a hypersurface defined by a function. The subject of embeddings is a very deep and subtle one. Even some very simple sounding questions (e.g., "For what k does P^n embed in R^{n+k}?") are very difficult. Because the theory of embedding is actually a part of differential topology (not differential geometry), we will only give the following theorem (whose proof is in Guillemin and Pollack [1974]) and make some comments. (One word of caution: the definition of embedding in differential topology is slightly different than that in differential geometry—in the former one also assumes that M has the subspace topology.)

THEOREM 5.7 (Whitney Embedding Theorem, 1937). Every n-manifold embeds in $\mathbf{R}^{2n+1}$.

A surface is a 2-manifold which is embedded in $\mathbf{R}^3$. It can be shown that P^2 cannot be embedded in $\mathbf{R}^3$ but can be embedded in $\mathbf{R}^4$ so there are 2-manifolds that are not surfaces. (In Problem 5.6 we give an embedding of P^2 into $\mathbf{R}^5$.) An *immersion* $\Phi: M \longrightarrow N$ is a differentiable map with $(\Phi_*)_p$ one-to-one at each point (so that an embedding is a one-to-one immersion). Although P^2 cannot be embedded in $\mathbf{R}^3$, it can be immersed in $\mathbf{R}^3$ (as Boy's surface, see Hilbert and Cohn-Vossen [1952]).

PROBLEMS

***5.1.** Prove Proposition 5.3.

5.2. If p is the north pole $(0, 0, 1)$ and $\Phi: S^2 \longrightarrow P^2$ as in Problem 3.2, find $(\Phi_*)_p$ as a matrix. Show it is nonsingular.

5.3. Prove if $I: M \longrightarrow M$ is given by $I(x) = x$, then $(I_*)_p =$ identity for all $p \in M$.

5.4. A differentiable map $\Phi: M \longrightarrow N$ is a *diffeomorphism* if there is a differentiable map $\Psi: N \longrightarrow M$ such that $\Phi \circ \Psi = I$ and $\Psi \circ \Phi = I$.
- (a) If Φ is a diffeomorphism, prove that $(\Phi_*)_p$ is an isomorphism for all $p \in M$.
- (b) If there is a diffeomorphism between M and N then $\dim M = \dim N$.

5.5. Let G be a Lie group and $g \in G$. Define $L_g: G \longrightarrow G$ by $L_g(x) = gx$ (multiplication in the group).
- (a) If $\tilde{X}_e \in T_eG$, prove that the field of vectors X given by $X_g = (L_g)_{*e}(\tilde{X}_e)$ is actually a vector field. X is called a *left-invariant vector field* on G. (e denotes the identity of the group G.)
- (b) Show that any Lie group is parallelizable. (See Problem 4.6.)

5.6. Let $\Phi: P^2 \longrightarrow \mathbf{R}^5$ be defined as follows: Let $l \in P^2$; then $l = [(a^1, a^2, a^3)]$ for some $a = (a^1, a^2, a^3) \in \mathbf{R}^3$ with $(a^1)^2 + (a^2)^2 + (a^3)^2 = 1$. (There are two such points a.) Let $\Phi(l) = ((a^1)^2, (a^2)^2, a^1a^2, a^1a^3, a^2a^3)$. Prove that Φ is an embedding of the projective plane P^2 into $\mathbf{R}^5$. (Does this help you visualize P^2?)

7–6. LINEAR CONNECTIONS ON MANIFOLDS

In Chapter 4 we saw how important it was to differentiate a vector field along a curve α. This really meant that given a vector field Y we could differentiate Y in the direction of the tangent to α. This led naturally to the definition of a parallel vector field and gave a necessary and sufficient condition for a curve to be a geodesic (that the tangent vector field be parallel along the curve). In this section we formalize the notion of differentiating a vector field with respect to a vector field by defining a linear connection and we present some examples and technical results in preparation for the discussion of parallel vector fields and geodesics in the next section.

DEFINITION. A *linear connection* on M is a function $\nabla: \mathfrak{X}(M) \times \mathfrak{X}(M) \to \mathfrak{X}(M)$ (which we write $\nabla_X Y$) such that for $X, Y, Z \in \mathfrak{X}(M)$, $r \in \mathbf{R}$ and $f \in \mathfrak{F}(M)$

(a) $\nabla_X(Y+Z) = \nabla_X Y + \nabla_X Z$ and $\nabla_X rY = r\nabla_X Y$;

(b) $\nabla_{X+Y} Z = \nabla_X Z + \nabla_Y Z$ and $\nabla_{fX} Y = f\nabla_X Y$; and

(c) $\nabla_X fY = (Xf)Y + f\nabla_X Y$.

Note that Condition (c) makes sense both technically and philosophically. Technically, $Xf \in \mathfrak{F}(M)$ so that $(Xf)Y$ makes sense and the right-hand side of (c) is a vector field. Philosophically, (c) is just a "derivative of the product" rule. ∇ is called a linear connection, but $\nabla_X Y$ should be read as the *covariant derivative* of Y with respect to (or in the direction of) X. It can be shown that any manifold admits many linear connections (Hicks [1965]).

If we are in a chart (U, ϕ) so that $\{(\partial/\partial x^i)_p | i = 1, \ldots, n\}$ is a basis for T_pM for all $p \in U$, then any vector field X can be expressed locally in the form $\sum X^i (\partial/\partial x^i)$, where $X^i \in \mathfrak{F}(U)$. Because of the various linearity properties of ∇ and the product rule, the behavior of ∇ is completely determined by the values of $\nabla_{\partial/\partial x^i}(\partial/\partial x^j)$. These values must be expressible as linear combinations of the $\partial/\partial x^k$ with the coefficients in $\mathfrak{F}(U)$.

DEFINITION. Let ∇ be a connection on M and let (U, ϕ) be a proper coordinate chart. The *Christoffel symbols* of ∇ with respect to (U, ϕ) are the functions $\Gamma_{ij}{}^k \in \mathfrak{F}(U)$ defined by

$$\nabla_{\partial/\partial x^i}\left(\frac{\partial}{\partial x^j}\right) = \sum_k \Gamma_{ij}{}^k \frac{\partial}{\partial x^k}. \tag{6-1}$$

Note that nothing in this definition allows us to conclude that $\Gamma_{ij}{}^k$ and $\Gamma_{ji}{}^k$ are equal. In fact this is often not true.

We will show in the next section that (in the presence of a Riemannian metric) these Christoffel symbols play the same role that those of Chapter 4 did. Condition (c) of the definition of $\nabla_X Y$ shows why $\Gamma_{ij}{}^k$ had such a strange transformation law (see Problem 4.11 of Chapter 4). Note that if M can be covered with one chart, then giving the $\Gamma_{ij}{}^k$ with respect to this chart defines a unique linear connection. We will use this approach in Example 6.2.

EXAMPLE 6.1 (Flat Euclidean Space). Let $M = \mathbf{R}^n$. If $Y \in \mathfrak{X}(\mathbf{R}^n)$, then $Y = \sum f^i e_i$ for some $f^i \in \mathfrak{F}(\mathbf{R}^n)$. We define the flat connection on $\mathbf{R}^n$ by $\nabla_X Y = \sum_{i=1}^{n} (Xf^i)e_i$. That this defines a linear connection is the content of Problem 6.1. Note that $\nabla_{\partial/\partial x^i}(\partial/\partial x^j) = 0$ for all i and j so that $\Gamma_{ij}{}^k = 0$ for all i, j, and k. $\nabla_X Y$ is the usual directional derivative of a vector-valued function.

EXAMPLE 6.2 (Hyperbolic (or Poincaré) Upper Half Plane). Let

$$M = H = \{(x, y) \in \mathbf{R}^2 \mid y > 0\}.$$

Define a linear connection on H by setting all $\Gamma_{ij}{}^k$ to zero except that $\Gamma_{12}{}^1 = \Gamma_{21}{}^1 = -\Gamma_{11}{}^2 = \Gamma_{22}{}^2 = -1/y$. This gives a connection on H because H can be covered by one coordinate chart.

Before giving another example we shall need some technical results. We shall fix a manifold M with linear connection ∇.

LEMMA 6.3. If $X, Y \in \mathfrak{X}(M)$ and $X_p = Y_p$ for some $p \in M$, then $(\nabla_X Z)_p = (\nabla_Y Z)_p$ for all $Z \in \mathfrak{X}(M)$.

Proof: If $W = X - Y$, then $W_p = 0$. Thus, in some coordinate chart about p, $W = \sum f^i\,(\partial/\partial x^i)$ and $f^i(p) = 0$ for all i. This means that for any $Z \in \mathfrak{X}(M)$, $(\nabla_W Z)_p = \sum f^i(p)(\nabla_{\partial/\partial x^i} Z)_p = 0$. ∎

What Lemma 6.3 means is that $(\nabla_X Y)_p$ depends only on the value of X at the point p. The Y dependence is not so simple. It depends on the value of Y along any curve which "fits" X as we see in Lemma 6.6.

DEFINITION. Z is a *vector field along the curve* $\alpha: I \longrightarrow M$ if Z assigns to each $t \in I$ an element $Z_{\alpha(t)} \in T_{\alpha(t)}M$ such that $t \longrightarrow Z_{\alpha(t)}(f)$ is a differentiable real-valued function of t for each $f \in \mathfrak{F}(M)$.

DEFINITION. Let $\alpha: I \longrightarrow M$ be a curve. The *tangent vector field* to α, T_α, is given by $(T_\alpha)_{\alpha(t)} = (\alpha_*)_t(d/dt)$, where t is the natural coordinate for I.

If there is no ambiguity, we write T for T_α. Note that T_α is an example of a vector field along α. In local coordinates about $\alpha(t_0)$ we have

$$T_{\alpha(t)} = \sum_i \frac{d\alpha^i}{dt}(t)\left(\frac{\partial}{\partial x^i}\right)_{\alpha(t)} \tag{6-2}$$

(see Equation (5-3)) just as in the Euclidean case. This is the reason one often sees $d\alpha/dt$ used instead of T_α. We have no concept of length at this point so we cannot talk about T_α being a unit vector. Note T_α is a function of t, not $\alpha(t)$. This was called the velocity vector field in Chapter 2.

DEFINITION. Let $\alpha: I \longrightarrow M$ and $Y \in \mathfrak{X}(M)$ be given. We shall now define the *covariant derivative of* Y *with respect to* α, $\nabla_T Y$. Let t_0 be given and let X be any vector field on M such that $X_{\alpha(t_0)} = T_{\alpha(t_0)}$. (Such an X

exists by Problem 4.2.) $\nabla_T Y$ is the vector field along α defined by $(\nabla_T Y)_{\alpha(t_0)} = (\nabla_X Y)_{\alpha(t_0)}$. (This is well defined by Lemma 6.3.)

EXAMPLE 6.4. Let M be flat Euclidean 2-space, $\alpha(t) = (\cos t, \sin t)$ for $0 < t < 2\pi$ and $Y = ye_1 - xe_2$. Since $T = -\sin t\, e_1 + \cos t\, e_2$, if we let $X = -ye_1 + xe_2$, then $X_{\alpha(t)} = T_{\alpha(t)}$. Therefore,

$$\nabla_T Y = \nabla_X Y = Xye_1 - Xxe_2 = xe_1 + ye_2$$

so that $(\nabla_T Y) = \cos t\, e_1 + \sin t\, e_2$.

LEMMA 6.5. If $f \in \mathfrak{F}(M)$ and $\alpha: I \longrightarrow M$ are such that $f \circ \alpha$ is constant, then $T_\alpha f = 0$.

Proof: $T_\alpha f = \alpha_*(d/dt)f = (d/dt)(f \circ \alpha) = 0$. ∎

LEMMA 6.6. Let $\alpha : I \longrightarrow M$ be a curve and $Y, Z \in \mathfrak{X}(M)$ such that $Y_{\alpha(t)} = Z_{\alpha(t)}$ for all $t \in I$. Then $\nabla_{T_\alpha} Y = \nabla_{T_\alpha} Z$ along α.

Proof: Let $W = Y - Z$ so that $W_{\alpha(t)} = 0$. We now show that $\nabla_T W = 0$. In local coordinates about some point on the curve we may write $W = \sum_i f^i (\partial/\partial x^i)$, where $f^i(\alpha(t)) = 0$. Therefore using Property (c) of the definition of ∇, we have

$$\nabla_T W = \sum T_\alpha f^i \frac{\partial}{\partial x^i} + \sum f^i \nabla_T \left(\frac{\partial}{\partial x^i}\right).$$

Since $f^i \circ \alpha$ is constant (it is actually zero), the first term of the right-hand side is zero by Lemma 6.5. Since f^i is zero at $\alpha(t)$, the second term is zero when evaluated along $\alpha(t)$. ∎

Three remarks about Lemma 6.6 are in order: (1) the conclusion $\nabla_T Y = \nabla_T Z$ is only valid at points along the curve α, (in fact, it does not make sense at any other point); (2) an examination of the proof shows that if $t_0 \in I$ and $Y_{\alpha(t)} = Z_{\alpha(t)}$ for t in some interval about t_0, then $(\nabla_T Y)_{\alpha(t_0)} = (\nabla_T Z)_{\alpha(t_0)}$; (3) if Y is a vector field along α, then we may define $\nabla_{T_\alpha} Y$ to be a vector field along α by picking any $\tilde{Y} \in \mathfrak{X}(M)$ such that $\tilde{Y}_{\alpha(t)} = Y_{\alpha(t)}$ in some neighborhood of t_0 in I and setting

$$(\nabla_T Y)_{\alpha(t_0)} = (\nabla_T \tilde{Y})_{\alpha(t_0)}.$$

This is well defined by Lemma 6.5.

We shall now assume that M is a submanifold of N and that M has been identified as a subset of N and T_pM as a subset of T_pN. If N has a linear connection $\bar{\nabla}$ and $X, Y \in \mathfrak{X}(M)$, then we may extend X and Y to vector fields on N which agree (at least locally) with X, Y; hence $\bar{\nabla}_X Y$ makes sense for $X, Y \in \mathfrak{X}(M)$. (It is well defined by Lemmas 6.3 and 6.6.) Since $\bar{\nabla}$ obeys the rules for a linear connection, the overly optimistic reader is tempted to think that $\bar{\nabla}$ is a linear connection on M. Alas, this is not true because $\bar{\nabla}_X Y$ need not be in $\mathfrak{X}(M)$ (although it must be in $\mathfrak{X}(N)$, of course). Specifically look at the case when $M = S^1$, $N = \mathbf{R}^2$, $X = -ye_1 + xe_2$, and $Y = ye_1 - xe_2$. Since S^1 is the hypersurface of $\mathbf{R}^2$ defined by $f(x, y) = x^2 + y^2 - 1$ and $\operatorname{grad} f = 2xe_1 + 2ye_2$, Proposition 5.6 shows that both X and Y are in $\mathfrak{X}(S^1)$. However, in Example 6.4, we computed $\bar{\nabla}_X Y$ to be $xe_1 + ye_2$ which is not in $\mathfrak{X}(S^1)$. However, guided by Gauss's formula (Equation (4-8) of Chapter 4) we can prove the following proposition:

PROPOSITION 6.7. Let $M = M_f$ be the hypersurface defined by f, let $\bar{\nabla}$ be the linear connection of flat Euclidean $(n + 1)$-space, and write $N = \operatorname{grad} f / |\operatorname{grad} f|$. If $X, Y \in \mathfrak{X}(M)$, then

$$\nabla_X Y = \bar{\nabla}_X Y - \langle \bar{\nabla}_X Y, N \rangle N \tag{6-3}$$

defines a linear connection on M.

Proof: We only show that $\nabla_X Y \in \mathfrak{X}(M)$ and leave the rest as Problem 6.2. $\nabla_X Y \in \mathfrak{X}(M)$ if and only if $\langle \nabla_X Y, N \rangle = 0$ (by Proposition 5.6), but this is clear. ▮

It will be shown in Section 7-8 (Problem 8.4) that

$$\langle \bar{\nabla}_X Y, N \rangle = -\langle Y, \bar{\nabla}_X N \rangle.$$

If we write down explicitly what $\bar{\nabla}_X N$ is, we find that $\bar{\nabla}_X N = -L(X)$ is our old friend the Weingarten map (if $n = 2$)! Now write out Equation (6-3) in local coordinates. It should be Gauss's formulas (4-8) of Chapter 4.

PROBLEMS

6.1. Prove that flat $\mathbf{R}^n$ (Example 6.1) is a manifold with linear connection.

***6.2.** Prove Proposition 6.7.

6.3. Let H be the hyperbolic upper half plane (Example 6.2) and let $\alpha(t) = (t, 1)$. If

$$Y(t) = \sin t \left(\frac{\partial}{\partial x^1}\right)_{\alpha(t)} + \cos t \left(\frac{\partial}{\partial x^2}\right)_{\alpha(t)},$$

prove that $\nabla_{T_\alpha} Y = 0$.

6.4. Prove that Equation (6-3) yields Gauss's formulas when $n = 2$.

7–7. PARALLEL VECTOR FIELDS AND GEODESICS ON A MANIFOLD WITH A LINEAR CONNECTION

In this section we define the notions of parallel vector fields, parallel translation, and geodesics on a manifold with linear connection. We show that parallel translation is the "global" version of covariant differentiation and also compute the geodesics of the classical geometries, R^2, H, and S^2. We fix a manifold M and linear connection ∇.

What should we take for the definition of a vector field Y being parallel along a curve α? For a surface it was that the derivative of Y was normal to the surface. Equation (6-3) tells us that this happens precisely when $\nabla_{T_\alpha} Y = 0$.

DEFINITION. Let Y be a vector field along α. Y is *parallel along* α if $\nabla_{T_\alpha} Y = 0$.

EXAMPLE 7.1. For flat Euclidean space a vector field Y is parallel along α if and only if $Y_{\alpha(t)} = (\alpha(t); a^1, \ldots, a^n)$ for some constants $a^i \in \mathbf{R}$. (See Problem 7.1.)

A review of Section 4-6 will enable you to guess all of the results (and proofs) of this section (except Proposition 7.6).

THEOREM 7.2 (T. Levi-Civita, 1917). Let $\alpha\colon [c, d] \to M$, $p = \alpha(c)$ and $\tilde{Y}_p \in T_pM$. Then there is a unique vector field Y parallel along α such that $Y_{\alpha(c)} = \tilde{Y}_p$.

Proof: The proof proceeds exactly like that of Theorem 6.7 of Chapter 4. Assume that we are in a chart (U, ϕ) about p and that $Y = \sum Y^i(t)(\partial/\partial x^i)_{\alpha(t)}$ is a vector field along α, where the Y^i are differentiable. Y is parallel along α if and only if

$$0 = \nabla_T Y = \sum_i \left[\left(\frac{dY^i}{dt}\right)\left(\frac{\partial}{\partial x^i}\right) + Y^i \nabla_T \frac{\partial}{\partial x^i}\right]$$

or

$$0 = \sum \left[\left(\frac{dY^k}{dt}\right) + Y^i \frac{d\alpha^j}{dt} \Gamma_{ji}{}^k\right] \frac{\partial}{\partial x^k} \tag{7-1}$$

where $\alpha(t) = \phi^{-1}(\alpha^1(t), \ldots, \alpha^n(t))$. Thus Y is parallel along α if and only if the Y^i solve the initial value problem

$$\begin{cases} \dfrac{dY^k}{dt} + \sum Y^i \dfrac{d\alpha^j}{dt} \Gamma_{ji}{}^k = 0 \qquad k = 1, \ldots, n \\ Y^k(c) = \tilde{Y}^k. \end{cases} \tag{7-2}$$

An application of Picard's Theorem finishes the theorem in a neighborhood of $\alpha(c)$. Now repeat this process until we get to $\alpha(d)$. ▮

EXAMPLE 7.3. Let H be the hyperbolic upper half plane of Example 6.2 and let $\alpha(t) = (t, 1)$ with $c = 0$ (and $d > 0$) so that $p = \alpha(0) = (0, 1)$. Let $\tilde{Y}_p = (\partial/\partial x^2)_p$ so that in the notation of Equation (7-2), $\tilde{Y}^1 = 0$ and $\tilde{Y}^2 = 1$. Equation (7-2) becomes

$$\begin{cases} \dfrac{dY^1}{dt} - Y^1 \dfrac{d\alpha^2}{dt} \dfrac{1}{\alpha^2(t)} - Y^2 \dfrac{d\alpha^1}{dt} \dfrac{1}{\alpha^2(t)} = 0 \\ \dfrac{dY^2}{dt} + Y^1 \dfrac{d\alpha^1}{dt} \dfrac{1}{\alpha^2(t)} - Y^2 \dfrac{d\alpha^2}{dt} \dfrac{1}{\alpha^2(t)} = 0 \end{cases}$$

or

$$\frac{dY^1}{dt} - Y^2 = 0 \quad \text{and} \quad \frac{dY^2}{dt} + Y^1 = 0. \tag{7-3}$$

This coupled system (7-3) can be solved by obtaining $(d^2Y^2/dt^2) + Y^2 = 0$ so that $Y^2(t) = c_1 \sin t + c_2 \cos t$. Since $Y^2(0) = \tilde{Y}^2 = 1$ we see that $c_2 = 1$. Going back to Equation (7-3) yields $Y^1(t) = -c_1 \cos t + \sin t$. Since $Y^1(0) = 0$, $c_1 = 0$, and so

$$Y(t) = Y^1(t)(\partial/\partial x^1)_{\alpha(t)} + Y^2(t)(\partial/\partial x^2)_{\alpha(t)}$$

or

$$Y(t) = \sin t(\partial/\partial x^1)_{\alpha(t)} + \cos t(\partial/\partial x^2)_{\alpha(t)}.$$

(See Figure 7.5.)

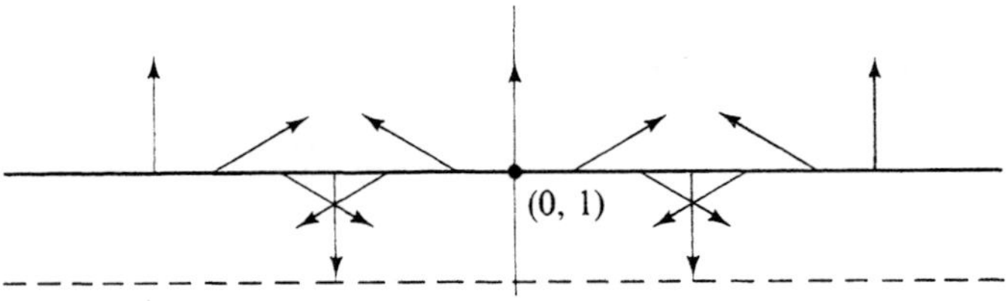

FIGURE 7.5

DEFINITION. Let α be a given curve. We define *parallel translation along* α, $P^\alpha: T_{\alpha(t_1)}M \longrightarrow T_{\alpha(t_2)}M$ by $P^\alpha(\tilde{Y}_{\alpha(t_1)}) = Y_{\alpha(t_2)}$ where Y is the unique parallel vector field along α such that $Y_{\alpha(t_1)} = \tilde{Y}_{\alpha(t_1)}$.

PROPOSITION 7.4. For each curve α, $P^\alpha: T_{\alpha(t_1)}M \longrightarrow T_{\alpha(t_2)}M$ is an isomorphism.
Proof: Problem 7.2. ▮

EXAMPLE 7.5. Let $M = H$ be the hyperbolic half plane and let $\alpha(t) = (t, 1)$. Example 7.3 shows that if $t_1 = 0$ then

$$P^\alpha\left(\frac{\partial}{\partial x^2}\right) = \sin t_2 \left(\frac{\partial}{\partial x^1}\right)_{(t_2,1)} + (\cos t_2)\left(\frac{\partial}{\partial x^2}\right)_{(t_2,1)}$$

It should be pointed out that Example 7.3 was picked with malice aforethought; that is, we picked it carefully so that we could actually solve the initial value problem (7-2). In practice it is often quite difficult to actually solve this initial value problem.

The next proposition shows either that parallel translation is a global version of covariant differentiation or covariant differentiation is the infinitesimal version of parallel translation, depending on your viewpoint. We feel that it gives a better intuitive feeling for covariant differentiation.

PROPOSITION 7.6. Let $P^{\alpha(t)}: T_{\alpha(0)}M \longrightarrow T_{\alpha(t)}M$ be parallel translation along α (with respect to the linear connection ∇). Then

$$(\nabla_{T_\alpha}Y)_{\alpha(0)} = \lim_{t\to 0} \frac{(P^{\alpha(t)})^{-1}(Y_{\alpha(t)}) - Y_{\alpha(0)}}{t}$$

Proof: Let Z_s be the unique parallel vector field along α guaranteed by Theorem 7.2 such that $(Z_s)_{\alpha(0)} = (P^{\alpha(s)})^{-1}(Y_{\alpha(s)})$. In a chart we write

$$(Z_s)_{\alpha(t)} = \sum Z_s^i(t)(\partial/\partial x^i)_{\alpha(t)},$$
$$T_\alpha = \sum (d\alpha^i/dt)(\partial/\partial x^i)_{\alpha(t)}$$

and

$$Y_{\alpha(t)} = \sum Y^i(t)(\partial/\partial x^i)_{\alpha(t)}.$$

Since Z_s is parallel along α, we apply Equation (7-2) to obtain

$$\text{(7-4)} \qquad \frac{dZ_s^k}{dt} + \sum_{i,j} \frac{d\alpha^i}{dt} Z_s^j(t)\Gamma_{ij}^k(\alpha(t)) = 0, \qquad k = 1, \ldots, n.$$
$$Z_s^k(s) = Y^k(s).$$

Applying the Mean Value Theorem to the function $Z_s^k(t)$ on the interval $0 \le t \le s$ we see that $Z_s^k(s) = Z_s^k(0) + sZ_s^{k\prime}(\xi_k)$ for some $0 < \xi_k < s$. Thus the kth component of

$$\frac{(P^{\alpha(s)})^{-1}(Y_{\alpha(s)}) - Y_{\alpha(0)}}{s}$$

is

$$\frac{Z_s^k(0) - Y^k(0)}{s} = \frac{Z_s^k(s) - sZ_s^{k\prime}(\xi_k) - Y^k(0)}{s}$$
$$= -Z_s^{k\prime}(\xi_k) + \frac{Z_s^k(s) - Y^k(0)}{s}.$$

Using Equation (7-4) this becomes

$$\text{(7-5)} \qquad \sum_{i,j} \frac{d\alpha^i}{dt}(\xi_k) Z_s^j(\xi_k)\Gamma_{ij}^k(\alpha(\xi_k)) + \frac{Y^k(s) - Y^k(0)}{s}.$$

Letting $s \to 0$ in Formula (7-5) and remembering that $\xi_k \to 0$ as $s \to 0$, we get the kth component of

$$\lim_{s \to 0} \frac{(P^{\alpha(s)})^{-1}(Y_{\alpha(s)}) - Y_{\alpha(0)}}{s}$$

is

$$\sum \frac{d\alpha^i}{dt}(0) Z_0^j(0)\Gamma_{ij}^k(\alpha(0)) + \frac{dY^k}{dt}(0)$$
$$= \sum_{i,j} \frac{d\alpha^i}{dt}(0) Y^j(0)\Gamma_{ij}^k(\alpha(0)) + \frac{dY^k}{dt}(0)$$
$$= k\text{th component of } \nabla_{T_\alpha} Y \text{ by Equation (7-1).} \quad \blacksquare$$

A direct translation of the definition of geodesic applies in this setting. Unlike Chapter 4, we do not assume T_α has length one because we can't! There is no notion of length of vectors on a manifold with linear connection.

DEFINITION. A curve α on M is a *geodesic* (with respect to ∇) if $\nabla_{T_\alpha} T_\alpha = 0$.

Note that applying (7-2) in the case $Y = T_\alpha$ yields

$$\frac{d^2\alpha^k}{dt^2} + \sum_{i,j=1}^{n} \Gamma_{ij}{}^k \frac{d\alpha^i}{dt}\frac{d\alpha^j}{dt} = 0 \qquad \text{for } k = 1, \ldots, n \tag{7-6}$$

as the equation of a geodesic. From this we conclude

THEOREM 7.7. Let M be a manifold with linear connection ∇. Let $p \in M$ and $X_p \in T_pM$. Then there is an $\epsilon > 0$ and a unique geodesic α such that $\alpha\colon (-\epsilon, \epsilon) \longrightarrow M$, $\alpha(0) = P$ and $(d\alpha/dt)(0) = X_p$.

We shall finish this section by actually computing the geodesics of the hyperbolic upper half plane. We leave the geodesics of flat Euclidean space as a problem (see Problem 7.1).

EXAMPLE 7.8. Let H be the hyperbolic upper half plane (Example 6.2). With the given Christoffel symbols, a geodesic $\alpha(t) = (\alpha^1(t), \alpha^2(t))$ must satisfy

$$\begin{aligned} &\frac{d^2\alpha^1}{dt^2} - \frac{2}{\alpha^2(t)}\frac{d\alpha^1}{dt}\frac{d\alpha^2}{dt} = 0 \\ &\frac{d^2\alpha^2}{dt^2} + \frac{1}{\alpha^2(t)}\left(\frac{d\alpha^1}{dt}\right)^2 - \frac{1}{\alpha^2(t)}\left(\frac{d\alpha^2}{dt}\right)^2 = 0, \end{aligned} \tag{7-7}$$

which results from Equation (7-6). Recall that

$$\tanh t = \frac{e^t - e^{-t}}{e^t + e^{-t}} \quad \text{and} \quad \operatorname{sech} t = \frac{2}{e^t + e^{-t}}$$

are the hyperbolic tangent and hyperbolic secant. Solutions of (7-7) are of the following form:

$$\alpha^1(t) = a + b \tanh(rt) \quad \text{and} \quad \alpha^2(t) = b \operatorname{sech}(rt) \tag{7-8}$$

or

$$\alpha^1(t) = c \quad \text{and} \quad \alpha^2(t) = de^{pt}, \tag{7-9}$$

where $a, b, c, d, p, r \in \mathbf{R}$.

Those geodesics which are in the form (7-9) are straight lines perpendicular to the x-axis. Those geodesics which are in the form (7-8) are upper semicircles of radius $|b|$ with center at $(a, 0)$ since

$$(\alpha^1(t) - a)^2 + (\alpha^2(t) - 0)^2 = b^2.$$

See Figure 7.6.

It is important to examine the hyperbolic upper half plane in regard to Euclid's Fifth Postulate. H is a space on which classical hyperbolic geometry lies. Note that if we use line to mean geodesic and parallel lines to mean they never meet (when infinitely extended), then given a line in H and a point not

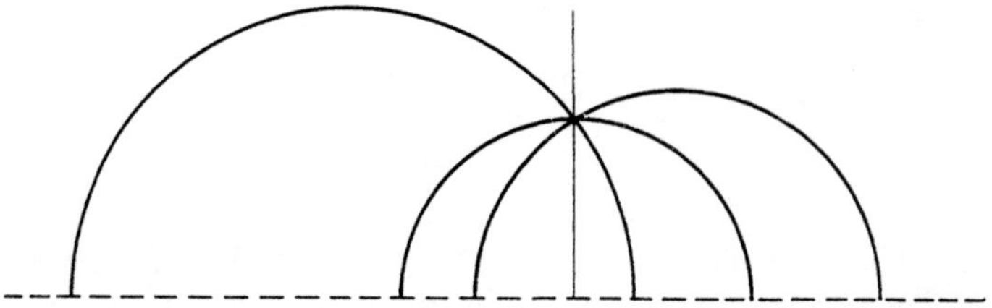

FIGURE 7.6

on the line there are infinitely many lines through the point which are parallel to the given line.

PROBLEMS

***7.1.** Prove Example 7.1 and find the geodesics.

***7.2.** Prove Proposition 7.4.

7.3. Let M^n be a parallelizable manifold with parallelization

$$X_1, \ldots, X_n \in \mathfrak{X}(M).$$

(See Problem 4.6.) Show there is a linear connection ∇ on M such that each X_i is parallel along any curve. (*Hint:* Analyze the example of flat Euclidean space.)

7.4. Let $M = \mathbf{R}^2$ and define ∇ on $\mathbf{R}^2$ by $\Gamma_{jk}{}^i = 0$ except $\Gamma_{21}{}^1 = \Gamma_{12}{}^1 = 1$. Find all geodesics through $p = (a, b) \in \mathbf{R}^2$. (Note that the geodesics of this example are not the same as the geodesics of flat $\mathbf{R}^2$.)

7.5. Use the connection of Proposition 6.7 to find all the geodesics on $S^2 \subset \mathbf{R}^3$.

***7.6.** Let $\{W_1, W_2, \ldots, W_k\}$ be parallel vector fields along α. Show that the W_i are linearly independent at each point of α if and only if they are linearly independent at one point of α.

7.7. Verify that Equations (7-8) and (7-9) do indeed give the solution of Equation (7-7).

7–8. RIEMANNIAN METRICS, DISTANCE, AND CURVATURE

In this final section we define a Riemannian metric and a Riemannian manifold. We show that every Riemannian manifold has a natural linear connection associated to it (called the *Riemannian connection*) so that the notion of geodesic makes sense on a Riemannian manifold. We then put this natural connection on a Riemannian manifold and discuss the distance-minimizing properties of geodesics. (Compare with Theorem 5.9 of Chapter 4.) We also define several different kinds of curvature on a Riemannian manifold.

A Riemannian metric is defined as a differentiable assignment of an inner product on T_pM for each $p \in M$. More formally

DEFINITION. A *field of metrics* g on a manifold M is an assignment of a map $g_p : T_pM \times T_pM \to \mathbf{R}$ to each $p \in M$ such that for all $X_p, Y_p, Z_p \in T_pM$ and $r \in \mathbf{R}$:

(a) $g_p(X_p + Y_p, Z_p) = g_p(X_p, Z_p) + g_p(Y_p, Z_p)$ and

$$g_p(rX_p, Y_p) = rg_p(X_p, Y_p);$$

(b) $g_p(X_p, Y_p) = g_p(Y_p, X_p)$; and

(c) $g_p(X_p, X_p) \geq 0$ with $g_p(X_p, X_p) = 0$ if and only if $X_p = 0$.

Note that Conditions (a) and (b) imply that

$$g_p(X_p, Y_p + Z_p) = g_p(X_p, Y_p) + g_p(X_p, Z_p)$$

and

$$g_p(X_p, rY_p) = rg_p(X_p, Y_p).$$

Thus a metric is an inner product on T_pM because it is symmetric (b), bilinear (a), and positive definite (c). If g is a field of metrics on M and X, $Y \in \mathfrak{X}(M)$, then $g(X, Y)$ is a real-valued function on M, whose value at p is $g_p(X_p, Y_p)$.

DEFINITION. A *Riemannian metric* on a manifold M is a field of metrics g such that $g(X, Y) \in \mathfrak{F}(M)$ for all $X, Y \in \mathfrak{X}(M)$. A *Riemannian manifold* is a manifold M together with a fixed Riemannian metric.

It can be shown that every manifold possesses a Riemannian metric (see Hicks [1965]). Note that in geometry the term metric does not refer to a distance function but rather to an inner product. We shall see below that there is a natural distance function associated with a Riemannian metric.

As in Chapter 4, we may define n^2 real-valued functions g_{ij} on a coordinate chart (U, ϕ) by

$$g_{ij}(p) = g_p\left(\left(\frac{\partial}{\partial x^i}\right)_p, \left(\frac{\partial}{\partial x^j}\right)_p\right)$$

EXAMPLE 8.1. Let $N = \mathbf{R}^n$ and let $g = \langle\ ,\ \rangle$ (the usual inner product) so that

$$g_p(X_p, Y_p) = \sum a^i b^i,$$

where $X_p = (p; a^1, \ldots, a^n)$ and $Y_p = (p; b^1, \ldots, b^n)$. Clearly $g_{ij}(p) = \delta_{ij}$ (the Kronecker delta).

EXAMPLE 8.2. Assume that M is a submanifold of a manifold N and that M (resp. T_pM) has been identified as a subset (resp. subspace) of N (resp. T_pN). If N has a Riemannian metric g^N and $X_p, Y_p \in T_pM \subset T_pN$, then $g_p{}^N(X_p, Y_p)$ makes sense and we may define $g_p{}^M(X_p, Y_p) = g_p{}^N(X_p, Y_p)$. This is exactly the way in which we made a surface in $\mathbf{R}^3$ into a Riemannian manifold.

DEFINITION. The Riemannian metric g^M defined in Example 8.2 above is called the *Riemannian metric induced by* g^N (or the *induced metric*).

EXAMPLE 8.3. Let $M = H$ be the hyperbolic upper half plane of Example 6.2. If $X_p, Y_p \in T_pH$, then $\langle X_p, Y_p\rangle$ will denote the dot product of Example 8.1. We define, at the point $p = (x, y) \in H$,

$$g_p(X_p, Y_p) = \frac{\langle X_p, Y_p\rangle}{y^2}. \tag{8-1}$$

g is a Riemannian metric on H. This example is what we were alluding to in Problem 2.4 of Chapter 6. g is *not* an induced metric.

We are now in the pleasant situation of having an entire chapter for motivation. We will examine some of the results of Chapter 4 and see which carry over to the more general setting of Riemannian manifolds. Assume M is a manifold with a fixed Riemannian metric.

DEFINITION. Suppose that $A\colon T_pM \longrightarrow T_qM$ is a linear transformation. A is an *isometry* if for all $X_p, Y_p \in T_pM$, $g_p(X_p, Y_p) = g_q(AX_p, AY_p)$.

DEFINITION. Let ∇ be a linear connection on the Riemannian manifold M. ∇ is called *metrical* if for all $X, Y, Z \in \mathfrak{X}(M)$

$$Xg(Y, Z) = g(\nabla_X Y, Z) + g(Y, \nabla_X Z). \tag{8-2}$$

A word is in order about what Equation (8-2) means. $g(Y, Z)$ is a function so $Xg(Y, Z)$ makes sense (and is a function). Since $\nabla_X Y$ and $\nabla_X Z \in \mathfrak{X}(M)$, the right-hand side is also a function. Intuitively, there are two ways to view (8-2). The first comes by reading $g(Y, Z)$ as "Y dot Z." From this viewpoint Equation (8-2) reads exactly like the formula for the derivative of a dot product where the derivative on the left is a directional derivative and those on the right are covariant derivatives. The second way to view (8-2) is geometrically via the following proposition.

PROPOSITION 8.4. Let ∇ be a linear connection on a Riemannian manifold M. ∇ is a metrical connection if and only if for every curve α on M, parallel translation along α is an isometry.

Proof: First assume that ∇ is metrical and let α be a curve in M. If $\tilde{Y}, \tilde{Z} \in T_{\alpha(0)}M$, let Y and Z be their parallel translates along α so that $Y_{\alpha(t)} = P^{\alpha(t)}(\tilde{Y})$ and $Z_{\alpha(t)} = P^{\alpha(t)}(\tilde{Z})$. We must show that $g_{\alpha(0)}(\tilde{Y}, \tilde{Z}) = g_{\alpha(t)}(Y, Z)$ for each t in the domain of α. We shall do this by showing the function $f(t) = g_{\alpha(t)}(Y, Z)$ is constant. By Equation (8-2) we have

$$\frac{df}{dt} = T_\alpha(f) = T_\alpha g_{\alpha(t)}(Y, Z) = g_{\alpha(t)}(\nabla_T Y, Z) + g_{\alpha(t)}(Y, \nabla_T Z) = 0 + 0 = 0,$$

since both Y and Z are parallel along α. Hence f is constant.

Now assume P^α is an isometry for each curve α. We shall verify Equation (8-2). Let $X_p \in T_pM$ and let α be the unique geodesic such that $\alpha(0) = p$ and $T_{\alpha(0)} = X_p$, as guaranteed by Theorem 7.7. Let $\{\tilde{W}_1, \tilde{W}_2, \ldots, \tilde{W}_n\}$ be an orthonormal basis of T_pM and let W_i be the parallel translate of $\tilde{W}_i$ along α. The uniqueness of a parallel translate (Proposition 7.2) guarantees that the W_i are linearly independent at each point of α (see Problem 7.6). Hence $\{W_1, \ldots, W_n\}$ is a basis at each point of α. By hypothesis,

$$\delta_{ij} = g_p(W_i, W_j) = g_{\alpha(t)}(W_i, W_j) \tag{8-3}$$

and the W_i give an orthonormal basis of the tangent space at each point of α.

Let $Y, Z \in \mathfrak{X}(M)$. There are differentiable functions $Y^i(t)$ and $Z^j(t)$ such that $Y_{\alpha(t)} = \sum Y^i(t)(W_i)_{\alpha(t)}$ and $Z_{\alpha(t)} = \sum Z^j(t)(W_j)_{\alpha(t)}$. By Equation (8-3),

$$g_{\alpha(t)}(Y, Z) = \sum Y^i Z^j g_{\alpha(t)}(W_i, W_j) = \sum Y^i Z^j \delta_{ij} = \sum Y^i Z^i.$$

Then

$$T_\alpha g(Y, Z) = \sum T_\alpha(Y^i Z^i). \tag{8-4}$$

Since the W_i are parallel along α,

$$\nabla_T Y = \sum T(Y^i)W_i + Y^i \nabla_T W_i = \sum T(Y^i)W_i.$$

Thus

$$\begin{aligned} g(\nabla_T Y, Z) + g(Y, \nabla_T Z) &= \sum T(Y^i)Z^j g(W_i, W_j) + \sum Y^i T(Z^j) g(W_i, W_j) \\ &= \sum T(Y^i Z^j) g(W_i, W_j) = \sum T(Y^i Z^i). \end{aligned}$$

If we combine this result with Equation (8-4) and evaluate at $t = 0$ (remembering that $T_{\alpha(0)} = X_p$), we obtain

$$X_p g(Y, Z) = g(\nabla_{X_p} Y, Z) + g(Y, \nabla_{X_p} Z). \tag{8-5}$$

Hence ∇ is metrical. ∎

Definition. A linear connection ∇ is *torsion-free* (or *symmetric*) if

$$\nabla_X Y - \nabla_Y X = [X, Y] \qquad \text{for all } X, Y \in \mathfrak{X}(M).$$

This condition is not as intuitive as the property of being metrical. In particular, it has nothing to do with the torsion of a curve. The reason for the terminology "symmetric" is the following lemma, whose proof we leave as Problem 8.2.

LEMMA 8.5. ∇ is torsion-free if and only if in every coordinate chart,

$$\Gamma_{ij}{}^{k} = \Gamma_{ji}{}^{k} \qquad \text{for all } 1 \leq i, j, k \leq n.$$

THEOREM 8.6 (Fundamental Lemma of Riemannian Geometry). Let M be a Riemannian manifold with a Riemannian metric g. Then there is a unique torsion-free metrical linear connection ∇ on M.

Proof: A straightforward calculation using the metrical and torsion-free conditions shows that if ∇ exists, then

$$g(\nabla_X Y, Z) = \tfrac{1}{2}(Xg(Y, Z) - Zg(X, Y) + Yg(Z, X) - g(X, [Y, Z]) + g(Z, [X, Y]) - g(Y, [Z, X])) \tag{8-6}$$

for all $X, Y, Z \in \mathfrak{X}(M)$. Since the right-hand side is totally determined by g, the value of $\nabla_X Y$ is determined. Hence ∇ is unique if it exists. Equation (8-6) is equivalent to the equation

$$\Gamma_{ij}{}^{k} = \frac{1}{2} \sum g^{lk} \left(\frac{\partial g_{li}}{\partial u^{j}} - \frac{\partial g_{ij}}{\partial u^{l}} + \frac{\partial g_{jl}}{\partial u^{i}} \right), \tag{8-7}$$

where (g^{lk}) is the inverse of (g_{ij}) (Lemma 8.5 has been used). This latter formula should be familiar as the intrinsic formula for the $\Gamma_{ij}{}^{k}$ (see Equation (4-11) of Chapter 4) and is the actual motivation behind Equation (8-6).

On the other hand, Equation (8-6) may be used to define a connection ∇: $(\nabla_X Y)_p$ is the unique element of T_pM whose inner product with an arbitrary Z_p is given by the right-hand side of (8-6) evaluated at p. A tedious calculation shows that this does give a linear connection and that it is metrical and torsion-free. ∎

In Problem 8.3 you will verify that (8-7) follows from (8-6). Note that (8-7) can be used to find the Christoffel symbols for this connection.

DEFINITION. The linear connection ∇ guaranteed by Theorem 8.6 is called the *Riemannian connection* of the Riemannian manifold.

EXAMPLE 8.7. The Riemannian connection of Example 8.1 gives flat Euclidean space (Example 6.1). The Riemannian connection of Example 8.2 is given in Problem 8.7. The Riemannian connection of the hyperbolic upper half plane is given in Example 6.2 as may be seen by computing the $\Gamma_{ij}{}^{k}$ via Equation (8-7).

We now put a notion of distance on a Riemannian manifold and discuss the length-minimizing properties of geodesics (with respect to the Riemannian connection). Again we fix a manifold M and a Riemannian metric g and let ∇ be the Riemannian connection. We shall assume that M is path-connected (i.e., if $p, q \in M$, then there is a curve $\alpha : [a, b] \longrightarrow M$ such that $\alpha(a) = p$ and $\alpha(b) = q$). We shall not prove any of the results in the rest of this section but merely give references.

DEFINITION. If $\alpha : [a, b] \longrightarrow M$ then the *length of* α is

$$|\alpha| = \int_a^b \sqrt{g_{\alpha(t)}(T_{\alpha(t)}, T_{\alpha(t)})}\, dt.$$

If $p, q \in M$, the *distance from p to q* is $d(p, q) = \inf |\alpha|$, where the infimum is taken over all broken C^∞ curves joining p to q.

PROPOSITION 8.8. The Riemannian manifold M with the distance defined above is a metric space.

The symmetry and the triangle inequality for d are very easy. The hard part of Proposition 8.8 is showing that $d(p, q) = 0$ implies $p = q$. This requires Theorem 8.11 below. Note that we have assumed that M is a metric space to start with. Is the distance defined by g the same as the original metric? The answer is an emphatic no. The upper half plane is a metric space (as a subset of $\mathbf{R}^2$). The metric space structure of H coming from a Riemannian metric is entirely different (see Problem 8.8). However, we can show the topologies are the same. (If you don't know what a topology is, then skip the next result.)

PROPOSITION 8.9. The topology of a Riemannian manifold M as a metric space is the same as the topology induced by the metric space structure coming from the Riemannian metric g (Hicks [1965, p. 70]).

Because we are assuming that M is a Riemannian manifold it makes sense to talk about a parametrization by arc length.

DEFINITION. The curve α is *parametrized by arc length* if $g(T_\alpha, T_\alpha) = 1$.

We see that the analogue of Theorem 5.9 of Chapter 4 is true even in this very general setting. A reference for the next four results is Milnor [1963, pp. 55–67] or Hicks [1965, Chapter 9].

THEOREM 8.10. Suppose $\alpha : [a, b] \longrightarrow M$ is a curve parametrized by arc length. If $|\alpha| \leq |\beta|$ for all broken C^∞ curves β such that $\beta(a) = \alpha(a)$ and $\beta(b) = \alpha(b)$, then α is a geodesic.

This says that any curve which minimizes distance is a geodesic. The converse is false as S^2 with the induced metric shows. (Great circles are geodesics, as usual.) However, locally things are better (compare with Theorem 5.12 of Chapter 4).

THEOREM 8.11 (J. H. C. Whitehead, 1932). For every point p in a Riemannian manifold M there is a neighborhood U such that
(a) any two points q and r in U may be joined by a geodesic α whose image lies in U;
(b) the geodesic α is the unique geodesic joining q to r which has length $d(q, r)$; and
(c) there is a local coordinate chart (U, ϕ) such that the geodesics through p take the form $\alpha(t) = \phi^{-1}(a^1 t, \ldots, a^n t)$ for some constants $a^1, \ldots, a^n \in \mathbf{R}$.

As mentioned in Chapter 4 it need not be possible to join any two points by a geodesic but:

THEOREM 8.12 (Hilbert, 1902). If a connected Riemannian manifold M is complete as a metric space with the metric induced by g, then any two points can be joined by a geodesic which minimizes distance between the points.

Hilbert's Theorem has a partial converse:

THEOREM 8.13 (Hopf-Rinow, 1931). If every geodesic on M may be extended indefinitely (i.e., is defined on all of $\mathbf{R}$), then any two points may be joined by a geodesic of minimal length and M is complete as a metric space under the distance determined by g.

We now turn to the notion of curvature. It is not even clear what the definition of curvature should be on a Riemannian manifold. We cannot use the Weingarten map L because there is not one on an arbitrary Riemannian manifold (there is no concept of a normal vector). It is Gauss's *Theorema Egregium* that tells us what to do. Indeed, that is the whole point of the *Theorema Egregium.*

DEFINITION. The *Riemann-Christoffel curvature tensor of type* (1, 3) is the map $R: \mathfrak{X}(M) \times \mathfrak{X}(M) \times \mathfrak{X}(M) \longrightarrow \mathfrak{X}(M)$ by

$$R(X, Y)Z = \nabla_X \nabla_Y Z - \nabla_Y \nabla_X Z - \nabla_{[X,Y]} Z.$$

In local coordinates we have

$$R\left(\frac{\partial}{\partial x_j}, \frac{\partial}{\partial x^k}\right) \frac{\partial}{\partial x^i} = \sum R_i{}^l{}_{jk} \frac{\partial}{\partial x^l},$$

where $R_i{}^l{}_{jk}$ is defined by Equation (9-1) of Chapter 4 (see Problem 8.9). Note that if $X = \partial/\partial x^i$ and $Y = \partial/\partial x^j$, then $R(X, Y)Z$ measures how much the "equality of mixed partials" fails to hold for covariant derivatives.

A simple calculation shows that $R \equiv 0$ in flat Euclidean space, which intuitively does not curve. On the other hand, R is nonzero in S^n (as we have actually computed in the case of S^2 in Chapter 4). As a hypersurface in R^{n+1}, S^n does have an intuitive notion of curvature. Thus in some sense R does measure curvature, or lack of flatness. (It can be shown that a manifold with $R \equiv 0$ cannot be distinguished *locally* from $\mathbf{R}^n$ geometrically—about each point there can be found a coordinate chart in which g_{ij} is identically equal to δ_{ij}.) However, we would prefer to have a numerical invariant for curvature. There are several ways to obtain a numerical invariant from the Riemann-Christoffel tensor. We shall give the definition of three: sectional, Ricci, and scalar curvatures. Each is defined in terms of the Riemann-Christoffel curvature tensor of type (1, 3) or type (0, 4).

DEFINITION. The *Riemann-Christoffel tensor of type* (0, 4) is the map R: $\mathfrak{X}(M) \times \mathfrak{X}(M) \times \mathfrak{X}(M) \times \mathfrak{X}(M) \longrightarrow \mathfrak{F}(M)$ given by

$$R(X, Y, Z, W) = g(R(X, Y)Z, W).$$

A straightforward calculation shows that the values of $R(X, Y)Z$ and $R(X, Y, Z, W)$ at $p \in M$ depend only on the values of X_p, Y_p, Z_p, and W_p and not on the vector fields themselves.

In surface theory we studied curvature by breaking the surface up into curves. Analogously, we can try to break a manifold up into two-dimensional submanifolds, or we may look at two-dimensional subspaces of T_pM.

DEFINITION. Let Π be a two-dimensional subspace of T_pM. The *sectional curvature* of Π is $K_p(\Pi) = R(X, Y, Y, X)(p)$, where $\{X_p, Y_p\}$ is an orthonormal basis of Π.

For surfaces in $\mathbf{R}^3$, $K_p(\Pi)$ is exactly the same as the Gaussian curvature at p.

The other two kinds of curvature, the Ricci and scalar curvatures, are quite frequently used in physics (especially relativity) as well as mathematics.

If $X_p, Y_p \in T_pM$, we may define a map $\Xi_p(X_p, Y_p): T_pM \longrightarrow T_pM$ by $\Xi_p(X_p, Y_p)V_p = R(V_p, X_p)Y_p$. Problem 8.10 shows that $\Xi_p(X_p, Y_p)$ is a linear transformation for each p and $X_p, Y_p \in T_pM$ and so we may take the trace of $\Xi_p(X_p, Y_p)$.

DEFINITION. The *Ricci curvature tensor* S is an assignment, to each $p \in M$, of a function S_p: $T_pM \times T_pM \longrightarrow \mathbf{R}$ defined by

$$S_p(X_p, Y_p) = \text{trace}(\Xi_p(X_p, Y_p)).$$

DEFINITION. If M^n is a Riemannian manifold with Ricci tensor S, then the *scalar curvature* of M at p is $\sum_{i=1}^{n} S_p((X_i)_p, (X_i)_p)$, where $\{(X_1)_p, \ldots, (X_n)_p\}$ is any orthonormal basis of T_pM.

It should be pointed out that the applications of the above to relativity theory is not Riemannian geometry. In relativity theory one does not have a Riemannian metric but rather a *nondegenerate* metric; Condition (c) of the definition is replaced by

(c′) $\quad g(X_p, Y_p) = 0 \quad$ for all $Y_p \in T_pM$ if and only if $X_p = 0$.

This lack of positive definiteness does create certain problems in translating Riemannian geometry into relativity theory.

One area of research in differential geometry revolves around the effect of assumptions about curvature on the topology of the manifold. The results in this area are fascinating (and quite difficult). Most of them require more topology than we can assume, so we will only mention a few of them and refer the interested reader to Kobyashi and Nomizu [1963, especially p. 294]. Readers who have gone through Chapter 6 have seen some results of this type.

There are many very simple sounding questions which are extremely difficult to answer. For example, there are very few examples known of compact manifolds with every where positive sectional curvature. In fact, it is not even known what the spaces with constant positive sectional curvature $1/a^2$ are. However, we do have the following.

THEOREM 8.14. Every connected complete Riemannian manifold M of even dimension which has constant positive sectional curvature $1/a^2$ is either a sphere of radius a or a suitable projective space.

If we are willing to assume that M is simply connected, then there is the following classification.

THEOREM 8.15. If M is a complete, connected, simply connected Riemannian manifold of constant sectional curvature c, then M is either
(a) a sphere of radius $\sqrt{1/c}$ $(c > 0)$;
(b) flat Euclidean space $(c = 0)$; or
(c) hyperbolic space form (see Problem 8.13) $(c < 0)$.

In Section 6-5 we saw that the Euler characteristic χ was $(1/2\pi)$ times the integral of the Gaussian curvature so that $K > 0$ implies $\chi > 0$. There is an analogous formula in higher even dimensions but the following conjecture, with the appropriate meaning for Euler characteristic, has remained unsolved for over fifty years:

CONJECTURE 8.16. If M is a compact, orientable, even-dimensional manifold with everywhere positive sectional curvature, then M has positive Euler characteristic.

PROBLEMS

8.1. If g^0 and g^1 are Riemannian metrics on M, prove that $tg^0 + sg^1$ is also a Riemannian metric on M if t and s are both positive.

***8.2.** Prove Lemma 8.5.

***8.3.** Prove that Equation (8-7) follows from Equation (8-6).

***8.4.** Prove that the Riemannian connection of Example 8.1 is flat Euclidean space. Verify that $\langle \bar{\nabla}_X Y, N \rangle = -\langle Y, \bar{\nabla}_X N \rangle$ as claimed at the end of Section 7-6.

***8.5.** Prove that the Riemannian connection of the hyperbolic upper half plane (Example 8.3) is the linear connection given in Example 6.2.

***8.6.** Prove Proposition 8.8 assuming Theorem 8.11.

†8.7. (Riemannian Connection for Submanifolds). Let M be a submanifold of N, $M \subset N$ and write $\bar{g}$ for the Riemannian metric of N (and $\bar{\nabla}$ for the Riemannian connection of N). If $p \in M$, then the *normal space* to M at p is

$$\nu_p = \{X_p \in T_pN \mid g_p(X_p, Y_p) = 0 \text{ for all } Y_p \in T_pM\}.$$

Since $T_pN = T_pM \oplus \nu_p$ for each $p \in M$, we may talk about the tangential and normal components of an element of T_pN. If $X, Y \in \mathfrak{X}(M)$, then $V(X, Y)$ is (by definition) the normal component of $\bar{\nabla}_X Y$. (V is called the *second fundamental tensor* of M in N.) Define $\nabla_X Y$ by:

$$\nabla_X Y = \bar{\nabla}_X Y - V(X, Y). \tag{8-8}$$

(This is known as Gauss's formula.)

(a) Show that $V(X, Y) = V(Y, X)$ and $V(fX, Y) = fV(X, Y)$ for any $f \in \mathfrak{F}(M)$.

(b) Show that ∇ is a linear connection on M.

(c) Show that ∇ is the Riemannian connection for the induced metric on M.

(d) If $M = M_f$ is the hypersurface of $N = R^{n+1}$ defined by f, show that Equation (8-8) reduces to Equation (6-3).

***8.8.** Let $0 < \epsilon < 1$. Show that the distance between $p = (0, 1)$ and $q_\epsilon = (0, \epsilon)$ is $(1/\epsilon) - 1$ in the hyperbolic upper half plane. (*Hint:* Consider the curve in H, $\alpha(t) = (0, 1 - t)$ for $0 \leq t \leq 1 - \epsilon$.) Note that the Euclidean distance between p and q_ϵ is $1 - \epsilon$ (hence stays finite as $\epsilon \longrightarrow 0$) whereas the "origin" is infinitely far away from (0, 1) in the hyperbolic distance.

8.9. Show that, in local coordinates,

$$R\left(\frac{\partial}{\partial x^j}, \frac{\partial}{\partial x^k}\right)\frac{\partial}{\partial x^i} = \sum_l \left\{\frac{\partial \Gamma_{ik}{}^l}{\partial x^j} - \frac{\partial \Gamma_{ij}{}^l}{\partial x^k} + \sum_p (\Gamma_{ik}{}^p\Gamma_{pj}{}^l - \Gamma_{ij}{}^p\Gamma_{pk}{}^l)\right\}\frac{\partial}{\partial x^l}$$

for all $1 \leq i, j, k \leq n$. Compare with Equation (9-1) of Chapter 4.

***8.10.** Show that $\Xi_p(X_p, Y_p) : T_pM \longrightarrow T_pM$ is a linear transformation for each $p \in M$.

8.11. Compute the sectional, Ricci, and scalar curvatures of $\mathbf{R}^n$ and S^n.

8.12. Let M^2 be a surface in $\mathbf{R}^3$ with Gaussian curvature $K(p)$ and mean curvature $H(p)$ at p. Express the sectional, Ricci, and scalar curvature of M at p in terms of K and H. Are you surprised at the answer?

8.13. (Hyperbolic Space Form.) Let $H^n = \{x \in R^n \mid |x| < 1\}$. Define the Riemannian metric on H^n at $x \in H^n$ by

$$g_{ij}(x) = \frac{4a^2\delta_{ij}}{(1 - |x|^2)^2}.$$

We are taking the chart on H^n given by the identity map.

(a) Find the Riemannian connection of H^n. (*Answer:* $\Gamma_{ij}{}^k$ is zero unless at least two of the indices i, j, k are equal, and then $\Gamma_{ij}{}^i = \Gamma_{ji}{}^i = -\Gamma_{ii}{}^j = \Gamma_{jj}{}^j = 2x^j/(1 - |x|^2)$.

(b) Show that the sectional curvature of any plane at any point is $-1/a^2$. (*Hint:* Show $R_j{}^i{}_{kr} = 0$ unless $k = i$ and $r = j$, or $k = j$ and $r = i$.)

Comment: H^2 is not the hyperbolic upper half plane. It is the *Poincaré disk*. We choose H as the model for hyperbolic geometry rather than H^2 because the computation and visualization of geodesics on H is easier. It can be shown that H^2 and H are the same in the sense that they are isometric.

††8.14. Prove Conjecture 8.16 and win a prize.

Appendix

Historical Notes

The history of mathematics, in addition to being a fascinating field in its own right, is a very valuable aid in learning a new field of mathematics. For this reason we have included these historical notes. They are not meant to be a history of differential geometry or even a history of the results contained in this book. Such a history itself would be at least as long as this volume. For a more complete history see Struik [1933], [1967], or Coolidge [1940]. There is also an interesting chapter in Boyer [1968].

Problems in the differential geometry of plane curves have been studied since the invention of calculus, but we shall not concern ourselves with this aspect of the history. (One amusing fact: some historians date the beginning of differential geometry before the invention of the calculus!) The first major contributor to the subject was Leonhard Euler (1707–1783). In 1736, he introduced the intrinsic coordinates of a plane curve (arc length s and the radius of curvature ρ instead of the curvature k), and so began the study of intrinsic geometry. You may recall that the description of curvature as the rate of change of a particular angle was due to Euler. The original theory of curves of constant width is due to him. He also contributed to the theory of surfaces, especially (with John and Daniel Bernoulli (1667–1748) and (1700–1782)) some work on geodesics, and was the first to describe geodesics as the solutions to certain differential equations. Motivated by physical problems, he showed, in 1736, that a point mass constrained to lie on a surface and subject to no other forces must move along a geodesic.

A second major figure in the history of differential geometry was Gaspard Monge (1746–1818). He also was motivated by practical problems—in his case, questions of fortification. Monge started the theory of space curves in a paper he wrote in 1771 and published in 1785. His methods were very geometric and reflected his interest in partial differential equations. He published the first text on differential geometry in 1807. An indication of the importance of this book is the appearance of a fifth edition, thirty years after his death. Monge is remembered not only for his original contributions but also for the results of his teaching. Among his pupils were Pierre Laplace (1749–1827), Jean Meusnier (1754–1793), Joseph Fourier (1768–1830), Michel Lancret (1774–1807), André Ampère (1775–1836), Étienne Malus (1775–1812), Siméon Poisson (1781–1840), Charles Dupin (1784–1873), Victor Poncelet (1788–1867), and Olinde Rodrigues (1794–1851). Today the work of the Monge school is hard to read because they thought in terms of infinitesimals. Although Euler had introduced analytical methods to the study of space curves just before his death, these were not even used in Monge's text twenty-five years later. As an example of this awkward language, we note that Lancret defined the "première flexion" (curvature) and "seconde flexion" (torsion) as the differential of the angles between two consecutive normal or osculating planes. It was Augustin Cauchy (1789–1857), one of the founders of group theory and the theory of limits, who first expressed curvature and torsion in terms of finite quantities as we do today.

Even in the 1840s the theory of space curves was inelegant and difficult to find. To ameliorate this situation Barré de Saint Venant (1797–1886) wrote in 1846 a study on space curves. In this study, he included much historical material and collected the available results together in one work. He is responsible for the term "binormal" and the first proof of Lancret's Theorem. The theory of space curves finally became unified when F. Frenet (1816–1868) and Joseph Serret (1819–1885) working independently came up with what we call the Frenet-Serret equations in 1847 and 1851. Unfortunately, their work was not highly regarded when it appeared. Part of this was due to the lack of the language of linear algebra, which we have used extensively. For example, instead of computing the derivative of the normal with respect to arc length, they computed the derivatives of the direction cosines of the line in the direction of the normal. It was Gaston Darboux (1824–1917) who first unified the theory of curves with his concept of a moving frame ("trièdre mobile"). In this he was motivated by the theory of mechanics. This is the modern theory which in turn, under the guidance of Élie Cartan (1869–1951), gave valuable insight into the theory of manifolds.

The contributions of Carl Friedrich Gauss (1777–1855) can be found in his *Disquisitiones generales circa superficies curvas* of 1827, which has been translated into English (Gauss [1965]). Once you overcome the differences in mathematical language, this translation is not too difficult to read, especially with the aid of the section "How to Read Gauss" in Spivak [1970, vol. II].

Struik [1933, p. 164] notes that Gauss "became the teacher of the entire learned world." The reason for this goes beyond the fact that he proved new and startling results. He was responsible for an entirely new approach to differential geometry, one that has proved fruitful for over 150 years: intrinsic geometry based upon the first fundamental form. The idea that a surface could be written parametrically as a function of two variables, at least locally, was known to Euler, but Gauss emphasized that a surface should be described this way, rather than as a set of points in $\mathbf{R}^3$ whose coordinates satisfy some relation. The sphere (or Gauss) map was also known to Euler. Gauss's approach made heavy use of it. In fact, it is the first concept that he defines in his book. Rodrigues had earlier found the limit of the ratio between the area of a surface and the area of the corresponding region on the sphere (Proposition 8.6 of Chapter 4). Gauss actually was the first to recognize the importance of this limit and used it as the definition of the curvature of a surface at a point. This is a second important aspect of Gauss's contributions: his insight into what was useful and important.

One can say that geometers before Gauss viewed a surface as being made up of infinitely many curves, whereas Gauss viewed the surface as an entity in itself. To better understand this, think about computing the curvature of a surface in two ways. One way would be extrinsic: find the principal directions and then compute the product of the normal curvatures of the corresponding lines of curvature, an approach essentially known to Euler. The second way would be intrinsic: use the *Theorema Egregium.* The contrast between the extrinsic and the intrinsic calculations of curvature is a very deep philosophical one, of which Gauss was well aware. His statement of the *Theorema Egregium* (Gauss [1965, p. 20]) illustrates this, along with his interest in the surface as a whole: "If a curved surface is developed upon any other surface whatever, the measure of curvature in each point remains unchanged." (Here "developed" means mapped in a one-to-one, onto, distance-preserving fashion.) Although Gauss was making tremendous steps forward with his intrinsic geometry, it was not completely recognized at the time. While he was advocating the study of the Gaussian curvature because it was invariant under a development, a mathematician as good as Sophie Germain (1776–1831) was advocating the mean curvature as the main object of study. (This is understandable since it came up constantly in her study of elasticity.)

If there are two key ideas which have permeated the history of geometry from 1827 to the present, they are (1) the continuing influence of physical problems in directing various avenues of research and (2) the desire to both understand and enlarge upon the work of Gauss. These two ideas are not unrelated. Indeed, Gauss was surveying the Kingdom of Hanover from 1821 to 1825 and this practical work combined with his work as a theoretical geodesist motivated much of his work in differential geometry.

We shall not go into the applications of differential geometry to physics, mechanics, and the like except to mention those most responsible. Gabriel Lamé (1795–1871) worked in the theory of elasticity and heat, as did de Saint Venant. Ampère and Lazare Carnot (1753–1823) were guided by their interest in electricity. Many individuals, including Alexis Clairaut (1713–1765), Monge, Ampère, and Henri Poincaré (1854–1912), studied differential geometry to gain insight into the theory of partial differential equations. Malus came to differential geometry from optics (line congruences). Dupin was interested in the applications of differential geometry to mechanics. Albert Einstein (1879–1955) based his theory of general relativity on differential geometry in 1915. There are still applications of differential geometry being made in new areas. See *Newsweek*, January 19, 1976, pp. 54–55 for a nontechnical account of René Thom's catastrophe theory. (It is actually more topology than geometry.) Catastrophe theory is also covered in Zeeman [1976].

The idea of the surface itself being an important entity was taken up again in the brilliant address of Bernard Riemann (1826–1866) in 1854. A translation of this may be found in Spivak [1970, vol. II], which we urge the reader to try to read. We have already discussed a part of Riemann's philosophy in the introductions to Chapters 4 and 7. We mentioned the philosophical importance of his realization that space and geometry are different. Of equal importance is his realization that a "quadratic differential" (which we now call the Riemannian metric) is the structure to add to the notion of manifold. This gives an infinitesimal way to measure distance and an enormous amount of structure on manifolds. To appreciate Riemann's contribution in this area we must realize that geometers had been putting a Riemannian metric on surfaces in $\mathbf{R}^3$ (the induced metric) without realizing that it was an extra structure. Riemann realized that this notion was extraordinarily important and separated it out. There are now more general notions of geometry (e.g., linear connections or connections on a fiber bundle), but this does not diminish Riemann's contribution.

In the period between 1854 and 1900 geometers were mostly obsessed with the language of differential geometry rather than the subject itself. This tendency for symbolism culminated with the development of the tensor calculus by Gregorio Ricci (1853–1925) and his student Tullio Levi-Civita (1873–1941). It was upon Riemann's abstract view of space and this tensor calculus that Einstein based his theory of gravitation, commonly called general relativity. One of the important historical developments during this period was the use of group theory in differential geometry. This unifying concept was introduced by Felix Klein (1849–1925) and Sophus Lie (1842–1899). It was further developed by Cartan in a plethora of fascinating papers.

Unfortunately, much of Cartan's work is difficult to read because of the lack of a suitable theory of topology at the time he wrote. For a modern interpretation of Klein's ideas see Millman [1977].

For those readers who are interested in what has happened more recently, a survey by Chern [1946] is a good source. Since then there has been a movement to change the notation once again and get away from the local tensor notation of Ricci and Levi-Civita. This new invariant notation, which we used to some degree in Chapter 7, along with the definition of a linear connection due to Koszul, first appeared as recently as 1954 in an important paper by Katsumi Nomizu.

We conclude these notes and this book with a quote from Gauss [1965, p. 45] which is as valid now as it was in 1827:

> Although geometers have given much attention to general investigations of curved surfaces and their results cover a significant portion of the domain of higher geometry, this subject is still so far from being exhausted, that it can be well said that, up to this time, but a small portion of an exceedingly fruitful field has been cultivated.

Bibliography

In addition to those books and papers referred to in the text, we have included several other related books. Some words are in order relative to the level and difficulty of these books. Struik [1961], Laugwitz [1965], O'Neill [1966], Stoker [1969], Goetz [1970], and Do Carmo [1976] are all undergraduate texts. Struik and Laugwitz are classical in outlook while O'Neill is quite modern. Stoker, Goetz, Do Carmo, and O'Neill are at about the same level as this text, but Stoker and Do Carmo assume more analytical and topological background and O'Neill uses the language of differential forms. At the crossover to graduate level; we have Spivak [1970, 1975]. An introduction to differential geometry at the graduate level may be found in Hicks [1965] or Boothby [1975]. At a more advanced level is Bishop and Crittenden [1964], and finally Kobayashi and Nomizu [1963, 1969].

The books by Matsushima [1972] and Warner [1971] are about differentiable manifolds, not geometry (that is, the concept of a connection does not appear). They cover material basic to both differential geometry and differential topology. For the latter subject both Guillemin and Pollack [1974] and Milnor [1965] are excellent and readable.

Any advanced study of differential geometry requires some background in point set topology, as may be found in Munkres [1975], and in algebraic topology, some of which can be found in Singer and Thorpe [1967] and Massey [1967].

ALMGREN, F., *Plateau's Problem*. New York: Benjamin, 1966.

BERGER, M., *Lectures on Geodesics in Riemannian Geometry*, Tata Institute on Fundamental Research Lectures in Mathematics, No. 33. Bombay: Tata Institute, 1965.

BERTRAND, J., "La théorie des courbes a double courbure," *Journal de Mathématiques Pures et Appliquées*, **15** (1850), 332–350.

BIRKHOFF, G., and G. ROTA, *Ordinary Differential Equations* (2nd ed.). Waltham, Mass.: Blaisdell, 1969.

BISHOP, R., "There is More than One Way to Frame a Curve," *American Mathematical Monthly*, **82** (1975), 246–251.

———, and R. CRITTENDEN, *Geometry of Manifolds*. New York: Academic Press, 1964.

BLASCHKE, W., *Vorlesungen über Differentialgeometrie, I* (3rd ed.). Berlin: Springer, 1930.

BOOTHBY, W., *An Introduction to Differentiable Manifolds and Riemannian Geometry*. New York: Academic Press, 1975.

BORSUK, K., "Sur la courbure totale des courbes fermées," *Annales de la Société Polonaise de Mathématique*, **20** (1948), 251–265.

BOYER, C., *A History of Mathematics*. New York: Wiley, 1968.

BREUER, S., and D. GOTTLIEB, "Explicit Characterizations of Spherical Curves," *Proceedings of the American Mathematical Society*, **27** (1971), 126–127.

CHERN, S., "Some New Viewpoints in Differential Geometry in the Large," *Bulletin of the American Mathematical Society*, **52** (1946), 1–30.

———, "Curves and Surfaces in Euclidean Space," in *Studies in Global Geometry and Analysis*, ed. S. Chern, 16–56. Englewood Cliffs, N.J.: Mathematical Association of America and Prentice-Hall, 1967.

CODDINGTON, E., and N. LEVINSON, *Ordinary Differential Equations*. New York: McGraw-Hill, 1955.

COOLIDGE, J., *A History of Geometrical Methods*. Oxford, England: Oxford University Press, 1940.

DO CARMO, M., *Differential Geometry of Curves and Surfaces*. Englewood Cliffs, N.J.: Prentice-Hall, 1976.

FARY, I., "Sur la courbure totale d'une courbe gauche faisant un noeud," *Bulletin de la Société Mathématique de France*, **77** (1949), 128–138.

FENCHEL, W., "Uber Krümmung and Windung geschlossenen Raumkurven," *Mathematische Annalen*, **101** (1929), 238–252.

———, "On the Differential Geometry of Global Space Curves," *Bulletin of the American Mathematical Society*, **57** (1951), 44–54.

FULKS, W., *Advanced Calculus* (2nd ed.). New York: Wiley, 1969.

GAUSS, K., *General Investigations of Curved Surfaces*, tr. A. Hiltebertel and J. Morehead. Hewlett, N.Y.: Raven Press, 1965.

GLUCK, H., "Higher Curvatures of Curves in Euclidean Space, I, II," *American Mathematical Monthly*, **73** (1966), 699–704; **74** (1967), 1049–1056.

GOETZ, A., *Introduction to Differential Geometry*. Reading, Mass.: Addison-Wesley, 1970.

GUILLEMIN, V., and A. POLLACK, *Differential Topology*. Englewood Cliffs, N.J.: Prentice-Hall, 1974.

HALL, T., *Carl Friedrich Gauss, A Biography*, tr. A. Froderberg. Cambridge, Mass.: MIT Press, 1970.

HICKS, N., *Notes on Differential Geometry*. Princeton: van Nostrand, 1965.

HILBERT, D., *The Foundations of Geometry* (2nd ed.), tr. E. Townsend. Chicago: Open Court, 1921.

———, *Grundlagen der Geometrie* (8th ed.). Stuttgart: Teubner, 1956.

———, and S. COHN-VOSSEN, *Geometry and the Imagination*, tr. P. Nemenyi. New York: Chelsea, 1952.

HOFFMAN, K., and R. KUNZE, *Linear Algebra* (2nd ed.). Englewood Cliffs, N.J.: Prentice-Hall, 1971.

HOPF, H., "Uber die Drehung der Tangenten und Sehnen ebener Kurven," *Compositio Mathematica*, **2** (1935), 50–62.

JACKSON, S., "Vertices of Plane Curves," *Bulletin of the American Mathematical Society*, **50** (1944), 564–578.

KNESER, A., "Bemerkungen über die Anzahl de Extreme der Krümmung auf geschlossenen Kurven und über verwandte Fragen einer nicht-Euklidischen Geometrie," *Festschrift zum siebzigsten Geburstag von Heinrich Weber*, 170–180. Leipzig: Teubner, 1912.

KOBAYASHI, S., and K. NOMIZU, *Foundations of Differential Geometry, I, II*. New York: Interscience, 1963, 1969.

LAUGWITZ, D., *Differential and Riemannian Geometry*. New York: Academic Press, 1965.

MASSEY, W., *Algebraic Topology: An Introduction*. New York: Harcourt, Brace and World, 1967.

MATSUSHIMA, Y., *Differentiable Manifolds*. New York: Marcel Dekker, 1972.

MILLMAN, R., "Kleinian Transformation Geometry," *American Mathematical Monthly*, **84** (1977).

———, and A. STEHNEY, "The Geometry of Connections," *American Mathematical Monthly*, **80** (1973), 475–500.

MILNOR, J., "On the Total Curvature of Knots," *Annals of Mathematics*, **52** (1950), 248–257.

———, *Morse Theory*, Annals of Mathematics Studies No. 51. Princeton: Princeton University Press, 1963.

———, *Topology from the Differentiable Viewpoint*. Charlottesville, Va.: University Press of Virginia, 1965.

MISNER, C., K. THORNE, and J. WHEELER, *Gravitation*. San Francisco: Freeman, 1973.

MUKHOPADHYAYA, S., "New Methods in the Geometry of a Plane Arc," *Bulletin of the Calcutta Mathematical Society*, **1** (1909), 31–37.

MUNKRES, J., *Topology, A First Course*. Englewood Cliffs, N.J.: Prentice-Hall, 1975.

O'NEILL, B., *Elementary Differential Geometry*. New York: Academic Press, 1966.

RUTISHAUSER, H., and H. SAMELSON, "Sur le rayon d'une sphère dont la surface contient une courbe fermée," *Comptes Rendus Hebdomadaires des Séances de l'Académie de Science*, **227** (1948), 755–757.

SALKOWSKI, F., "Zur Transformation von Raumkurven," *Mathematische Annalen*, **66** (1909), 517–557.

SAMELSON, H., "Orientability of Hypersurfaces in R^n," *Proceedings of the American Mathematical Society*, **22** (1969), 301–302.

SCHERRER, W., "Eine Kennzeichnung der Kugel," *Vierteljahresschrift Naturforscher Gesellschaft in Zurich*, **85** (1940), 40–46.

SCHMIDT, E., "Uber das isoperimetrische Problem in Raum von n Dimensionen," *Mathematische Zeitschrift*, **44** (1939), 689–788.

SEGRE, B., "Sui circoli geodetici di una superficie a curvature totale contante, che contengono nell'interno una linea assegnata," *Bollettino della Unione Matematica Italiana*, **13** (1934), 279–283.

———, "Una nuova caratterizzazione della sfera," *Atti della Accademia Nazionale dei Lincei, Rendiconti*, **3** (1947), 420–422.

———, "Sulla torsione integrale delle curve chiuse sghembe," *Atti della Accademia Nazionale dei Lincei, Rendiconti*, **3** (1947), 422–426.

SINGER, I., and J. THORPE, *Lecture Notes on Elementary Topology and Geometry*. Glenview, Ill.: Scott, Foresman, 1967.

SPIVAK, M., *A Comprehensive Introduction to Differential Geometry, I, II, III, IV, V.* Boston: Publish or Perish, 1970, 1975.

STERNBERG, S., *Lectures on Differential Geometry*. Englewood Cliffs, N.J.: Prentice-Hall, 1964.

STOKER, J., *Differential Geometry*. New York: Interscience, 1969.

STRUIK, D., "Outline of a History of Differential Geometry," *Isis*, **19** (1933), 92–120; **20** (1933), 161–191.

———, *Lectures on Classical Differential Geometry* (2nd ed.). Reading, Mass.: Addison-Wesley, 1961.

———, *A Concise History of Mathematics* (3rd ed.). New York: Dover, 1967.

THOMAS, G., *Calculus and Analytic Geometry* (4th ed.). Reading, Mass.: Addison-Wesley, 1968.

VOSS, A., "Uber Kurvenpaare in Raume," *Sitzungsberichte, Akadamie der Wissenschaften zu München*, **39** (1909), 106.

VOSS, K., "Eine Bemerkung über die Totalkrümmung geschlossenen Raumkurven," *Archiv der Mathematik*, **6** (1955), 259–263.

WARNER, F., *Foundations of Differentiable Manifolds and Lie Groups*. Glenview, Ill.: Scott, Foresman, 1971.

WONG, Y., "A Global Formulation of the Condition for a Curve to Lie in a Sphere," *Monatschefte Für Mathematik*, **67** (1963), 363–365.

———, "On an Explicit Characterization of Spherical Curves," *Proceedings of the American Mathematical Society*, **34** (1972), 239–242.

ZEEMAN, E., "Catastrophe Theory," *Scientific American*, **234** (1976), 65–83.

INDEX

I.

Notational Index

II.

Topical Index